Reconsidering Resilience in African Pastoralism

Towards a Relational and Contextual Approach

Reconsidering Resilience in African Pastoralism

Towards a Relational and Contextual Approach

Edited by

Shinya Konaka

Greta Semplici

and

Peter D. Little

Kyoto University Press

TRANS
PACIFIC
PRESS

Published in 2023 jointly by:

Kyoto University Press
69 Yoshida Konoe-cho
Sakyo-ku, Kyoto 606-8315, Japan
Telephone: +81-75-761-6182
Fax: +81-75-761-6190
Email: sales@kyoto-up.or.jp
Web: http://www.kyoto-up.or.jp

Trans Pacific Press Co., Ltd.
2nd Floor, Hamamatsu-cho Daiya Building
2-2-15 Hamamatsu-cho, Minato-ku, Tokyo
105-0013, Japan
Telephone: +81-50-5371-9475
Email: info@transpacificpress.com
Web: http://www.transpacificpress.com

Distributors

USA and Canada
Independent Publishers Group (IPG)
814 N. Franklin Street
Chicago, IL 60610, USA
Telephone inquiries: +1-312-337-0747
Order placement: 800-888-4741 (domestic only)
Fax: +1-312-337-5985
Email: frontdesk@ipgbook.com
Web: http://www.ipgbook.com

Europe, Oceania, Middle East and Africa
EUROSPAN
Gray's Inn House,
127 Clerkenwell Road
London, EC1R 5DB
United Kingdom
Telephone: +44-(0)20-7240-0856
Email: info@eurospan.co.uk
Web: https://www.eurospangroup.com/

Japan
For purchase orders in Japan, please contact
any distributor in Japan.

China
China Publishers Services Ltd.
718, 7/F., Fortune Commercial Building,
362 Sha Tsui Road, Tsuen Wan, N.T.
Hong Kong
Telephone: +852-2491-1436
Email: edwin@cps-hk.com

Southeast Asia
Alkem Company Pte Ltd.
1, Sunview Road #01-27, Eco-Tech@Sunview
Singapore 627615
Telephone: +65 6265 6666
Email: enquiry@alkem.com.sg

ISBN 978-1-920850-10-4 (hardback)
ISBN 978-1-920850-06-7 (paperback)
ISBN 978-1-920850-07-4 (eBook)

Contents

List of Figures

List of Table

List of Photographs

Contributors

Shinya Konaka is Professor, School of International Relations, and Dean, Graduate School of International Relations, University of Shizuoka. He has been conducting field research among Maa-speaking pastoralists, mainly the Samburu, since 1992. He was awarded the 13th Okita Prize in International Development, the 8th Japan Consortium for Area Studies Award for Prominent Scholar, and the JASID Special Award, 2018. He is the author of *Localization of Humanitarian Assistance Frameworks for East African Pastoralists, African Study Monographs. Supplementary Issue*, 53 (coedited with Sun Xiaogang, 2017) and 'Livestock as interface: The case of the Samburu in Kenya' in *Anthropology of Things* (TransPacific Press 2018).

Email: africanlake@gmail.com

Peter D. Little is the Samuel Candler Dobbs Professor of Anthropology and Director of the Global Development Studies Program, Emory University. He has conducted studies of political ecology, pastoralism, poverty and inequality, informality, development, and statelessness in Kenya, Somalia, and Ethiopia. Currently, he is engaged in a comparative study of 'Changing Perceptions of Poverty and Well-being in Baringo County, Kenya and South Wollo, Ethiopia'. Dr. Little is the author of *Economic and Political Reform in Africa: Anthropological Perspectives* (Indiana University Press, 2013), *The Elusive Granary: Herder, Farmer, and State in Northern Kenya* (Cambridge University Press, 1999), *and Somalia: Economy without State* (Indiana University Press/ James Currey, 2003); and co-author of *Risk and Social Change in an African Rural Economy: Livelihoods in Pastoralist Communities* (Routledge, 2016).

E-mail: pdlittl@emory.edu

Greta Semplici is a Max Weber Fellow at the European University Institute. She has earned a D.Phil. (Ph.D.) from the Oxford Department of International Development. She conducted an ethnographic study on the understanding of resilience from the perspective of mobile pastoralists in Turkana County, in the arid lands of Northern

Kenya. Her current research interests lie at the interface between nomadism, mobility and migration, drawing lessons from the experiences of 'mobile peoples'. She is also interested in human-animal-environment relationships, and broader issues concerning development and international cooperation. Previously, Greta worked for FAO Somalia as Monitoring and Evaluation International Consultant and collaborated with LAMA Development and Cooperation Agency for research on social protection strategies in rural Malawi. She also held several RA positions with ODI (Overseas Development Institute), IMI (International Migration Institute) and EUI (European University Institute). She holds a B.A. in Development Economics and International Studies (laurea triennale) and M.Sc. in Development Economics (laurea specialistica) from the University of Florence. Greta spent the academic year 2021/22 working on her book project, provisionally titled, Moving Deserts: Resilience and Development in the Drylands.

Email: Greta.Semplici@eui.eu

Go Shimada is a Professor at the School of Information and Communication, Meiji University and Editor-in-Chief of *Journal of International Development Studies*. He holds a Ph.D. from Waseda University (2014) and MA from Manchester (1999). He is a founding board member of the Japanese Association for Development Economics (JADE). Previously, he was an Associate Professor at the University of Shizuoka, Visiting Scholar at Columbia University, First Secretary of the Permanent Mission of Japan to the United Nations, and Japan International Cooperation Agency (JICA). He is the author of 'Does environmental policy make African industry less competitive? The possibilities in green industrial policy' in *The Quality of Growth in Africa*, edited by Ravi Kanbur, Akbar Noman, and Joseph E. Stiglitz (2019), 350–372. New York: Columbia University Press. He is also the co-editor of *Workers, Managers, Productivity: Kaizen in Developing Countries* (Singapore, Palgrave Macmillan, co-edited with Akio Hosono and John Page).

Email: go_shimada@meiji.ac.jp

Tamara Enomoto is Professor at the Organization for the Strategic Coordination of Research and Intellectual Properties, Meiji University. She holds a Ph.D. from the University of Tokyo (2015), an M.A. in Conflict, Development and Security (2004) and an M.A. in Development Studies (2002) from the University of Leeds. From 2003 to 2015, she worked at Oxfam where she was in charge of policy, advocacy and research on humanitarian and arms control issues and coordinated humanitarian aid and advocacy projects. Her recent publications include 'Demarcating battle lines: Citizenship and agency in the era of misanthropy' in Ituhiro Hazama, Kiyoshi Umeya, and Francis B. Nyamnjoh (eds). *Citizenship in Motion: South African and Japanese Scholars in Conversation* (2019, Langaa RPCIG), 'Overcoming the dichotomy between Africa and the West: Norms and measures for arms transfers to non-state actors' in Mitsugi Endo, Michael Neocosmos and Ato Kwamena Onoma (eds). *African Politics of Survival Extraversion and Informality in the Contemporary World* (2020, Langaa RPCIG), *Weapon Taboos: Genealogies of Pariah Weapons* (2020, Nihon Keizai Hyouronsha, edited book in Japanese), and *The Arms Trade Treaty: The Self, Sovereignty, and Arms Transfer Control* (2020, Kouyou Shobou, book in Japanese).

E-mail: tamaraenomoto67@gmail.com

Toru Sagawa is Associate Professor at the Faculty of Letters, Keio University, Japan. He has conducted field research in East Africa. His recent works include 'Large-scale development projects, food security policy and livelihood of agro-pastoralists in southwestern Ethiopia', in M. Takahashi et al. (eds.) *Development and Subsistence in Globalising Africa: Beyond the Dichotomy* (2021, Langaa RPCIG), and 'Waiting on a friend: Hospitality and gift to the 'enemy' in the Daasanach', *Nilo-Ethiopian Studies* 23 (2019).

Email: waraji.1125@gmail.com

Rahma Hassan is a social-economic researcher from Kenya and a PhD fellow from the University of Nairobi and the University of Copenhagen, under the Rights and Resilience Project. She has worked in social development research, in government and research on health, gender, and governance. Currently, she is researching community land rights among pastoralist communities in Kenya.

Email: rahmahassan@uonbi.ac.ke

Itsuhiro Hazama is a Professor of Anthropology at Toyo University in the Faculty of Sociology. His research focuses on ethnography of human-animal interaction and citizenship studies. He has recently co-authored 'Naturalography of co-existence among East African pastoral societies: An introductory overview of Japanese scholarship' (*African Study Monographs* 40 (2–3): 45–75, 2019); and co-edited *Citizenship in Motion: South African and Japanese Scholars in Conversation* (Bamenda: Langaa Rpcig, 2019).

Email: hazama@toyo.jp

Giulia Gonzales is a Max Weber Fellow at the European University Institute (Florence) and is affiliated with PASTRES, Pastoralism, Uncertainty, Resilience. Her current project, 'Emerging (Dis)Connections. Crisis, (im)mobilities and citizenship: Navigating uncertainties among Kel Tamasheq in Bamako' emphasises the making and unmaking of connections by Kel Tamasheq within and beyond Bamako as ways to embrace current socio-economic and political uncertainties, and as processes of state-formation within a globalised world. She has conducted fieldwork in Bamako (Mali), Ouagadougou and the refugee camp of Sag-Nioniogo (Burkina Faso), exploring issues of ethnicity and nationalism, belonging, place-making, migration and displacement, and everyday politics. She has also worked on migration in Europe, focusing on receptions of health services by third-country communities (CeSPI), and connectivity and mobility (ERC TRAFIG). She has published on Nomadic Peoples, L'Ouest Saharien, and contributed to co-edited volumes.

Email: Giulia.Gonzales@eui.eu

Takuto Sakamoto is a Professor at the Graduate Program on Human Security, the Graduate School of Arts and Sciences, the University of Tokyo. Dr. Sakamoto has been conducting research on a broad array of subjects, including human security, global governance, state and conflict, and dryland pastoralism, all with a particular focus on Africa. He also has extensive methodological interests in data analysis and simulation. His major works include *Integration and Disintegration of Territorial Rule. Multi-Agent Simulation Analysis of the States in Northeast Africa* (Tokyo: Shoseki-kobo Hayama, 2011) (in Japanese) and '*Computational* *research on mobile pastoralism using agent-based modeling and satellite imagery'* (PLOS ONE, Vol. 11, No. 3, 2016).

E-mail: sakamoto@hsp.c.u-tokyo.ac.jp

Rumiko Murao is a researcher at the Research Institute of Humanity and Nature, Kyoto, Japan. She holds a Ph.D. from the University of Kyoto (2009) in Area Studies. Her recent publications include 'The endogenous reintegration of post-conflict Angola society,' *African Study Monographs* (in Press, 2022), 'Zambia' In K. Makino and E. Iwasaki (eds). *Global Social Welfare: Africa and Middle East.* Tokyo. Jyunpousha. pp. 225–247. (2020, in Japanese), 'The daily life strategies of small-scale farmers after a prolonged war: The long-term influence of humanitarian assistance.' *African Study Monographs Supplementary Issue*, 53: 103–116. (2017), 'Land use of Angolan immigrants in western Zambia: Rethinking the autonomy and coexisting of self-settled refugee communities in host countries.' In A. Takada and I. K. Nyamongo. (eds). *Mila Special Issue.* pp. 59–67. (2014), *Creativity of African Farmers.* Kyoto: Showado (2012. a monograph in Japanese)

Email: rumiko.murao@gmail.com

Ian Scoones is a Professor at the Institute of Development Studies at the University of Sussex. He leads the ERC Advanced Grant funded PASTRES programme (Pastoralism, Uncertainty and Resilience: Global Lessons from the Margins, www.pastres.org). He has worked on land, agriculture and livelihoods, including pastoralism, for over 35 years, mostly in sub-Saharan Africa (see ianscoones.net)

Email: I.Scoones@ids.ac.uk

Introduction: Rethinking Resilience in the Context of East African Pastoralism

Shinya Konaka, Peter D. Little, and Greta Semplici

The promise of resilience is compelling. In a world increasingly exposed to uncertainty (Scoones and Stirling 2020), resilience proposes itself as the governance of last resort, the last card to play before we reach the limits of our planet. The once stable, predictable environment of the Holocene has been replaced by the advance of the Anthropocene (Chandler et al. 2020). It thereby follows that the modernist pursuit of command-and-control of the environment, of human superiority over the non-human, of centralisation and forecasting, is no longer tenable as it harms more than it saves the planet. Resilience grew from the corners of ecology, engineering, and psychology to now be invoked against the challenges posed by the Anthropocene, as a type of governmental reform that embraces change, diversity, and surprise. In popular terms, resilience refers to the capacity to persist in the face of change, to continue developing with ever changing environments, to adapt and evolve with change (Folke 2016). It reflects the ability of people, communities, and societies to navigate uncertainty and complexity, to continue learning, self-organising, and becoming. Resilience promises us that we will survive and even prosper despite the grim realities of increasing disasters and crises that surround us (Tierney 2014). As it stands, resilience can be seen as "the key political category of our times" (Neocleous 2013: 3).

But resilience is a slippery concept. It pervades every domain of our lives despite, or perhaps because of, a widely acknowledged vagueness (Strunz 2011). Grove defines it as "an essentially contested concept" (2018: 31). With this book, we do not aim to solve the mystery of *what resilience is*; perhaps the opposite, as we will argue that we rather reject the impulse to operationalise the concept and achieve conceptual clarification to prioritise its *localisation* in context and practice. With this book, we pursue instead a far more grounded goal. We aim to interrogate the increasingly overused concept of resilience by examining its application to research on pastoral populations through a series of case studies from Africa, with a focus on East Africa. Starting from this ethnographically rich discussion, we hope to contribute some insights about *thinking-resilience*, how to reconfigure the concept of resilience to accommodate specific contexts, instead of the more standard *resilience-thinking* approach (Folke et al. 2010). In short, we want to show how to apply the concept of resilience to specific contexts.

The capacity to interface change and variability with livelihood adaptations is well known by pastoralist populations across the world. Variability managers and innovators, pastoralists maneuver through complex environments comprised of ephemeral resources, climatic variations, and increasing connectivity across scales (Nori 2019). The pastoral scholarship is loud in advocating the possibilities to learn from pastoralists about how to manage change, to inform knowledge and decision-making across different domains from the vast dryland stretches of Africa to the Asian highlands where collectively tens of millions of pastoralists reside. These vast territories are characterized by varied uncertainties – including those caused by climate and environmental change, financial and commodity markets, disease outbreaks, unstable infrastructures, misplaced migration policies and insecurity and conflict (Nori and Scoones 2019).

Pastoralism thus appears a fertile field for enquiring into the multiple facets of resilience, starting from the simple but fundamental premise that resilience is context-dependent and can only be understood in relation to a given context and locality (Carpenter et. al. 2001, Chandler and Coaffee 2017). As a concept, resilience was originally framed to be general, universal and scientifically neutral. However, since the 2000s, it has come to be framed as a highly context-dependent or context-sensitive concept and has been applied across a number of ecological and livelihood scenarios (Waller 2001, Ungar 2008, Fletcher and Sarkar 2013, Panter-Brick 2014, Weichselgartner and Kelman 2014, Davies et al. 2015, Chandler and Coaffee 2017, Panter-Brick et al. 2018). As Oliver-Smith (2017) puts it, 'like adaptation and vulnerability, resilience as a broad concept must be specified to be of use in research, policy, and practice.' In short, the concept needs to be *localized* to and critically assessed for each specific context of pastoralism in East Africa (see also Konaka 2017).

In this context, resilience – and its frequently presumed antonym 'vulnerability' – are commonly used terms in academic and non-academic narratives about pastoralists and the risks they confront, especially those related to climate change. This book addresses the ways in which anthropologists have studied the interactions between pastoral communities and outside actors (e.g., development and government agencies) under the guise of *building resilience*. We identify a paradox of representation as scholars portray pastoralists as one of the most resilient groups (Chatty 2007) while practitioners see them as among the most vulnerable. Increasingly, food insecurity, extreme poverty, low-intensity conflict, displacement and natural disasters that are impacting East African pastoralists frequently are attributed to global climate change, regardless of the empirical evidence. In particular, the need to 'build resilience' among East African pastoralists is frequently highlighted whenever humanitarian and developmental interventions to address these problems are considered (Pavanello 2009; Headey et al. 2012; Lind 2016; Cervigni and Morris 2016; Catley 2017; FAO 2018). Thus, the commanding presence of resilience discourses related to pastoralism and development summon researchers and policymakers alike to critically consider the applicability of resilience thinking to the practice and future of East African pastoralism.

In the remainder of this introduction, we provide an overview of the mainstream theoretical and methodological arguments related to resilience. This is followed by an examination of the concept of resilience in the context of pastoralism in East Africa. Finally, the different contributions of the volume are introduced.

An overview of theory and application

Resilience is both an interdisciplinary and an ordinary term that derives from the Latin word *resilire* or *resilio* (to recoil or leap [jump] back) (Masten and Gewirtz 2006; Klein, Nicholls and Thomalla 2003), which refers to the capacity to 'bounce back'. As a metaphor, *bouncing back* has long constrained the application of resilience thinking to refer to the return to an imagined equilibrium upon a perturbation. But the concept of resilience has greatly evolved since, and it reflects now, as we will see, a much richer idea encompassing change and transformability.

In practice, resilience has become a 'catch-all' phrase used to vaguely express a wide range of responses under many different contexts of risk and threat (Manyena 2006; Brand and Jax 2007; Chandler and Coaffee 2017). As an academic term, it is said to have originated in physics and mathematics (Van der Leeuw and Leygonie 2000; Norris et al. 2008; MacKinnon and Derickson 2012), although most applications beginning in the 1970s largely derive from the disciplines of ecology and psychology. The concept gradually spread to other disciplines and fields, such as economics, geography, anthropology, environmental sciences and others.

In terms of disciplinary interpretations and definitions, resilience is quite controversial, and numerous definitions have been offered and examined (Adger 2000; Manyena 2006; Brand and Jax 2007; Fletcher and Sarkar 2013; Southwick et al. 2014). In the field of ecology, it is widely recognised that Holling's (1973) seminal article entitled 'Resilience and stability of ecological systems' marked a radical change challenging, at that time, the dominant stable-equilibrium view of ecosystems (Folke 2016) and launched the beginning of ecological studies of resilience (Manyena 2006). In the article, Holling distinguishes 'stability' from 'resilience', with the former indicating the capacity of a system to return to equilibrium and the latter the capacity to maintain internal structure in a period of perturbation (Van der Leeuw and Leygonie 2000: 9). Later, Holling (1996) differentiated two aspects of resilience. One focuses on efficiency, constancy and predictability and is referred to as *engineering resilience*. The other highlights persistence, change and unpredictability and is referred to as *ecological resilience*. In making this distinction, Holling implies that engineering resilience leads to initial success in managing variability but over time it can diminish resilience and increase the vulnerability of ecosystems (Holling 1996: 38). As such, Holling (1996: 41) stressed the importance of flexible, diverse and redundant regulation for sustainable development.[1] Along similar lines, Chandler and Coaffee (2017) summarise how the paradigm of resilience thinking has shifted from the *homeostasis* approach of the earlier period to the *autopoiesis* approach (ie., a system capable of reproducing itself) of the more recent era.

The scope and study of resilience has been gradually expanded from its roots in the natural sciences to the study of human societies, where the focus has been on the 'resilience of socio-ecological systems' (often abbreviated as 'SES') (Adger 2000; Van der Leeuw and Leygonie 2000; Folke 2006; Walker et al. 2006). In contrast to the 'adaptability' concept described in earlier interpretations of resilience, which indicates the capacity to continue developing within the current stability domain or basin of attraction (Folke et al. 2010), SES studies introduce the concept of 'transformability' defined as 'the capacity to create a fundamentally new system when ecological, economic, or social structures make the existing system untenable' (Walker et al. 2004: 5). Furthermore, Holling and his colleagues elaborated the theory of adaption cycles, referred to as 'panarchy' (from *Pan*, the Greek god of nature and *archia*, meaning hierarchy/order) (Holling 2001; Gunderson and Holling 2002; Walker et al. 2004; Resilience Alliance 2010; Allen et al. 2014). This term is used to describe a model of hierarchically linked systems represented as adaptive cycles that interact across scales (Resilience Alliance 2010: 29).

Another important field that brought resilience to prominence is developmental psychology. In examining the lives of 'at-risk' children, pioneering psychologists recognised that some youngsters thrive amid adversity and eventually become well adjusted, healthy adults (Garmezy 1971; Werner and Smith 1982; Rutter 1987). Resilience was initially conceptualised in terms of personality traits or coping styles that allowed some children to progress along a positive developmental trajectory even when confronted with considerable adversity. It is noteworthy that

---

1 For pastoralists this supports the argument that a diversified herd of goats, sheep, cattle, and/or camels is less susceptible to losses from animal disease, market disruptions, and climate shocks (cf. Niamir-Fuller 1998). Under current economic conditions, some pastoralists with more diversified livelihoods also can be more resilient than specialized pastoralists (cf. Leach et al. 2010).

some psychologists shifted from studying resilience from an individual personality perspective to consider socio-cultural environments beyond the individual. One psychologist stated that 'the interaction between people and their environments is an important consideration when conceptualizing resilience' (Fletcher and Sarkar 2013: 15). Waller claimed that 'the study of resilience is evolving from static, individualistic conceptualizations to an appreciation of the complex relational and contextual aspects of positive adaptation in the face of adversity' (2001: 296).

Propositions for new frontiers of resilience

As already suggested, the field of resilience has generally evolved and enlarged. In many fields, we are now very far from the original propositions of resilience as 'the capacity of a system to absorb disturbance and reorganize while undergoing change so as to still retain essentially the same function, structure, identity, and feedbacks' (Walker et al. 2004: online). More dynamic approaches to resilience today are common in many academic and professional fields including international development and humanitarianism, disaster management, urban planning, and security approaches. However, resilience is not immune to criticism, and some scholars go as far as counting the number of days when it will disappear from the international stage (Chandler 2020). This is not the space to review the critiques of resilience, which are many and relevant (see for example: Brown 2015; Scott-Smith 2018; Cannon and Müller-Mahn 2010). Instead, we focus on three propositions that we feel ally with the burgeoning literature to remain vigilant to the ways that resilience often is employed. These are: (1) a localization of the concept, (2) a reconfiguration of its ontological framing, and (3) a movement away from its closed tendencies to consider multiple relations, including political.

A problem with conceptions of resilience thinking is that they tend to presuppose generality, universality and neutrality. These imaginaries of resilience aspire to compile a somewhat magic policy making bullet to be employed for all problems, as diverse as underdevelopment, conflict, and environmental crises (Chandler 2020). Such a tendency has exposed resilience to the critiques of serving the goal to consolidate neoliberal socio-ecological relations through elements of privatisation, individualisation, deregulation, marketisation, and normalisation of vulnerability (MacKinnon and Derickson 2013; Evans and Reid 2014; Duffield 2012). To quote Olsson et al., 'given the lack of attention to agency, conflict, knowledge and power, and its insensitivity to theoretical development in contemporary social science, resilience is becoming a powerful depoliticising or naturalising scientific concept and metaphor when used by political regimes' (2017: 59). In contrast, the reality is that resilience differs substantially from society to society, and from culture to culture. There is no one-size-fits-all model that can be universally applied.

Anthropological studies of resilience have tended to focus more on holism and local cultural perspectives than do those in ecology and psychology. According to Oliver-Smith, 'anthropological perspectives on adaptation, vulnerability, and resilience, grounded in a tradition of holistic research on local lifeways, often vary significantly from those used in the climate change literature and policy frameworks' (2017: 207). Another anthropologist, Crane, also distinguishes the anthropological from the ecological approach: 'The examination of the relationship between resilience as a quality of ecological systems and resilience as an experience within subjective and collective cultural frameworks will be a key challenge in making models more meaningful and useful to people who live within modeled systems' (Crane 2010: online). Our first proposition is in line with these claims, promoting a contextual approach to resilience. This framing is similar to Roe's (2020) approach, applied to pastoral contexts, where the 'real-time operation' of pastoralists can be understood as seeking the reliability of the system in order to minimise volatility and to insure its future. In this regard, the 'positionality' of the different stakeholders of resilience, including pastoralists, researchers and policymakers, must be contextually located before we can discuss the issue of resilience. In other words, any research on resilience should never go without asking the question 'resilience for whom and by whom' (Little and McPeak 2014; Chandler and Coaffee 2017).

As proposed by Chandler, universal resilience approaches reify a modernist construction which assumes the problems as external and solutions as internal and achievable by enabling and building local capacities. Therefore, they evade our co-responsibility in the production of the same problems to be faced (2020). It is no longer possible to presume a modernist world external to us and amenable to governing, there is no longer an inside (human) and outside (non-human, nature, environment), a local and a global, a traditional and a modern. Such dichotomies were once hallmarks of the modernist project of civilisation, progress, development, and are now confronted by the advancement of the Anthropocene which warns that 'the consequences of what we do stick with us' (Chandler 2020: 50). In terms of ontological boundaries, 'these "liberal" binaries are understood as legacies of modernist approaches, which are 'reductionist' – reducing the complexity of the world to separate and discrete objects rather than looking at the dynamic relationships involved' (Chandler 2014: 7). Our second proposition lies on what we could call an 'ontological framing' that questions scientific epistemologies of knowledge making and disrupts ontological dichotomies by adopting local epistemologies and assigning agency and creativity also to nature and the non-human world.

Bypassing such crude divisions implies leaving a world made of discrete and rigid entities to one of relations and openness. Hence, our third proposition is to employ a relational approach to resilience (Chander et al 2020; Schreiber 2021; Darnhofer et al 2016; Schluter et al 2019). There has recently been a shift to more interactional ways to comprehend socio-ecological relations, less focused on states and stability, on entities and their properties or behavioural capacities, and more on the relational networks from which entities emerge and are constantly modelled (Folke 2021; Hertz et al. 2020; West et al. 2020). From this perspective, resilience can no longer be described through a set of over-imposed principles but through the relational context which will impact the outcome of planned intervention. It is no longer the resilience of an entity that matters, but "the resilience of A to B" (Grove 2018: 8), where A and B themselves are not closed units, but assemblages that change over time and are heterogenous.

Only in this way, could resilience properly emphasise change and the wider patterns that enable or constrain change.

In this book we incorporate elements of these perspectives to the study of pastoralist resilience, including contextual, ontological, and relational perspectives (see Chandler 2014; Little and McPeak 2014; McPeak and Little 2017; Krätli 2017). It emphasises that: 'resilience has played a transitional role of shifting from modernist subject-centred perspective (of a strong subject) to a relational ontology of system dynamics (with a relational, embedded subject) through its focus on non-linearity (unexpected outcomes)' (Chandler 2014: 9). Our perspective, therefore, suggests that the complex and non-linear dynamics of pastoralist resilience are embedded in specific contexts and relations that privilege local epistemologies and negate modernist thinking and assumed dichotomies.

Resilience in the context of pastoralism

In a historical context, 'resilience thinking' about pastoralism, especially in terms of rangeland ecology, dates to the 1980s (Walker et al. 1981; Ellis and Swift 1988). Researchers from this period drew on the ideas of Holling and his colleagues. However, for general terminology, most researchers of pastoralism and rangeland ecology preferred the concept of 'disequilibrium' from what was termed the 'new ecological thinking', rather than resilience to theorise about pastoral ecologies and societies, especially how they respond to different ecological and human-induced disturbances (Ellis and Swift 1988; Behnke et al. 1993; Scoones 1995, 2004; McCabe 2004). Since the 2000s, researchers have applied the notion of resilience more explicitly in studies of pastoralism. In this respect, Niamir-Fuller (1998) published one of the first studies focused specifically on the concept of pastoralist resilience. She suggests that the mobility characteristics of pastoralism, comprising seasonal migration of herds and rotational grazing often across vast territories, are the main contributing factors explaining the resilience of pastoralist communities. Recent studies of pastoralism and resilience shifted the target from rangeland ecologies to consider more social and cultural factors. For example, Robinson and Berkes

(2010) applied the '*threshold*' concept of resilience theory to the analysis of the social-ecological system of the Gabra in north-central Kenya. Leslie and McCabe (2013), in turn, argue that '*response diversity*' (RD), a concept that originated in ecology, can be employed to understand the resilience of the Turkana (Kenya) and Maasai (Tanzania) social-ecological systems. In a more recent project, Anderson and Bolling (2017) analyse the environmental and social histories of the Lake Baringo-Bogoria Basin, Kenya by applying Gunderson and Holling's (2002) resilience cycle model of '*panarchy*'. This approach allowed the authors to examine the interactions of dynamic social and ecological relations in the basin over temporal and geographic scales. Although the application of resilience thinking for each of these scholars varies, the study of resilience in pastoralism has clearly entered a new phase of increased emphasis on the resilience concept and related theory.

The adaptation of 'resilience-thinking' to the field of pastoralism is significant. The 'contextual approach' to resilience that we advocate differs slightly from those mentioned above due to our greater attention to the dynamic nature of pastoralism itself. Instead of applying resilience as a general and neutral term to the field of pastoralism, resilience theory should be reformulated to account for the dynamic context of pastoralism, including its social dimensions. This does not mean a denial of resilience as a valuable concept – far from it. What we propose is that the combination of both 'resilience-thinking' (applying the resilience concept to specific contexts) and 'thinking-resilience' (reconfiguring the resilience concept to accommodate specific contexts) will enrich the study of resilience in pastoralism.

Our contextual approach to pastoralist resilience contributes three important perspectives. The first is to see resilience not as something pre-existing among pastoral communities but something constructed through interactions (relations) between the community and external actors. Roe's (1994) 'narrative policy' approach is helpful here by showing how policymakers (external actors) try to stabilise assumptions and concepts in situations of considerable uncertainty, interdependence and disagreement (1994: 34). While Roe did not specifically address resilience in this work, the concept can be added to the list of development policy narratives that policymakers apply to pastoral development. This perspective makes it possible to rethink resilience not as a mere academic concept but also as a policy narrative that finances, mobilises and even creates the lived realities of pastoralists and their communities.

It is also necessary to acknowledge that the ontology of resilience thinking is found among external actors and institutions rather than originating in pastoral communities. While it may be possible to search for equivalent vernacular word(s) among pastoral communities (Galvin et al. 2013; Liao and Fei 2015, 2016), resilience thinking is unlikely to be part of the knowledge systems of pastoralists. Rather, one should look for the origins of the resilience concept for pastoralists based on their interactions with external actors, including us – the researchers.

The second perspective that the contextual approach contributes to understandings of pastoralist resilience is the recognition that boundaries between pastoral communities and external actors cannot be, if they ever were, analytically demarcated. Instead, they are intricately intertwined in ways that defy rigid dichotomies. This complex interrelationship challenges the common narrative that local communities have been passively affected by external factors, including policy narratives, or that local actors are unable to effectively cope with these externalities. As Chandler puts it, 'resilience is thereby both about adapting to the external world and about being aware that in this process of adaptation the world is being reshaped' (2014: 7). If we question resilience under the current complex context of pastoralism, the complex interrelationships between inside (local) and external actors need to be carefully examined and simple dichotomies avoided.

The third framing highlights the glaring disconnect in the realities of resilience among pastoral communities and those of outside agents and events. This gap also can be found within certain pastoral communities differentiated by age, class, gender and other social factors. Accordingly, to generalise about 'pastoral resilience' (Little and McPeak 2017: 18) may mask important perspectives and experiences for sub-groups within the population of pastoral areas. The wide gap in the understanding of what resilience implies for local (inside) actors and for development agencies (external) and their policies and schemes has been noticed by some anthropologists (Crane 2010; Oliver-Smith 2017). Certainly, pastoral communities demonstrate resilience in their actions regardless of how it is interpreted and, thus, the concept has proved to be increasingly useful for current research on pastoralism in East Africa (Robinson and Berkes 2010; Anderson and Bolling 2017). In short, pastoralists are far more resilient than normally assumed by development and humanitarian agencies (Niamir-Fuller 1998; Fratkin and McCabe 1999; Fratkin 2001; McCabe 2002; Scoones 2004; Little and McPeak 2014; Catley 2017; Andeson and Bolling 2017). By acknowledging that resilience is constructed through interactions between pastoralists themselves and external actors, it is demonstrated that their lived realities and perceptions of resilience are not the same. Understanding the differences and glaring gaps between them can provide clues and a meaningful counter-narrative to guide more effective humanitarian and development planning than is usually the case.

An overview of the chapters

This book comprises twelve chapters which shed new theoretical and ethnographic light on various aspects of resilience and resilience-related phenomena in Africa, predominantly among African pastoralists, as well as agro-pastoralists, cultivators, and city dwellers. Although ethnographic research on East African pastoralists comprises the primary contents of the book, the authors hail from a variety of disciplines: anthropology, area and development studies, international relations, international politics, and development economics, as resilience requires an inter-disciplinary approach.

Part I addresses the political economy of resilience from a global perspective. It provides a bird's-eye view of the backdrop of resilience and resilience-related phenomena in Africa from the perspective of development economics and international politics. The chapter by Shimada raises a fundamental question for the book: does aid contribute to making Africa resilient? Using panel data from several African countries, the author assesses the damage caused by natural disasters due to climate change and the extent to which aid can offset such damage. Shimada found that official development assistance (ODA) for disaster prevention and mitigation, humanitarian and emergency ODA have small impact coefficients in reducing deaths, and that particularly food aid needs to be re-evaluated because of its crowding-out effect on domestic production. His insights remind us that aid and resilience in Africa is not a simple input-output issue; and that using a resilience perspective can illuminate answers to unforeseen questions.

Resilience and ODA is also discussed by Enomoto, but from the perspective of international politics. It examines how the dominant discourse in the development and humanitarian sector has been transformed since the mid-twentieth century, and attempts to understand how such transformations have led to the widespread use of the term 'resilience'. She then presents some of the criticisms of the term, and outlines the potentials and difficulties of employing such a concept in the current criticism. Her insights offer a cautious view on the use of the term, considering the difficulty of avoiding the trap of legitimising and normalising Northern-centric, Western-centric or racist structures and discourses underlying current humanitarian and development sectors. This chapter provides a critical view of resilience that acknowledges the power dynamics between the global north and south.

Shifting the perspective from macro to micro, Part II, 'Resilience through livelihood diversification,' marks a primary contribution of the book: ethnographic case studies on the resilience of African pastoralists. The view that pastoralism alone can ensure a pastoralist livelihood is clearly obsolete in contemporary settings. Pastoralism must be considered within broad relationships and contexts of wholistic livelihood making, of which pastoralism is only one part. The two chapters in Part II demonstrate aspects of East African pastoralists' resilience through livelihood diversification. In the chapter by Little, the concept of resilience is used to examine livelihood and asset diversification among Il Chamus of Baringo County, Kenya from 1980 to 2018. By addressing diversification trends among different groups of households, the chapter argues that both better-off and poor households, headed either by females or males, pursue non-pastoral strategies and assets, but opportunities differ greatly along gender and class lines. Lucrative diversification options are increasingly town-based and available only to a small percentage of better-off households, while for poor families they are a survival strategy as many are unable to rely on pastoralism for their livelihood. By highlighting the importance of identities and relationships sustained over time and across different rural and urban spaces, the chapter shows that being a 'resilient' Il Chamus means more than owning a large herd of livestock.

The chapter by Sagawa, in turn, also sheds light on livelihood diversification of East African pastoralists, albeit through a more cultural framework. Sagawa points out that pastoralists diversify their livelihood not only out of economic necessity, but also for cultural value and to maintain social relations. In turn, the diversification process affects their cultural values and social relations. Sagawa examines how the youth of Daasanach in southwestern Ethiopia legitimise their choice to be involved in fishing activities, which is considered a low status activity according to their cultural values. They recognised that highlanders were invading their territory and attempted to maintain a livestock-centred livelihood through fishing. Employing the relational approach, he concludes that as the social and political contexts surrounding pastoralism have changed, local evaluations and attitudes towards other activities have also been altered.

The aforementioned chapters focused primarily on the political economy of resilience. In Part III, we turn to the socio-cultural aspects of resilience. In particular, the role of identity as assuring socio-cultural coherence and endurance even when external environmental factors are incessantly shifting. Using a similar cultural framework to Sagawa, Semplici reconsiders an overlooked dimension of cultural resilience, 'resilience and identity', through an ethnographic examination of the '*raiya*' concept among Turkana herders in Northern Kenya. Semplici argues that a part of their resilience stems from a sense of belonging and solidarity centred on a collective identity built in opposition to urbanities along symbolic boundaries. Moreover, she demonstrates how such an identity remains flexible and responsive to change, disrupts dichotomies and weaves together different social worlds, such as rural and urban. She concludes that resilience is found in the capacity for change and in remaining open and flexible with regard to decisions and practices. It rests in being mobile, and thus serves to navigate change and different cultural environments.

Hassan's chapter examines resilience strategies adopted by Samburu women under changing land tenure and heightened subdivision in certain Samburu rangelands after the Kenyan land reforms of 2016, with a focus on their gender dimensions. She presents different resilience strategies adopted by women, which reflect the options available for pastoralist households and the role women play to adjust to disruptions and new lifestyles. By focusing on Samburu women, she highlighted how land tenure changes customary systems, and that resource pressures are related to the ways groups take uncertainty into account. She also observes that identity might play an important role in the inclusion of landless women in the community. Hassan demonstrates that the gendered nature of resilience reflects how various relationships among pastoralists fluctuate, responding to shifts in the community and land tenure. This finding constitutes an important contribution to the relational and contextual approach to resilience.

Changing our gaze from the daily livelihood of the internal community to the emergency period of nation-state intervention, Part IV focuses on displaced pastoralists and refugee pastoralists in East Africa during and after conflicts and subsequent state interventions. Conflict tremendously damages all aspects of livelihood of pastoralists, and in some cases wholly destroys pastoralism. Examining how pastoralists and ex-pastoralists have recovered is crucial for understanding pastoralist resilience. The first chapter by Konaka demonstrates the potential of a contextual and relational approach to resilience with reliability theories under the mid-and post-conflict settings. Konaka elaborates on the 'resilience of pastoralism' in a situation where survival is at stake. The chapter presents an ethnographic case study of a series of conflicts in Kenya between the Samburu and the Pokot that emerged in 2004. It explores the theoretical possibilities of Roe's reliability pastoralist theory. In the conflict case study, we can observe how clustered settlements and inter-ethnic mobile phone networks improvised real-time management of 'reliability professionals' (in this case, pastoralists). The case study illustrates that pastoralists' livelihoods were secured during these crises by an improvised real-time management effort which significantly disregarded outside stable norms and cultural values. This case illustrates how pastoralists dealt with the conflict as reliability professionals, as well as provides a new way of thinking about the 'resilience of pastoralism' beyond conventional thinking and disciplinary borders.

Another chapter by Konaka explores contextualising the concept of resilience in relation to the material culture of East African pastoralists and humanitarian assistance through the ethnographic case studies of the Samburu, Tugen, and Il Chamus in Northern Kenya – who were displaced after conflict with the Pokot in 2004. Using inventory survey data, the study identifies the possessions of displaced people. The survey also revealed the necessary items they carry when fleeing, which constitute a 'minimum set of possessions.' His observation that people who have lost everything move towards recovery by restoring a minimum set of possessions that has been regarded as a part of their bodies provides a starting point of resilience that cannot be overemphasised. The findings suggest the need to redefine the framework of humanitarian assistance for East African pastoralist communities whose lives are dominated by uncertainty and unpredictability, and to rethink humanitarian assistance that is more rooted in cultural dignity.

The subsequent chapter by Hazama focuses on pastoralist citizenship in Karimojong and Dodoth in Uganda, which widens the scope of resilience to include broader social science arguments. Karimojong and Dodos were oppressed by the Ugandan state, resulting in strong regional tensions. Established by displaced pastoralists, the collaborative camp is a substitute for the dysfunctional citizenship of the nation-state. Hazama outlines a multi-species ethnographic case of pastoralists. He considers resilience to be the ability to regain well-being – which may have developed uniquely in a human-domestic animal relationship that is found in pastoralist societies – by referring to how Karamoja pastoralists resist oppressive policies and organise citizenship-related practices. Hazama interprets animal behaviour to suggest that pastoralists recognise the political subjectivity of animals, thereby identifying them as co-citizens. The chapter concludes that the resilience of the pastoralist society is embedded in a man-animal identity and social relationship.

Finally, Part V seeks to broaden our scope from pastoralists to include farmers and city dwellers and their different relationship with pastoralists through a comparative perspective. The research site shifts from East Africa to West and Southern Africa. The chapter by Gonzales depicts a completely different conception of 'urbanization' or 'urban-rural relations' than what is conventionally used. She demonstrates how practices of mobility and immobility become resourceful strategies for a nomadic pastoralist population to embrace emerging possibilities in the urban space, drawing on case studies of several families of Kel Tamasheq (aka Tuareg) in northern Mali. To discuss mobile/immobile practices in Bamako, Gonzales analyses visiting patterns among families in urban areas through narratives of collective identity while referencing the concept of patience. Resilience is considered as the capacity to switch from being mobile to being immobile depending on available possibilities, interests, emotional attachments, and so on. Embedded as a feature of their collective nomadic, pastoralist, and Muslim identity, patience provides a continuity in representations of Kel Tamasheq collective identity across the urban-rural continuum despite switches from mobility to immobility.

The chapter by Sakamoto advances the conventional understanding of contemporary farmer-herder conflicts in the Sahel. He analyses processes and patterns of crop damage that involve pastoral Fulbe, Tuareg herders, and Hausa farmers in southwestern Niger. First, he estimates patterns of land cover and land use in the study area through the combined uses of field data and satellite images. The data reveal significant spatial constraints that herders now face with respect to access to resources. Second, he analyses GPS records of Fulbe pastoralists' daily herding activities. Through his analyses, he demonstrates that the daily movements of Fulbe and Tuareg pastoralists seem to be constantly adjusted, if not optimised, to ensure access to the patchy resources thinly spread over the area. Their grazing strategies avoid encroaching on cropland as much as possible, which is a sign of the resilience of local pastoralists. He suggests that, on a wider spatial scale, there are certain structural limitations to these efforts, such as 'hotspots' with relatively abundant vegetation resources and surface water which trigger farmer-herder conflict. His analysis also suggests that pastoralists' traditional coping mechanism, namely mobility, has been greatly disrupted under the weight of enormous social and ecological pressures. His findings remind us that pastoralism occupies just a small part of regional networks constituted by various actors and resilience must be considered in the broader context.

The final chapter does not focus on pastoralists; but instead provides for a comparison between pastoralists and ex-refugee farmers in Zambia. Murao clarifies how the refugee peasants – Mbunda – reconstructed their livelihood based on agriculture after the implementation of local integration and resettlement projects through which the Mbunda people were given legal status and land rights to live and cultivate land in Zambia. Murao focussed on the Mbunda reorganised their social relations both in the refugee settlements and in the resettlement schemes to examine the internal support that underpinned their resilience. In comparison with East African pastoral societies, African farmer societies have chiefs as traditional authorities who control land distribution. However, Mbunda operated a flexible land management system where access to land and labour was determined by the matrilineal kinship principle. They had not greatly depended on the livelihoods based on privatised land rights as residents because of their mutual relationships based on the matrilineal kin group. Mbunda's resilience has developed owing to internal relations of the former refugees under land privatisation for refugee schemes. Therefore, the resilience of African farmers to land privatisation may not be so different to African pastoralists as is normally assumed owing to the individual or chiefship dichotomy – as both Hassan and Murao identify communal ties, clanship, kinship acting as buffers against hardship.

We do not intend to summarise all the arguments of each chapter here. Instead, the epilogue by Scoones provides the theoretical perspective and elaboration of this book. Although the topics, approaches, and ethnographic locations of each chapter in this book vary, each provides new insights for reconsidering and reforging the concept of resilience. The catch-all, muddled usage of the resilience concept is to be avoided; resilience varies according to relationship, context, and location. That is precisely what the relational and contextual approach to resilience was originally intended to achieve. In fact, each chapter revealed various unconsidered aspects of resilience in Africa beyond universal and neutral models of resilience: political power, state intervention, economic inequality, gender, human-animal relations, technology, regional networks, farmer-herder relations, cultural identity, pastoralist's values, and so on. We hope that reconsidering the concept of resilience from various perspectives and research locales will lead to new pathways of understanding or practicing resilience based on the various lived realities and surrounding networks in this complex and precarious world.

Acknowledgments

This book builds on a panel titled 'Pastoralists and Resilience: Rethinking the Inside and Outside Perspectives of the Pastoral Communities' held at the IUAES 2019 inter-congress in Poznan convened by Shinya Konaka and Saverio Krätli. The authors are grateful to the conveners and participants in the panel.

The Introduction, chapters 4, 5, 6, 8, and 11 are revised from articles in the Special Issue on 'Rethinking Resilience in the Context of East African Pastoralism' of *Nomadic Peoples* 25 (2) edited by Shinya Konaka and Peter D. Little. The authors would like to thank the White Horse Press for permitting the articles to be reprinted.

The publication was supported by the JSPS KAKENHI Grants JP18H03606.

The authors' individual acknowledgements are included at the end of the respective chapters.

References

Adger, W.N. 2000. 'Social and ecological resilience: Are they related?'. *Progress in Human Geography* 24 (3): 347–364. https://doi.org/10.1191/030913200701540465

Allen, C.R., D.G. Angeler, A.S. Garmestani, L.H. Gunderson and C.S. Holling. 2014. 'Panarchy: Theory and application'. *Ecosystems* 17: 578–589. https://doi.org/10.1007/s10021-013-9744-2

Anderson. D. M. and M. Bolling. (eds). 2017. *Resilience and Collapse in African Savannahs: Causes and consequences of environmental change in East Africa.* London: Routledge.

Behnke, R.H. Jr., I. Scoones and C. Kerven. (eds). 1993. *Range Ecology at Disequilibrium: New models of natural variability and pastoral adaption in African savannas.* London: Overseas Development Institute.

Brand, S.B. and K. Jax. 2007. 'Focusing the meaning(s) of resilience: Resilience as a descriptive concept and a boundary object'. *Ecology and Society* 12 (1): 23. https://doi.org/10.5751/ES-02029-120123

Carpenter, S., B. Walker, J. M. Anderies, and N. Abel. 2001. 'From metaphor to measurement: Resilience of what to what?'. *Ecosystems* 4: 765–781. https://doi.org/10.1007/s10021-001-0045-9

Catley, A. 2017. *Pathways to Resilience in Pastoralist Areas: A synthesis of research in the Horn of Africa.* Boston: Feinstein International Center, Tufts University.

Cervigni, R. and M. Morris. (eds). 2016. *Confronting Drought in Africa's Drylands: Opportunities for enhancing resilience.* Washington: International Bank for Reconstruction and Development, The World Bank. https://doi.org/10.1596/978-1-4648-0817-3

Chandler, D. 2014. *Resilience: The governance of complexity.* London: Routledge. https://doi.org/10.4324/9781315773810

Chandler, D. and J. Coaffee. (eds). 2017 *The Routledge Handbook of International Resilience.* London: Routledge. https://doi.org/10.4324/9781315765006

Crane, T., A. 2010. 'Of models and meanings: Cultural resilience in social-ecological systems'. *Ecology and Society* 15 (4): 19. https://doi.org/10.5751/ES-03683-150419

Davies, J., L.W. Robinson and P. J. Ericksen. 2015. 'Development process resilience and sustainable development: Insights from the drylands of Eastern Africa'. *Society and Natural Resources* 28: 328–343. https://doi.org/10.1080/08941920.2014.970734

Ellis, J.E. and D.M. Swift. 1988. 'Stability of African pastoral ecosystems: Alternate paradigms and implications for development'. *Journal of Range Management* 41: 450–459. https://doi.org/10.2307/3899515

FAO (Food and Agriculture Organization of the United Nations). 2018. *Pastoralism in Africa's Drylands.* Rome: FAO.

Folke, C. 2006. 'Resilience: The emergence of a perspective for social–ecological systems analyses'. *Global Environmental Change* 16: 253–267. https://doi.org/10.1016/j.gloenvcha.2006.04.002

Folke, C., S.R. Carpenter, B. Walker, M. Scheffer, T. Chapin and J. Rockstrom. 2010. 'Resilience thinking: Integrating resilience, adaptability and transformability'. *Ecology and Society* 15 (4): 20. https://doi.org/10.5751/ES-03610-150420

Fletcher, D. and M. Sarkar. 2013. 'Psychological resilience: A review and critique of definitions, concepts, and theory'. *European Psychologist* 18 (1): 12–23. https://doi.org/10.1027/1016-9040/a000124

Fratkin, E. and J.T. McCabe. 1999. 'Introduction' to a Special Issue of *Nomadic Peoples* 'East African Pastoralism at the Crossroads'. *Nomadic Peoples* 3 (2): 5–15. https://doi.org/10.3167/082279499782409442

Fratkin, E. 2001. 'East African pastoralism in transition: Maasai, Boran, and Rendille cases'. *African Studies Review* 44 (3): 1–25. https://doi.org/10.2307/525591

Galvin, K., R. Reid, D. Nkedianye, J. Njoka, J.R. de Pinho, D. Kaelo, and P.K. Thornton. 2013. 'Pastoral transformations to resilient futures: Understanding climate from the ground up'. *Research Brief* April: 1–4.

Garmezy, N. 1971. 'Vulnerability research and the issue of primary prevention'. *American Journal of Orthopsychiatry* 41 (1): 101–116. https://doi.org/10.1111/j.1939-0025.1971.tb01111.x

Gunderson, L. and C.S. Holling. (eds). 2002. *Panarchy: Understanding transformations in human and natural systems.* Washington, DC: Island Press.

Headey, D., A.S. Taffesse and L. You. (eds). 2012. *Enhancing Resilience in the Horn of Africa: An exploration into alternative options.* Development Strategy and Governance Division. Washington: International Food Policy Research Institute.

Holling, C.S. 1973. 'Resilience and stability of ecological systems'. *Annual Review of Ecology and Systematics* 4: 1-23. https://doi.org/10.1146/annurev.es.04.110173.000245

———— 1996. 'Engineering resilience versus ecological resilience'. In National Academy of Engineering (ed.), *Engineering Within Ecological Constraints.* Washington, DC: The National Academies Press. pp. 31–44. https://doi.org/10.17226/4919

———— 2001. 'Understanding the complexity of economic, social and ecological systems'. *Ecosystems* 4: 390–405. https://doi.org/10.1007/s10021-001-0101-5

Klein, R.J.T., R.J. Nicholls and F. Thomalla, 2003. 'Resilience to natural hazards: How useful is this concept?' *Environmental Hazards* 5: 35–45. https://doi.org/10.1016/j.hazards.2004.02.001

Konaka, S. 2017. 'Introduction: The articulation-sphere approach to humanitarian assistance to East African pastoralists'. *African Study Monographs, Supplementary Issue* 53: 1–17. https://doi.org/10.14989/218918

———— 2018. 'Livestock as interface: The case of the Samburu in Kenya'. In I. Tokoro and K. Kawai. (eds). *An Anthropology of Things.* Melbourne: Trans Pacific Press. pp. 241–257.

Krätli, S. 2017. 'Pastoral localization of humanitarian aid: The need to re-qualify the pastoral context'. *African Study Monographs, Supplementary Issue* 53: 141–146.

Leach, M., I. Scoones and A. Stirling. 2010. *Dynamic Sustainabilities: Technology, environment, social justice.* London: Routledge. https://doi.org/10.4324/9781849775069

Leslie, P. and J.T. McCabe. 2013. 'Response diversity and resilience in social-ecological systems'. *Current Anthropology* 54 (2): 114–143. https://doi.org/10.1086/669563

Liao, C. and D. Fei. 2015. 'Resilience of what to what? Evidence from pastoral contexts in East Africa and Central Asia'. *Resilience* 4 (1): 14–29. https://doi.org/10.1080/21693293.2015.1094167

———— 2016. 'The operationalization of resilience in pastoral regions in East Africa and Central Asia'. In D. Chandler and J. Coaffee. (eds). *The Routledge Handbook of International Resilience.* London: Routledge. pp. 198–210.

Lind. J. 2016. *Changes in the Drylands of Eastern Africa: Implications for resilience-strengthening efforts.* Sussex: Institute of Development Studies, University of Sussex.

Little, P.D. and J.G. McPeak. 2014. 'Resilience and pastoralism in Africa South of the Sahara'. In S. Fan, R. Pandya-Lorch and S. Yosef. (eds). *Resilience for Food and Nutrition Security.* Washington, DC: International Food Policy Research Institute. pp. 75–82.

MacKinnon, D. and K.D. Derickson. 2012. 'From resilience to resourcefulness: A critique of resilience policy and activism'. *Progress in Human Geography* 37 (2): 253–270. https://doi.org/10.1177/0309132512454775

Manyena, S.B. 2006. 'The concept of resilience revisited'. *Disasters* 30: 433–450. https://doi.org/10.1111/j.0361-3666.2006.00331.x

Masten, A.S. and A.H. Gewirtz. 2006. 'Resilience in development: The importance of early childhood'. In R.E. Tremblay, R.G. Barr and R. DeV. Peters. (eds). *Encyclopedia of Early Childhood Development.* Montreal: Centre of Excellence for Early Childhood Development. pp. 1–6.

Masten A.S. 2014. 'Global perspectives on resilience in children and youth'. *Child Development* 85: 6–20. https://doi.org/10.1111/cdev.12205

McCabe, J.T. 2002. 'The role of drought among the Turkana of Kenya'. In A. Oliver-Smith and S. Hoffman. (eds). *Culture and Catastrophe.* Santa Fe: School of American Research. pp. 213–236.

———— 2004. *Cattle Bring Us to Our Enemies: Turkana ecology, politics, and raiding in a disequilibrium system.* Michigan: University of Michigan Press. https://doi.org/10.3998/mpub.23477

McPeak, J.G. and P.D. Little. 2017. 'Applying the concept of resilience to pastoralist household data'. *Pastoralism: Research, Policy and Practice* 7 (14): 1–18. https://doi.org/10.1186/s13570-017-0082-4

Niamir-Fuller, M. 1998. 'The resilience of pastoral herding in Sahelian Africa'. In F. Berkes and C. Folke. (eds). *Linking Social and Ecological Systems: Management Practices for Building Resilience.* Cambridge: Cambridge University Press. pp. 250–284.

Norris, F.H., S.P. Stevens, B. Pfefferbaum, K.F. Wyche and R.L. Pfefferbaum. 2008. 'Community resilience as a metaphor, theory, set of capacities, and strategy for disaster readiness'. *American Journal of Community Psychology* 41: 127–150. https://doi.org/10.1007/s10464-007-9156-6

Oliver-Smith, A. 2017. 'The concepts of adaptation, vulnerability, and resilience in the anthropology of climate change'. In H. Kopina and E. Shoreman-Ouimet. (eds). *Routledge Handbook of Environmental Anthropology.* London: Routledge. pp. 116–136. https://doi.org/10.4324/9781315768946-17

Olsson, L., A. Jerneck, H. Thorén, J. Persson and D. O'Byrne. 2017. 'A social science perspective on resilience'. In D. Chandler, and J. Coaffee. (eds). *The Routledge Handbook of International Resilience.* London: Routledge. pp.49-62.

Panter-Brick, C. 2014. 'Health, risk, and resilience: Interdisciplinary concepts and applications'. *Annual Review of Anthropology* 43: 431–48. https://doi.org/10.1146/annurev-anthro-102313-025944

Panter-Brick, C., K. Hadfield, R. Dajani, M. Eggerman, A. Ager and M. Ungar. 2018. 'Resilience in context: A brief and culturally grounded measure for Syrian refugee and Jordanian host-community adolescents'. *Child Development* 89 (5): 1803–1820. https://doi.org/10.1111/cdev.12868

Pavanello, S. 2009. *Pastoralists' Vulnerability in the Horn of Africa: Exploring political marginalization, donors' policies, and cross-border issues, literature review.* London: Humanitarian Policy Group (HPG) Overseas Development Institute.

Resilience Alliance. 2010. *Assessing Resilience in Social-ecological Systems: Workbook for practitioners.* Version 2.0. Halifax: Resilience Alliance.

Robinson, L.W. and F. Berkes. 2010. 'Applying resilience thinking to questions of policy for pastoralist systems: Lessons from the Gabra of Northern Kenya'. *Human Ecology* 38: 335–350. https://doi.org/10.1007/s10745-010-9327-1

Roe, E. 1994. *Narrative Policy Analysis: Theory and practice.* Durham: Duke University Press. https://doi.org/10.1515/9780822381891

————— 2020. *A New Policy Narrative for Pastoralism? Pastoralists as reliability professionals and pastoralist systems as infrastructure,* STEPS Working Paper 113. Sussex: STEPS Centre.

Rutter, M. 1987. 'Psychosocial resilience and protective mechanisms'. *American Journal of Orthopsychiatry* 57 (3): 316–331. https://doi.org/10.1111/j.1939-0025.1987.tb03541.x

Scoones, I. (ed.). 1995. *Living with Uncertainty: New directions in pastoral development in Africa.* London: International Institute for Environment and Development. https://doi.org/10.3362/9781780445335

Scoones, I. 2004. 'Climate change and the challenge of non- equilibrium thinking'. *IDS Bulletin* 35 (3): 114–119. https://doi.org/10.1111/j.1759-5436.2004.tb00144.x

Southwick, S.M., G.A. Bonanno, A.S. Masten, C. Panter-Brick and R. Yehuda. 2014. 'Resilience definitions, theory, and challenges: Interdisciplinary perspectives'. *European Journal of Psychotraumatology* 5: 25338. https://doi.org/10.3402/ejpt.v5.25338

Ungar, M. 2008. 'Resilience across cultures'. *British Journal of Social Work* 38: 218–235. https://doi.org/10.1093/bjsw/bcl343

Van der Leeuw, S.E. and C.A. Leygonie 2000. 'A Longterm Perspective on Resilience in Socionatural Systems'. Paper Presented at the Workshop on System Shocks–System Resilience, Abisko. pp. 22–26.

Waller, A. 2001. 'Resilience in ecosystemic context: Evolution of the concept'. *American Journal of Orthopsychiatry* 71 (3): 290–297. https://doi.org/10.1037/0002-9432.71.3.290

Walker, B.H., D. Ludwig, C.S. Holling and R.M. Peterman. 1981. 'Stability of semi-arid savanna grazing systems'. *Journal of Ecology* 69: 473–498. https://doi.org/10.2307/2259679

Walker, B.H., C.S. Holling, S.R. Carpenter and A. Kinzig, 2004. 'Resilience, adaptability and transformability in social-ecological systems'. *Ecology and Society* 9 (2): 5. https://doi.org/10.5751/ES-00650-090205

Walker, B.H., J.M. Anderies, A.P. Kinzing and P. Ryan. 2006. 'Exploring resilience in social-ecological systems through comparative studies and theory development: Introduction to the Special Issue'. *Ecology and Society* 11 (1): 12. https://doi.org/10.5751/ES-01573-110112

Werner, E.E. and R.S. Smith. 1982. *Vulnerable but Invincible: A study of resilient children*. New York: McGraw-Hill.

Weichselgartner, J. and I. Kelman. 2015. 'Geographies of resilience: Challenges and opportunities of a descriptive concept'. *Progress in Human Geography* 39 (3): 249–267. https://doi.org/10.1177/0309132513518834

Political
Economy of
Resilience
from a Global
Perspective

CHAPTER

01

Does Aid Make Africa Resilient? Disasters' impacts on economic growth, agriculture, and conflicts[1]

Go Shimada

Introduction

The resilience of African societies today is crucially dependent on climate change because it causes unpredictable disasters, such as drought. Those disasters disrupt their lives, for instance, decreasing agricultural production and killing nomadic livestock. To cope with the disruptions, building African climate resilience is essential.

Every year, irregular and extreme weather events are reported worldwide. In 2021, Europe experienced an oppressive heatwave and was hit by devastating floods in July (Centre for Research on the Epidemiology of Disasters, 2021). The magnitude of destruction was severe in Germany and Belgium, causing several deaths. In Asia, tropical cyclones have become stronger (Peduzzi et al., 2012). Also, Japan has recorded torrential rainfall, flooding, and landslides more often than before.

Climate change is a worldwide phenomenon, but African countries are disproportionately punished, even though they contributed the least

---

1 This chapter is an edited version of the following open-access paper:

Shimada, Go. 2022. 'The Impact of climate-change-related disasters on Africa's economic growth, agriculture, and conflicts: Can humanitarian aid and food assistance offset the damage?' *International Journal of Environmental Research and Public Health* 19 (1): 467. https://doi.org/10.3390/ijerph19010467.

towards greenhouse gas emissions compared with developed countries (IPCC 2022; Shimada 2019). As the UNFCCC (United Nations Framework Convention on Climate Change) Earth Summit in Rio de Janeiro in 1992 discussed, Africa has common but differentiated responsibilities (CBDR). Therefore, do donor countries contribute sufficient aid to African countries to promote Africa's adaptive capacity to become resilient? To answer this question, we must understand the extent of the collateral damage of climate change in Africa. This study analyses the damage caused by climate-related natural disasters, such as floods, droughts, and storms, and the extent to which aid can offset damage.[2] The remainder of this paper is organised as follows. Section 1 examines the current damage trends caused by climate-related natural disasters. Section 2 presents the study methods and data. Section 3 discusses the results of analysis. Based on the results, the discussion is presented in Section 4. Finally, conclusions are drawn from the results and discussion.

Background: Climate change and natural disasters in Africa

This section reviews existing literature on the impact of climate change and related natural disasters, focusing on the impacts on GDP per capita, agricultural production, poverty and conflicts, and government effectiveness (Field and Barros, 2014; Bernard et al., 2013; Borras et al., 2022).

GDP per capita

Using satellite images, Tellman et al. (2021) found that most of the population was exposed to floods between 2000 and 2015, especially in Asia and sub-Saharan Africa. The authors also argued that their projections for 2030 indicate that a larger number of people will be exposed to flood threats. Gross domestic product (GDP) per capita helps measure the

---

2 The phenomenon of extreme temperature is not included in this study, as they are not frequent in Africa.

economic and social impact of climate-related natural disasters. There are several studies on the overall impact of natural disasters on economic growth but not specific to climate-related disasters (Kahn 2005; Strömberg 2007; Noy and Vu 2010; E.A. Cavallo and Noy 2009; E. Cavallo et al. 2013). While there is no consensus regarding its impact on economic growth, some found a significantly negative long-term effect (Noy 2009; Shimada 2012, 2015; Benson and Clay 2006; Shimada 2017; Rasmussen 2004). Conversely, some studies show a positive impact on economic growth as disasters promote the "Schumpeterian creative destruction" process (Albala-Bertrand 1993; Dacy and Kunreuther 1969; Toll and Leek 1999; Skidmore and Toya 2002; Sawada, Bhattcharyay, and Kotera 2011). According to this view, disasters promote innovation and investment, destroying existing practices, products, or services. Other research reveals mixed results on economic growth due to natural disasters (Charvériat 2000; Hochrainer 2009). Therefore, it remains necessary to clarify the effect of climate-related disasters in Africa.

Agricultural production and aid

The socioeconomic impact on agriculture and food security have been identified as critical sectors in the contemporary era of change (Vermeulen et al., 2012). Lesk, Rowhani, and Ramankutty (2016) found that droughts have reduced national cereal production by 9–10% in terms of impact on crop production. However, they found no statistically significant effect of floods. Nonetheless, the question arises if this is truly a global phenomenon or specific to Africa.

Africa is particularly vulnerable to climate change, as the region's adaptive capacity has certain constraints (Connolly-Boutin and Smit 2016; Shimada and Motomura 2017; Higuchi and Shimada 2019; Hosono, Page, and Shimada 2020; Shimada and Sonobe 2021). Notably, For African farmers, many of whom are small-scale farmers, obtaining the skills and knowledge to adapt to the damage caused by global warming and natural disasters is a difficult challenge (Phiiri, Egeru, and Ekwamu 2016). As there is limited empirical data, this study examines the impact of climate-related disasters on African agriculture because this sector is more critical for people's livelihoods than other sectors. Therefore, if disasters decrease agricultural production, poverty increases in rural and urban areas.

In this regard, the literature does not consider social variables. Agricultural production is not determined solely by disasters. Other factors, such as farmers' human capital and market demand for agricultural commodities, are essential production factors (Schultz 1964, 1961). The prior literature does not control these social variables and may produce biased results. Therefore, this study estimates the impact of climate-related disasters by controlling for these factors.

A critical control variable is disaster relief provided by international donors, including the United Nations (UN), World Bank, and bilateral governments. Abundant literature is available on disaster relief based on case studies (Cuny 1994). In contrast, there are few quantitative studies, especially on how disaster relief mitigates damage or contributes towards recovery (Raschky and Schwindt 2009; De Juan, Pierskalla, and Schwarz 2020). Raschky and Schwindt (2009) found that increased foreign aid resulted in higher fatality rates. According to De Juan, Pierskalla, and Schwarz (2020), international aid increases social strife rather than decreasing it, promoting a new conflict over the distribution of resources.

Some studies have shown that aid increased after a disaster (Strömberg 2007; Becerra, Cavallo, and Noy 2014).[3] However, none applied an aid disaggregation approach to examine the impact of different aid types. These studies only used aggregated aid data. This is challenging because aid has a distinct impact. For instance, aid for primary education shows different results from assistance provided for infrastructure or humanitarian food aid. If this study uses aggregated aid data, it may mislead us to incorrect policy recommendation. For instance, assistance for infrastructural growth tends to expand in budgets compared to agricultural development aid. It is essential to distinguish between various aid types, but there is a

---

3 There are some studies on the impact of domestic government aid to mitigate the damage, but not international aid (See for example, Andor, Osberghaus, and Simora 2020).

research gap in this regard. There is no study on whether humanitarian aid, food aid, and other assistance forms have mitigated the damage caused by natural disasters.

Impact on poverty and armed conflict

Regarding the impact on poverty, Kahn (2005) found that the Gini coefficient is positively correlated with disaster-related deaths. Barrett (2010) discussed how food insecurity is associated with sudden catastrophe-like disasters and chronic poverty. Damage due to disasters is of two types: direct and indirect. Notably, in direct damage, the disaster itself kills people. In indirect damage, there are cases where people die not because of the disaster itself but due to displacement (e.g., losing their job after the disaster) and resultant poverty. Therefore, it is essential to measure the impact of such indirect consequences.

There is no consensus in the literature on the natural disaster–conflict nexus. Some studies found a link, whereas others did not. For instance, O'Loughlin, Linke, and Witmer (2014) studied the link between climate variability and armed conflicts and found that, in general, extremely high temperatures are associated with greater conflict levels. They also found that the link varies depending on the conflict type and different subregions of Africa. Marshall B. Burke et al. (2009) found a strong historical association between civil war and temperature in Africa, indicating that by 2030, armed conflict is likely to increase by approximately 54%. However, Buhaug (2010) argued against this nexus, reporting that the incidence of armed conflict has declined in Africa since 2002 despite rising temperature levels. In response, Burke et al. (2010) stated that there are some econometric issues in Buhaug (2010).[4]

These studies examined the temperature-conflict nexus. However, this nexus does not have a direct causal relationship. There is an indirect link between temperature and conflict. High temperatures cause crop damage when temperatures are above 30 °C or the average temperature is above

---

4 Burke acknowledges that the climate-conflict still stands, but the nexus has weakened since 2002.

25 °C for a prolonged period. Tolerance to high temperatures varies among crops. Therefore, high temperatures would have some social impact, such as reducing income. Consequently, such outcomes potentially lead to conflicts. The literature does not consider the indirect causal relationship and treats the temperature-conflict link as a black box. The impact of high temperatures on vegetation is beyond the scope of this study; this research focuses on the impact of disasters on conflicts.

Government effectiveness

Strömberg (2007) studied whether government effectiveness is essential for dealing with disasters. The study used the government effectiveness index, an indicator produced by the World Bank, to test its importance of governance effectiveness. Government effectiveness is essential during disasters. In a crisis, the government needs to handle everything quickly and within a limited period. Even in developed countries, handling crises after disasters is challenging, and sometimes governments fail to cope with them. However, this is likely more difficult for developing countries. This raises the question: how important is government effectiveness in Africa? Strömberg found that government effectiveness reduces the number of people killed by natural disasters globally, not specifically in Africa. This study examines the importance of government effectiveness in mitigating damage caused by climate-related disasters.

This section has reviewed current trends in climate-related natural disasters in Africa. It examined the literature on the impact of disasters on GDP, agriculture, poverty and conflict, and government effectiveness.

First, based on the identified research gaps, the next section examines the impact of climate-related natural disasters and international aid on GDP per capita. Second, the disaster-agriculture nexus is discussed, focusing on major crops. Third, the impact on poverty and conflict is tested. Finally, factors contributing to decreasing (or increasing) the death toll due to natural disasters were tested. One of the factors examined is international aid extended to African countries.

Methods and Data

This section describes the analytical framework for empirical analysis. This study tested three aspects of climate-related disasters: (1) the effect on GDP per capita and agricultural production, (2) the impact on poverty and conflict, and (3) factors contributing towards reducing the impact of disasters.

Research gaps were identified based on the literature review in Section 1. There is no consensus on the impact of natural disasters on GDP. There is also no empirical analysis focusing on the impact of climate-related natural disasters on agricultural production in Africa. Even the global agricultural production literature does not control for other socio-economic conditions. Agricultural production is a part of economic activities; therefore, there is a need to control these variables. Otherwise, there is a potential for result bias. Therefore, it is important to understand the impact of climate-related natural disasters on GDP and agricultural production, considering other socio-economic factors.

The following formula is used to estimate the impact on GDP per capita growth, following the model used by Skidmore and Toya (2002).

$$\Delta(\tfrac{Y}{P})_{i,t} = \alpha_{i,t} + \beta_1\Delta(\tfrac{Y}{P})_{i,t-1} + \beta_2 Dis_{i,t} + \beta_3 Aid_{i,t} + \beta_4 Gov_{i,t} + \beta_5 X_{i,t} + \varepsilon_{i,t} \quad (1)$$

Y denotes GDP, and P represents the population. Therefore, $\Delta(\tfrac{Y}{P})_{i,t}$ is GDP per capita growth. i is a country index to capture country-specific effects, and t is the time (year) index. The lagged GDP per capita growth ($\Delta(\tfrac{Y}{P})_{i,t-1}$) is included because the previous year's growth trend greatly affects the current year's economic activities. $Dis_{i,t}$ is a measure of the impact of disasters specific to country i at time t. This study used the number of people affected by disasters for this variable. This is because these data represent the impact of disasters. The number of occurrences does not necessarily equate to the impact. If a disaster occurs in an uninhabited area, then the impact on human activities is limited, as people do not live there. Aid denotes international aid. This is global official development assistance (ODA) data and not a specific country's ODA. This study uses a different type of aid data, as impact varies depending on the aid type.

For instance, the impact of cereal food aid is not the same as medical aid. Therefore, it is necessary to disaggregate aid data. This study uses the following data: aid for agriculture, humanitarian aid, and cereal food aid. $Gov_{i,t}$ denotes government expenditure. $X_{i,t}$ is the other control variables, including the following variables: education, fertility rate, and government effectiveness index. The education variable represents human capital (Schultz 1961, 1988). This study uses the government effectiveness index developed by the World Bank, which measures the quality of public services, infrastructure, and civil service based on the World Bank's survey (The World Bank 2021).

The following analytical framework will be used to estimate the impacts on agricultural production, reformulating equation 1 to focus on agriculture.

$$Agr_{i,t} = \alpha_{i,t} + \beta_1 Dis_{i,t} + \beta_2 \Delta(\tfrac{Y}{P})_{i,t} + \beta_3 Aid_{i,t} + \beta_4 X_{i,t} + \varepsilon_{i,t} \ (2)$$

$Agr_{i,t}$ denotes the variable for agricultural production. $Dis_{i,t}$, $\Delta(\tfrac{Y}{P})_{i,t}$, and $Aid_{i,t}$ are as per equation (1) above. $\Delta(\tfrac{Y}{P})_{i,t}$ is included because agricultural production is affected by economic activities. $X_{i,t}$ denotes other control variables, such as inequality in educational attainment and government effectiveness. The former represents human capital.

As this estimation uses panel data, three methods are used to assess the results.[5] They estimate fixed effects (FE), random effects (RE), and pooling. Among the three estimation methods, the most appropriate estimation method is determined by the results of the F-test, Hausman test, and the Breusch–Pagan test. The method determined for each model is reported at the bottom of the result tables.

Table 1.1 presents the descriptive statistics of the data used in the empirical study. This dataset covers all African countries (including sub-Saharan and North African countries). This is unbalanced panel data. The period varies depending on the data, and there are gaps in the annual data. This dataset was constructed using the following four datasets.

---

5 This study used Stata/SE 17.0 for the estimation.

First, the Emergency Events Database (EM-DAT) was used for natural disaster-related data (EM-DAT 2021). The EM-DAT is an international disaster database widely used to analyse natural disasters and has been managed by the Centre for Research on the Epidemiology of Disasters (CRED) from 1988 until today. For disasters to be recorded as an extreme event, one of the following criteria must be met: (1) 10 or more people reportedly killed, (2) 100 or more people affected, (3) a declaration of a state of emergency, and (3) a call for international assistance.

Second, the WDI (World Development Indicators) dataset is compiled by the World Bank (The World Bank 2021). Third, the Food and Agriculture Organisation Corporate Statistical Database (FAOSTAT) is a dataset on agricultural production provided by the Food and Agriculture Organisation (FAO) (Food Agriculture Organisation of the United Nations 2021). Finally, the V-dem is used, produced by the Varieties of Democracy Project on government effectiveness and educational inequality data (Coppedge et al. 2019).

Five types of variables are used as measures of disaster impact. The four variables represent the number of people affected by the following: (1) climate-related disasters (aggregate variable), (2) droughts, (3) floods, and (4) storms. The variable of climate-related disasters is the aggregate variable of droughts, floods, and storms. This study also used the number of deaths caused by climate-related disasters. In addition, this study uses two other variables: corruption control and the local government index, as measures of government effectiveness. If corruption control is inadequate, the government's effectiveness is considered low and would affect recovery implementation after a disaster. For international aid, seven different types of data were prepared to examine how aid works to mitigate the impact of natural disasters. Aid for agricultural development works differently from humanitarian aid. For agricultural production, the agricultural production index is the aggregate index. Production data for major cereals, such as maise, sorghum, and millet, were used to examine the impact of crops.

Table 1.1 Descriptive statistics

Variable	Obs.	Mean	Std. Dev.	Min.	Max.	Year	Data Source
Number of people affected by climate-related disasters	3,394	171,620.9	906,675.7	0	23,000,000	1900-2021	EM-DAT
Number of people affected by drought	3,143	0.0140703	0.080764	0	1	1900-2021	EM-DAT
Number of people affected by flood	3,143	0.0024037	0.0179437	0	0	1927-2021	EM-DAT
Number of people affected by storm	3,143	0.0012416	0.0250523	0	1	1948-2021	EM-DAT
Number of deaths by climate-related disasters	3,394	267.0533	6475.562	0	300,000	1900-2021	EM-DAT
Control of corruption	1,185	-0.542972	0.687297	-1.869	2	1996-2019	V-Dem
Local Government Index	3,224	0.4376833	0.3243581	0	1	1900-2020	V-Dem
Educational inequality, Gini	2,283	60.83778	22.18283	11.875	99.804	1927-2010	V-Dem
Net ODA	2,286	10.48105	11.82156	-0.251879	147	1960-2011	WDI
Humanitarian ODA	503	66,800,000	171,000,000	1,387	1,380,000,000	2002-2011	WDI
ODA for reconstruction relief and rehabilitation	296	5,796,040	13,200,000	-52185	96,900,000	2002-2011	WDI
ODA for agriculture	513	29,000,000	40,400,000	7,069	387,000,000	2002-2011	WDI
Emergency ODA	493	64,100,000	165,000,000	1,387	1,280,000,000	2002-2011	WDI
ODA for disaster prevention and preparedness	249	1,209,090	2,266,356	-75420	16,700,000	2002-2011	WDI
Cereal food aid	1,309	67,786.64	165,895	0	1,900,805	1988-2012	WDI
Agriculture Production Index	2,583	70.33316	28.66801	13.42	193	1961-2011	WDI
Cereal production index	2,512	82.68087	77.53881	5.79	1,925	1961-2011	WDI
Maise production (ton)	1,950	584,838.4	1,210,479	4	10,500,000	1961-2019	FAOSTAT
Sorghum production (ton)	1,522	245,611.4	506,910.9	0	5,265,580	1961-2019	FAOSTAT
Millet production (ton)	1,323	131,864.3	262,178.4	54	1,878,527	1961-2019	FAOSTAT
Rice production (ton)	1,669	372,630.1	938,768.1	0	7,253,373	1961-2019	FAOSTAT
Wheat production (ton)	1,105	607,994	1,521,939	0	9,607,736	1961-2019	FAOSTAT
Barley production (ton)	552	449,506.7	749,312.6	100	3,831,130	1961-2019	FAOSTAT
Fonio production (ton)	381	36,381.73	83,020.09	100	530,227	1961-2020	FAOSTAT
Poverty gap at the urban poverty line (%)	76	11.90395	9.049087	1.8	40	1961-2011	WDI
Poverty gap at the rural poverty line (%)	77	22.07273	9.499953	3.6	53	1961-2012	WDI
Battle-related deaths (number of people)	278	1,411.522	4,618.351	0	50,293	1961-2013	WDI

Results

This section reviews the current trends of natural disasters in Africa before analysing the impacts of natural disasters. Figure 1.1 presents the number of people affected by droughts, floods, and storms caused by climate change. The data used the Emergency Events Database (EM-DAT), mentioned in the previous section (EM-DAT, 2021). As a measure of natural disasters, Figure 1.1 uses the number of people affected rather than the number of disasters because this is a better measure of disaster severity. Each disaster is different in scale and impact. If a disaster occurs in a remote mountainous area, the social impact is less, but the same type of event would have a devastating effect in an urban area.

Furthermore, an essential point must be considered when interpreting these numbers. The disaster number, such as the number of people affected by a disaster may be underreported compared with the actual situation (Strömberg 2007). For instance, some authoritarian African regimes may underreport the damage caused by disasters to avoid being criticised for their response. Given African governments' capacity for data authenticity, there may be misreporting across countries and over time (Jerven 2013).[6]

In such cases, there is a risk of underestimating the impact of climate-related disasters.

Figure 1.1 has two vertical axes because the damage caused by droughts is much larger than that caused by floods and storms. The left axis represents the number of people affected by droughts, and the right axis indicates the number of people displaced by floods and storms. The worst drought affected more than 35,000 people in 1999. In 2021, many African countries suffered from droughts, including Somalia, Ethiopia, Madagascar, Angola, and South Africa.[7]

---

6 Because some authoritarian regimes in Africa may under-report the damage caused by disasters, the actual situation might be more severe than reported here. One example of manipulated figures like this is Nigerian population data, which may lack accuracy. This is because population estimates also have political significance and the census directly impacts electoral representation. Nigeria is a country where political power is highly contested, and state oversight is weak.

7 These latest figures are not presented in Figure 2.1, which only covers the period 1950–2020.

Generally, in Africa the impact of floods and storms is much less than the impact of drought. Due to the nature of disasters, the three lines representing the different types of natural disaster fluctuate significantly. For instance, there was a massive drought in 1999, mainly in East Africa (Kenya and Ethiopia), but there was no drought in the preceding or subsequent year. The damage caused by droughts became much greater in the late 1960s, and they seem to occur quite frequently now. These days, droughts have begun to occur once every two years. At the same time, the damage caused has also become increasingly severe. A similar pattern can be observed for floods and storms. During the late 1990s, there was an unprecedented increase in the number of people affected by floods and storms.

To clearly understand the long-term trend of natural disasters, fitted lines for droughts, floods, and storms are drawn in Figure 1.1. These three fitted lines show a clear upward trend, especially due to droughts, followed by floods and storms. Figure 1.1 confirms that the damage caused

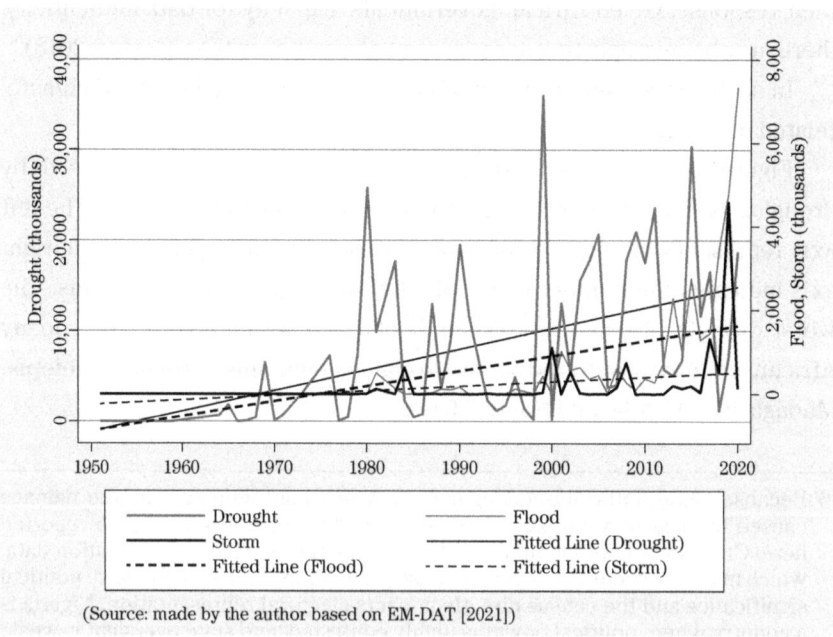

(Source: made by the author based on EM-DAT [2021])

Figure 1.1 The total number of people affected by droughts, floods, and storms in Africa 1950–2020.

by climate-related disasters has increased rapidly since the late 1960s, primarily during droughts. Therefore, the next question is how disaster-related damage affects Africa's socio-economic development and growth.

Table 1.2 presents the estimated coefficients in the impact on GDP per capita growth. The four models used different variables. Models 1 and 2 used aggregate disaster measures (number of people affected by climate-related disasters), and Models 3 and 4 used disaggregated disaster variables (droughts, floods, and storms). As the dataset is unbalanced panel data with gaps, the N used for estimation differs depending on the variable used.

Table 1.2 Economic growth impacts by climate-related disasters and aid

Dependent variable	GDP per capita growth (annual %)			
	(1)	(2)	(3)	(4)
Lagged GDP per capita growth	0.1694751***	0.1543027**	0.16712***	0.1519377**
	(2.80)	(2.49)	(2.75)	(2.44)
Education (+15 years old)	-1.228877	-1.139392	-1.25669	-1.172042
	(-0.56)	(-0.52)	(-0.58)	(-0.53)
Fertility rate	1.35599	2.05509	1.479018	2.180705
	(0.66)	(0.97)	(0.71)	(1.02)
HDI	-3.463914	-.953797	-2.788904	-.2467038
	(-0.17)	(-0.04)	(-0.13)	(-0.01)
Number of people affected by climate-related disasters	-8.057152**	-8.057298**		
	(-2.07)	(-2.05)		
Number of people affected by drought			-7.632749*	-7.614573*
			(-1.87)	(-1.85)
Number of people affected by flood			-13.04042	-13.28393
			(-0.99)	(-1.01)
Number of people affected by storm			0.00000684	0.00000685
			(1.05)	(1.05)
Government expenditure	0.1529741	0.1360603	0.1624341	0.1460737
	(1.53)	(1.31)	(1.62)	(1.40)
Humanitarian ODA	-0.0000000007		-0.000000001	
	(-0.14)		(-0.14)	
Emergency ODA		-0.000000001		-0.000000001
		(-0.180)		(-0.18)
Constant	1.043053	-3.935066	.1118494	-4.888921
	(0.06)	(-0.22)	(0.01)	(-0.28)
N	272	264	272	264
Type of regression	FE	FE	FE	FE

Note: Numbers in parentheses are t-values. ***, **, and * indicate statistical significance at the 1%, 5%, and 10% levels, respectively.

Models 1 and 2 show that climate-related natural disasters significantly lowered GDP per capita growth. However, as Models 3 and 4 showed, the impact differs depending on the natural disaster type. Only droughts had a statistically significant negative impact, as opposed to floods and storms. This indicates the importance of any drought policy to sustain GDP per capita growth in Africa.

Models 1 and 3 examined the impact of humanitarian ODA on GDP per capita. How do global efforts mitigate the negative impact of natural disasters? Similar to Models 2 and 4, which examined the emergency aid, all these variables became insignificant. This is reasonable considering that the amount of aid in these categories is too small to capture the impact compared with a country's GDP. Therefore, the study next examines aid impact focusing on crops because the agricultural sector is particularly vulnerable to natural disasters, affecting economic growth and damaging crop production.

Table 1.3 Impact on agricultural production by climate-related disasters and aid

Dependent variable	Agriculture Production Index			
	(1)	(2)	(3)	(4)
GDP per capita growth	-0.0003642	-0.1086933	-0.115295	-0.032736
	(-0.00)	(-0.95)	(-0.99)	(-0.10)
Number of people affected by climate-related disasters	-24.41415***	-23.02236***	-21.84345**	-35.02979**
	(-2.74)	(-2.58)	(-2.41)	(-2.06)
Educational inequality, Gini	-4.608486***	-4.224185***	-4.253704***	-6.260741***
	(-19.84)	(-15.48)	(-15.29)	(-6.89)
Humanitarian ODA	-0.0000000052			
	(-0.47)			
ODA for agriculture		0.000000075***	0.000000076***	
		(2.65)	(2.66)	
ODA for disaster prevention & preparedness (lagged)				-0.0000008
				(-0.95)
Local Government Index			-8.470382	3.579998
			(-1.35)	(0.14)
Constant	319.1347***	297.3906***	303.5181***	397.2466***
	(29.24)	(22.74)	(21.33)	(9.78)
N	370	379	370	120
Type of regression	FE	FE	FE	FE

Note: Numbers in parentheses are t-values. ***, **, and * indicate statistical significance at the 1%, 5%, and 10% levels, respectively.

Table 1.3 examines the impact on agricultural production. The number of people affected by natural disasters, an aggregate variable, is used to measure disaster impact. This variable is statistically significant, and the negative coefficient is also substantial. This result indicates that climate-related disasters have a significant negative impact on agricultural production. Educational inequality is also negative, indicating that human capital is important for agricultural production. As the educational Gini coefficient reflects income inequality, the widening rich-poor gap negatively impacts agricultural production. This is consistent with two studies (Schultz 1964, 1961). Therefore, human capital is required to cope with disasters caused by climate change, which are likely to increase in the future.

According to Models 1 and 4, humanitarian aid and ODA for disaster prevention and preparedness are not statistically significant. In contrast, according to Models 2 and 3, agricultural aid is positive. These results are consistent with the expected outcome, as humanitarian aid and disaster prevention assistance do not support agricultural production. However, this analysis shows that disasters substantially negatively impact agricultural production. Therefore, agricultural aid has become even more important during the age of climate change, specifically to fight against the long-term negative consequences. However, it is necessary to note that the coefficient of agricultural aid is small, indicating that agricultural aid alone is not enough to mitigate the negative impact of climate-related disasters.

Table 1.4 investigates the impact of climate-related disasters on agriculture. Compared to Table 1.3, the impact of different types of aid is tested in Table 1.4. Models 1, 2, and 3 examine emergency aid, agricultural aid, and cereal food aid, respectively.[8] Emergency aid is not significant,

---

8 The sectoral data used for this analysis is from OECD (Organisation for Economic Co-operation and Development) through WDI (World Development Indicators). These data are based on the distribution of bilateral ODA commitments and aggregates of individual projects notified under the Creditor Reporting System of the OECD, supplemented by reporting on the sectoral distribution of technical cooperation and actual disbursements of food and emergency aid. As this is aggregated data, it includes various types of aid such as cash transfers (conditional or not conditional), universal or targeted, temporary or recurrent.

similar to humanitarian aid. This is expected because emergency aid does not aim at agricultural development. The impact on agricultural aid and cereal food aid is opposite, wherein agricultural aid is positive, but cereal food aid is negative due to a substitutional effect. This is in line with the argument by Anderson (1999). The inflow of cereals from foreign countries seems to have a crowding-out effect on domestically produced cereals. Even if the coefficient is positive for agricultural aid, it is very small compared to the coefficient of disasters. In other words, agricultural aid does not compensate for the negative impact of disasters. The results for the impact of disasters and educational inequality are the same as those in Table 1.3, further confirming the earlier results.

Table 1.4 Impact on agricultural production by climate-related disasters and aid

Dependent variable	Agriculture Production Index		
	(1)	(2)	(3)
GDP per capita growth	-0.0106388	-0.1173564	0.3344108***
	(-0.09)	(-1.01)	(5.03)
Number of people affected by climate-related disasters	-23.4604***		
	(-2.58)		
Number of people affected by flood		-17.48007	-9.036425
		(-0.41)	(-0.30)
Number of people affected by drought		-24.51023***	-12.51029*
		(-2.61)	(-1.94)
Number of people affected by storm		-0.000009	-0.000007
		(-0.93)	(-1.57)
Educational inequality, Gini	-4.64529***	-4.230962***	-2.568219***
	(-19.48)	(-15.09)	(-26.68)
Local Government Index	-8.113644	-8.025702	20.9569***
	(-1.30)	(-1.28)	(6.93)
Emergency ODA	-0.000000003		
	(-0.775)		
ODA for Agriculture		0.00000008***	
		(2.70)	
Cereal food aid			-0.0000148***
			(-3.49)
Constant	326.9576***	302.2468***	206.6968***
	(26.54)	(21.07)	(37.24)
N	355	370	958
Type of regression	FE	FE	FE

Note: Numbers in parentheses are t-values. ***, **, and * indicate statistical significance at the 1%, 5%, and 10% levels, respectively.

Tables 1.5 and 1.6 examine the impact on a crop-by-crop basis. The difference between these two tables is that Table 1.5 includes ODA for agriculture, whereas Table 1.6 examines the impact of cereal aid. This analysis shows that the impact varies by crop. For instance, droughts negatively impact maise. Storms reduce rice and fonio production because, at the heading stage, strong winds can topple the panicles of rice and fonio. Floods also reduce fonio production. However, this result is not robust, as these crops are statistically insignificant (Table 1.6). Therefore, the impact of disasters on crops varies depending on the vegetation type.

Agricultural ODA positively impacts maise, sorghum, millet, and rice, but the impact differs by crop. Furthermore, the coefficient is very

Table 1.5 Impact on cereal production by climate-related disasters and aid

Dependent variable	(1) Cereal production index	(2) Maise production	(3) Sorghum production	(4) Millet production	(5) Rice production	(6) Wheat production	(7) Barley production	(8) Fonio production
GDP per capita growth	.3519966	1591.614	-1304.271	830.9917	-1334.704	9,952.268	24,650.14	-2,459.17
	(1.22)	[0.41]	(-0.74)	(0.62)	[-0.45]	[1.06]	[0.80]	(-1.21)
Educational inequality, Gini	-7.099192***	-32,610.23***	-5,121.155	-989.7303	-8121.831	-8,631.816	-1,084.82	-12,828.91***
	(-10.44)	[-5.19]	(-1.58)	(-0.40)	[-1.37]	[-0.54]	[-0.05]	(-8.28)
Number of people affected by flood	-4.82892	-910,600.5	27,443.77	-35,073.24	-223,888.4	1,576,617	-1,673,180	-745,509.2***
	(-0.05)	[-0.46]	(0.03)	(-0.06)	[-0.16]	[0.19]	[-0.07]	(-2.92)
Number of people affected by drought	-51.96536**	-1,027,551***	-39,551.82	-102,533.6	-46,956.45	91,261.4	-10,5139.6	118,431.8
	(-2.27)	[-3.75]	(-0.33)	(-0.94)	[-0.19]	[0.15]	[-0.09]	(0.46)
Number of people affected by storm	0.000000779	-0.1388403	-0.0418895	-0.1643853	-0.66323***	-0.0056365	0.0449533	-51.86224*
	(0.03)	[-0.64]	(-0.44)	(-0.62)	[-2.77]	[-0.01]	[0.01]	(-1.92)
ODA for Agriculture	0.000000116*	0.0040394***	0.001325***	0.000796***	0.0034716***	0.0011092	0.0014411	-0.0003032
	(1.65)	[5.45]	(3.95)	(3.19)	[5.11]	[0.53]	[0.42]	(-2.39)
Constant	433.7546***	2,230,139***	461,934.9***	208,877.2	866,423.4**	1,128,003	400,714.1	1,043,952***
	(13.19)	[5.79]	(2.82)	(1.55)	[2.43]	[1.32]	[0.36]	(8.78)
N	373	247	324	185	332	158	62	45
Type of regression	FE	FE	RE	FE	RE	RE	RE	FE

Note: Numbers in brackets are z-values, and in parentheses are t-values. ***, **, and * indicate statistical significance at the 1%, 5%, and 10% levels, respectively.

small. More importantly, it can be observed in Table 1.6 that the impact of cereal aid is different from agricultural ODA. Overall, cereal aid has a negative impact, but this influence differs depending on the crop. Maise, sorghum, rice, and wheat are all negatively affected, but millet production increases marginally. Although the reason for increased millet production is unclear, millet may be used as a substitute for cereals such as wheat, whose production decreased. In many cases, cereal aid is provided during humanitarian crises, including natural disasters. However, as Anderson (1999) discussed, cereal aid can negatively affect production because cereals provided through food aid have a crowding-out effect on domestic production.

Table 1.6 Cereal food aid and cereal production

Dependent variable	(1) Cereal production index	(2) Maise production	(3) Sorghum production	(4) Millet production	(5) Rice production	(6) Wheat production	(7) Barley production	(8) Fonio production
GDP per capita growth	0.9851934***	6623.565***	2369.321	1372.234	-645.8084	15,459.38***	42,489.02***	-775.6227
	(4.12)	[2.74]	(2.18)	(1.62)	[-0.42]	[2.75]	[4.30]	(-0.64)
Educational inequality, Gini	-1.605586***	-20,892.1***	-4,520.574***	-4,523.103***	-11,114.77***	-10,901.36**	10845.18	-6,237.34***
	(-4.97)	[-10.16]	(-4.98)	(-5.85)	[-4.99]	[-2.16]	[1.36]	(-7.90)
Number of people affected by flood	54.60008	-55925.41	-5490.607	-1908.055	-518772.2	-1131083	3,794,894	-313,625.6
	(0.50)	[-0.07]	(-0.02)	(-0.01)	[-0.64]	[-0.57]	[0.32]	(-1.02)
Number of people affected by drought	-31.99213	-552,229.1***	-61,476.59	-74,580.37	-1235.48	48,256.81	64,481.36	86,418.55
	(-1.39)	[-3.75]	(-1.04)	(-1.21)	[-0.01]	[0.15]	[0.13]	(0.77)
Number of people affected by storm	-0.000008	-0.119833	-0.0115932	0.000002	0.0225578	0.0035995	-5.263566	-66.31653
	(-0.50)	[-1.37]	(-0.33)	(0.00)	[0.25]	[0.02]	[-0.90]	(-1.54)
Cereal food aid	-0.000033***	-0.94005***	-0.073275**	0.1050056*	-1.45745***	-1.6366***	-0.0561999	0.1900641
	(-2.14)	[-10.41]	(-2.00)	(1.74)	[-15.80]	[-8.37]	[-0.30]	(0.71)
Constant	177.215***	1782212***	458177.8***	403646.2***	1150479***	1287005***	-120938.3	535114.8***
	(10.77)	[10.82]	(9.45)	(9.17)	[5.83]	[3.85]	[-0.25]	(8.68)
N	952	631	540	457	598	401	165	115
Type of regression	FE	FE	RE	FE	RE	RE	RE	FE

Note: Numbers in brackets are z-values, and in parentheses are t-values. ***, **, and * indicate statistical significance at the 1%, 5%, and 10% levels, respectively.

This study also examined non-cereal crops, such as bananas, cassava, tea, and coffee (refer to Annex). It can be observed that only coffee is negatively and strongly affected by droughts because water stress affects coffee production due to water availability sensitivity.

As discussed above, natural disasters negatively impact agricultural production. The next question is what the consequences of such effects are. Table 1.7 examines the impact of climate-related disasters on poverty and conflicts because reduced agricultural production indicates lower income for farmers. This would probably impact poverty and conflicts. As observed in Section 1 (Literature review), previous studies examine the temperature-conflict link without considering the internal mechanism. As discussed, using disaster data, this study examined the internal nexus between climate change and armed conflicts.

Model 1 examined the impact of aggregate climate-related natural disasters in rural areas and shows that poverty does not increase in rural areas. The growth in GDP per capita reduces poverty in rural areas. However, this situation contrasts in urban areas (Model 2), where climate-related natural disasters increase poverty. It seems there are two reasons behind this. One is a robust mutual support system in rural areas that are not affected by climate change. This system makes people in rural areas more resilient than those in the urban area providing support to the people affected by natural disasters. The other is that some people migrate from rural to urban areas searching for jobs after natural disasters, where the resilient system is weaker compared with rural areas. Therefore, rather than rural areas, poverty increases in urban areas.

Model 2 used aggregated data on disasters. Then, Model 3 analyses the impact of different disaster types disaggregating the disasters to flood, drought and storm. The focus is on urban areas. The results show that climate variability that leads to extreme events such as droughts causes a substantial increase in poverty in Africa.

Another possible impact of agricultural production decrease is conflict. Models 4 and 5 studied the impact of natural disasters on the number of battle-related deaths. Using aggregate natural disaster data, Model 4 shows that while growth in GDP per capita reduced battle-related deaths, natural disasters significantly increased the death toll. Model 5 used disaggregated data on disasters, then, this model also confirmed this aspect. Among the types of natural disasters, droughts exacerbated battle-related deaths in Africa.

As these results indicate, natural disasters reduce agricultural production and trigger poverty in urban areas. Furthermore, these disasters and increasing poverty in cities increase people's discontent. It then ends up leading to increase civil wars and international conflicts.

Finally, Table 1.8 examines the various factors that contributed to reducing the number of deaths caused by climate-related natural disasters. This analysis aims to explore how African countries can respond and mitigate the adverse effects of climate warming.

Model 1 examined HDI (Human Development Index), government effectiveness, and corruption control. Corruption control is a proxy for government transparency. Government effectiveness is strongly significant

Table 1.7 Poverty and battle

Dependent variable	Poverty in rural area (1)	Poverty in urban area (2)	Poverty in urban area (3)	Battle related deaths (4)	Battle related deaths (5)
GDP per capita growth	-0.5020925*	-0.0326405	-0.1644819	-87.08594**	-87.32139**
	[-1.84]	[-0.15]	(-0.63)	(-2.37)	(-2.40)
Number of people affected by climate-related disasters	9.285749	15.6348**		11,854.74***	
	[1.12]	[2.34]		(2.93)	
Number of people affected by flood			-64.6435		12,956.34
			(-1.50)		(0.57)
Number of people affected by drought			20.16717**		13,122.96***
			(-2.57)		(3.89)
Number of people affected by storm			-79.01091		-.0149493
			(-0.88)		(-0.28)
Constant	23.394***	11.288***	12.342***	1,142.126***	1,105.389***
	[16.03]	[8.01]	(12.38)	(3.91)	(3.77)
N	77	76	75	260	260
Type of regression	RE	RE	FE	FE	FE

Note: Numbers in parentheses are *t*-values. ***, **, and * indicate statistical significance at the 1%, 5%, and 10% levels, respectively.

in reducing the number of disaster-related deaths. The ODA for disaster prevention and preparedness has also become positive. However, the coefficient is not necessarily large enough to mitigate the impact on deaths. Therefore, the government's capacity building is critical for mitigating the impact of natural disasters. Model 2 includes the regulatory quality of government policies, but the results do not change.

Model 4 tests humanitarian aid, and Model 5 includes emergency aid. Both are statistically significant in reducing the number of deaths caused by natural disasters. This is an excellent outcome for international donors, but it is necessary to note that the coefficients are very small. In other words, to mitigate damage caused by increasing climate-related disasters, these aids are inadequate for coping with the damage. From this analysis, it can be concluded that government effectiveness is key to reducing the number of deaths caused by natural disasters.

Table 1.8 Factors contributed to reduce the number of deaths by disasters

Dependent variable	Total number of deaths by climate-related disasters			
	(1)	(2)	(3)	(4)
HDI (Human Development Index)	-504.342	-621.4799	-349.3779*	-354.2997*
	(-1.31)	(-1.50)	(-1.68)	(-1.68)
Gov. effectiveness	-131.8227**	-141.7638**	-54.9951*	-55.76964*
	(-2.37)	(-2.48)	(-1.77)	(-1.76)
Control of corruption	36.64476	20.4265	31.70747	31.43229
	(0.60)	(0.32)	(0.96)	(0.93)
Regulatory quality		49.01479	14.14797	13.79879
		(0.76)	(0.48)	(0.46)
ODA for disaster prevention and preparedness	-0.000008***	-0.0000073***		
	(-2.78)	(-2.65)		
Humanitarian ODA			-0.0000002***	
			(-4.04)	
Emergency ODA				-0.0000002***
				(-4.16)
Constant	187.7321	254.3052	181.7999*	181.3178*
	(1.05)	(1.28)	(1.74)	(1.73)
N	234	234	343	334
Type of regression	FE	FE	FE	FE

Note: Numbers in brackets are z-values, and in parentheses are t-values. ***, **, and * indicate statistical significance at the 1%, 5%, and 10% levels, respectively.

Discussion and Conclusions

Unlike previous studies, this study controls for social variables and examines the crop-by-crop impact. Using panel data from African countries, this study found the following four aspects:

1. Climate-related natural disasters negatively impact per capita GDP growth and agricultural production. The impact is severe on cereal production, especially droughts (maise) and storms (rice and fonio).
2. While ODA for agriculture has a slightly positive impact, cereal aid food negatively impacts cereal production (maise, sorghum, rice, wheat).
3. Climate-related disasters, primarily droughts, increase poverty in urban areas and increase battle-related deaths. This result supports the argument by Marshall B. Burke et al. (2009).
4. Finally, government effectiveness is key to determining the number of deaths caused by climate-related disasters.

There are several important policy implications. First, these findings show that climate change has severe consequences not only for the development of African countries but also for armed conflicts. As mentioned earlier, Africa is the least responsible for the increase in greenhouse emissions. Therefore, donor countries must initiate quick action to assist African countries and help them cope with climate change, especially climate-related natural disasters. Among natural disasters that are closely associated with human life, droughts have severe consequences on the following socio-economic activities: GDP per capita growth, agricultural production, poverty, and armed conflicts. International donors must focus on developing measures to prevent droughts and assist African countries' adaptive strategies to combat climate change. Further, as Chandler et al. (2020) discussed, it is also essential to find and promote alternative agricultural and food production systems which are less fuel intensive.

Second, government effectiveness is key to coping with climate-related disasters. International aid must be provided to improve government effectiveness. This is clear, as the coefficient of the impact to reduce the number of deaths was small on ODA for disaster prevention and preparedness, humanitarian ODA, and emergency ODA.

Third, there is a need to review if cereal aid is beneficial for African countries as it reduces cereal production, possibly due to a crowding-out effect on domestic production.

As discussed in Section 2, disaster data may be underreported. Most authoritarian governments do not overemphasise the damage caused by natural disasters. These governments have a strong incentive to underreport damages. Thus, our assessment likely underestimates the true damage level. Therefore, it is necessary to plan policies and measures that consider this possibility. Another underreporting issue is how the African community has been coping with climate-related natural disasters. The following chapters describe and discuss their effort under severe conditions. These aspects should not remain underreported.

Annex 1.1 Non-cereal crops

Dependent variable	(1) Banana production	(2) Cassava production	(3) Tea production	(4) Coffee production
GDP per capita growth	-3964.216*	-13413.36	493.2195	87.8662
	(-1.87)	[-1.51]	[0.98]	[0.26]
Educational inequality, Gini	-35370.73***	-123926.1***	-1405.38	623.709
	(-7.14)	[-5.87]	[-1.61]	[1.15]
Number of people affected by flood	-172524.4	4,010,153	-157,908.5	53,130.28
	(-0.14)	[0.69]	[-0.60]	[0.26]
Number of people affected by drought	117,739.9	-1717949	33,912.43	-81737.55**
	(0.59)	[-1.59]	[1.46]	[-2.08]
Number of people affected by storm	0.0329793	-1.00887	-0.0004288	0.0114696
	(0.25)	[-1.63]	[0.04]	[0.55]
ODA for Agriculture	0.0002185	0.0047701**	0.0001	-0.0000291
	(0.45)	[2.15]	[1.54]	[-0.35]
Constant	2,052,232***	9,049,522***	106,277.7**	2,775.313
	(8.55)	[6.25]	[1.89]	[0.10]
N	197	188	80	144
Type of regression	FE	RE	RE	FE

Note: Numbers in brackets are z-values, and in parentheses are t-values. ***, **, and * indicate statistical significance at the 1%, 5%, and 10% levels, respectively.

References

Albala-Bertrand, J.M. 1993. *Political Economy of Large Natural Disasters: With* Special Reference to Developing Countries. Oxford: Oxford University Press.

Anderson, M.B. 1999. *Do No Harm: How aid can support peace-or war.* Boulder, Colorado: Lynne Rienner Publishers.

Andor, M.A., D. Osberghaus, and M. Simora. 2020. 'Natural disasters and governmental aid: Is there a charity hazard?' *Ecological Economics* 169: 106534.

Barrett, C.B. 2010. 'Measuring food insecurity.' *Science* 327 (5967): 825–8. https://doi.org/10.1126/science.1182768. https://www.ncbi.nlm.nih.gov/pubmed/20150491.

Becerra, O., E. Cavallo, and Ilan Noy. 2014. 'Foreign aid in the aftermath of large natural disasters.' *Review of Development Economics* 18 (3): 445–460.

Benson, C., and E.J. Clay. 2006. 'Disasters, vulnerability and the global economy: implications for less-developed countries and poor populations.' In C.S. Galbraith and C.S. Stiles. (eds). *Developmental Entrepreneurship: Adversity, risk, and isolation.* Bingley: Emerald Group Publishing Limited. pp. 115–145.

Bernard, B., K. Vincent, M. Frank, and E. Anthony. 2013. 'Comparison of extreme weather events and streamflow from drought indices and a hydrological model in River Malaba, Eastern Uganda.' *International Journal of Environmental Studies* 70 (6): 940–951.

Borras Jr, S.M., I. Scoones, A. Baviskar, M. Edelman, N.L. Peluso, and W. Wolford. 2022. 'Climate change and agrarian struggles: an invitation to contribute to a JPS Forum.' *The Journal of Peasant Studies* 49 (1): 1–28.

Buhaug, H. 2010. 'Climate not to blame for African civil wars.' *Proceedings of the National Academy of Sciences* 107 (38): 16477–16482.

Burke, M.B., E. Miguel, Shanker Satyanath, John A Dykema, and David B Lobell. 2010. 'Climate robustly linked to African civil war.' *Proceedings of the National Academy of Sciences* 107 (51): E185–E185.

Burke, M.B., E. Miguel, Shanker Satyanath, John A. Dykema, and David B. Lobell. 2009. 'Warming increases the risk of civil war in Africa.' *Proceedings of the National Academy of Sciences* 106 (49): 20670–20674. https://doi.org/10.1073/pnas.0907998106. https://www.pnas.org/content/pnas/106/49/20670.full.pdf.

Cavallo, E.A, and I. Noy. 2009. 'The economics of natural disasters: a survey.' https://publications.iadb.org/en/publication/economics-natural-disasters-survey.

Cavallo, E., S. Galiani, I. Noy, and J. Pantano. 2013. 'Catastrophic natural disasters and economic growth.' *Review of Economics and Statistics* 95 (5): 1549–1561.

Centre for Research on the Epidemiology of Disasters. 2021. 'Extreme weather events in Europe.' *CRED Crunch* (64). https://www.cred.be/publications.

Chandler, D., K. Grove, and S. Wakefield. 2020. *Resilience in the Anthropocene: Governance and politics at the end of the world. Routledge research in the Anthropocene.* Abingdon, Oxon: Routledge.

Charvériat, C. 2000. 'Natural disasters in Latin America and the Caribbean: An overview of risk.' IDB Working Paper No. 364.

Connolly-Boutin, Liette., and B. Smit. 2016. 'Climate change, food security, and livelihoods in sub-Saharan Africa.' *Regional Environmental Change* 16 (2): 385–399.

Coppedge, M., J. Gerring, C.H. Knutsen, S.I. Lindberg, J. Teorell, D. Altman, M. Bernhard, M.S. Fish, A. Glynn, A. Hicken, A. Lührmann, K.L. Marquardt, K. McMann, P. Paxton, D. Pemstein, B. Seim, R. Sigman, S.-E. Skaaning, J. Staton, A. Cornell, L. Gastaldi, H. Gjerløw, V. Mechkova, J. von Römer, A. Sundström, E. Tzelgov, L. Uberti, Y. Wang, T. Wig, and D. Ziblatt. 2019. *V-Dem Codebook v9.* edited by Varieties of Democracy (V-Dem) Project.

Cuny, Frederick C. 1994. *Disasters and Development.* Dallas: Intertect Press.

Dacy, D.C., and H. Kunreuther. 1969. *Economics of Natural Disasters; Implications for Federal policy.* New York: Free Press.

De Juan, A., J Pierskalla, and E. Schwarz. 2020. 'Natural disasters, aid distribution, and social conflict: Micro-level evidence from the 2015 earthquake in Nepal.' *World Development* 126: 104715.

EM-DAT. 2021. The Emergency Events Database Belgium: Université catholique de Louvain (UCL) - CRED, D. Guha-Sapir - www.emdat.be, Brussels, Belgium.

Field, C.B., and V.R. Barros. 2014. *Climate change 2014: Impacts, adaptation and vulnerability: Regional aspects.* Cambridge: Cambridge University Press.

Food Agriculture Organization of the United Nations. 2021. FAOSTAT statistical database. Rome: FAO.

Higuchi, Y., and G. Shimada. 2019. 'Industrial policy, industrial development, and structural transformation in Asia and Africa.' In K. Otsuka and K. Sugihara. (eds). *Paths to the Emerging State in Asia and Africa.* Singapore: Springer. pp. 195–218.

Hochrainer, S. 2009. 'Assessing the macroeconomic impacts of natural disasters: Are there any?' *World Bank policy research working paper* (4968).

Hosono, A., J. Page, and G. Shimada (eds). 2020. *Workers, Managers, Productivity: Kaizen in developing countries.* Singapore: Palgrave Macmillan.

IPCC. 2022. *Climate Change 2022: Impacts, Adaptation, and Vulnerability. Contribution of Working Group II to the Sixth Assessment Report of the Intergovernmental Panel on Climate Change.* Edited by D.C. Roberts H.-O. Pörtner, M. Tignor, E.S. Poloczanska, K. Mintenbeck, A. Alegría, M. Craig, S. Langsdorf, S. Löschke, V. Möller, A. Okem, and B. Rama. Cambridge: Cambridge University Press.

Jerven, M. 2013. *Poor numbers.* Ithaca: Cornell University Press.

Kahn, M.E. 2005. 'The death toll from natural disasters: The role of income, geography, and institutions.' *Review of economics and statistics* 87 (2): 271–284.

Lesk, C., P. Rowhani, and N. Ramankutty. 2016. 'Influence of extreme weather disasters on global crop production.' *Nature* 529 (7584): 84–7. https://doi.org/10.1038/nature16467. https://www.ncbi.nlm.nih.gov/pubmed/26738594.

Noy, I. 2009. 'The macroeconomic consequences of disasters.' *Journal of Development Economics* 88 (2): 221–231.

Noy, I., and T.B. Vu. 2010. 'The economics of natural disasters in a developing country: The case of Vietnam.' *Journal of Asian Economics* 21 (4): 345–354.

O'Loughlin, J.A., M. Linke, and F.D.W. Witmer. 2014. 'Effects of temperature and precipitation variability on the risk of violence in sub-Saharan Africa, 1980–2012.' *Proceedings of the National Academy of Sciences* 111 (47): 16712–16717.

Peduzzi, P., B. Chatenoux, H. Dao, A. De Bono, C. Herold, J. Kossin, F. Mouton, and O. Nordbeck. 2012. 'Global trends in tropical cyclone risk.' *Nature climate change* 2 (4): 289–294.

Phiiri, G.K., A. Egeru, and A. Ekwamu. 2016. 'Climate change and agriculture nexus in sub-Saharan Africa: The agonizing reality for smallholder farmers.' *International Journal of Current Research and Review* 8 (2): 57.

Raschky, P., and M. Schwindt. 2009. 'Aid, natural disasters and the Samaritan's dilemma.' *World Bank Policy Research Working Paper* (4952).

Rasmussen, T. 2004. 'Macroeconomic implications of natural disasters in the Caribbean.' *IMF Working Paper* WP/04.224.

Sawada, Y., R. Bhattcharyay, and T. Kotera. 2011. *Aggregate impacts of natural and man-made disasters: A quantitative comparison.* Research Inst. of Economy, Trade and Industry.

Schultz, T.W. 1961. 'Investment in human capital.' *The American Economic Review* 51 (1): 1–17.

———— 1964. *Transforming Traditional Agriculture.* Chicago: University of Chicago.

———— 1988. *On Investing in Specialized Human Capital to Attain Increasing Returns. The state of development economics: Progress and perspectives.* Oxford: Basil Blackwell.

Shimada, G. 2012. 'The macroeconomic impacts of natural disasters: A case study of Japan.' *Journal of the Graduate School of Asia-Pacific Studies* 24: 121–137.

———— 2015. 'The role of social capital after disasters: An empirical study of Japan based on Time-Series-Cross-Section (TSCS) data from 1981 to 2012.' *International Journal of Disaster Risk Reduction* 14: 388–394. https://doi.org/10.1016/j.ijdrr.2015.09.004.

———— 2017. 'A quantitative study of social capital in the tertiary sector of Kobe – Has social capital promoted economic reconstruction since the Great Hanshin Awaji Earthquake?' *International Journal of Disaster Risk Reduction* 22: 494–502. https://doi.org/10.1016/j.ijdrr.2016.10.002.

————— 2019. 'Does Environmental Policy Make African Industry Less Competitive? - The Possibilities in Green Industrial Policy.' In R. Kanbur, A. Norman, and J. E. Stiglitz.(eds). *The Quality of Growth in Africa*. New York: Colombia University Press. pp. 350–372.

Shimada, G., and M. Motomura. 2017. 'Building resilience through social capital as a counter-measure to natural disasters in Africa: A case study from a project in pastoralist and agro-pastoralist communities in Borena, in the Oromia Region of Ethiopia.' *African Study Monographs. Supplementary issue.* 53: 35–51.

Shimada, G., and T. Sonobe. 2021. 'Impacts of management training on workers: Evidence from Central America and the Caribbean region.' *Review of Development Economics* 25 (4): 1492–1514. https://doi.org/10.1111/rode.12773.

Skidmore, M., and H. Toya. 2002. 'Do natural disasters promote long-run growth?' *Economic Inquiry* 40 (4): 664–687.

Strömberg, D. 2007. 'Natural disasters, economic development, and humanitarian aid.' *Journal of Economic Perspectives* 21 (3): 199–222. https://doi.org/10.1257/jep.21.3.199. https://pubs.aeaweb.org/doi/pdfplus/10.1257/jep.21.3.199.

Tellman, B., J.A. Sullivan, C. Kuhn, A.J. Kettner, C.S. Doyle, G.R. Brakenridge, T.A. Erickson, and D.A. Slayback. 2021. 'Satellite imaging reveals increased proportion of population exposed to floods.' *Nature* 596 (7870): 80–86. https://doi.org/10.1038/s41586-021-03695-w.

The World Bank. 2021. World development indicators. Washington, DC: The World Bank.

Toll, R., and F. Leek. 1999. 'Economic analysis of natural disasters.' In Thomas Downing, Alexander Olsthoorn, and Richard S.J. Tol. (eds). *Climate Change and Risk*. London: Routledge. pp. 308–327.

Vermeulen, S.J., P.K. Aggarwal, A. Ainslie, C. Angelone, B.M. Campbell, A.J. Challinor, J.W. Hansen, J.S.I. Ingram, A. Jarvis, P. Kristjanson, C. Lau, G.C. Nelson, P.K. Thornton, and E. Wollenberg. 2012. 'Options for support to agriculture and food security under climate change.' *Environmental Science & Policy* 15 (1): 136–144. https://doi.org/10.1016/j.envsci.2011.09.003.

02

Genealogies of Resilience in the Development and Humanitarian Sector: Potentials and Difficulties

Tamara Enomoto

Introduction

The term 'resilience' has been widely used in research and policy debate in the development and humanitarian sector since the 2000s. Government agencies, international organisations, non-governmental organisations (NGOs) and researchers have promoted resilience as a key concept in the sector, seeking to analyse the factors that may drive resilience and to devise measures to improve it. Some expect that the idea of resilience may have the potential to overcome existing problems of international policy debate, for example, the tendency to assume that people in the global South are powerless and helpless (Almedom and Tumwine 2008; FAO 2016).

Focusing on this sector, this chapter introduces the backdrop against which the term started to gain wide circulation and reflects on key criticisms of the resilience concept. As has been widely pointed out, the concept was mainstreamed in the sector in the 2000s due to the transformation of the dominant discourse in policy debate in the sector by the end of the 1990s. This chapter will examine how the dominant discourses in the sector were transformed since the mid-twentieth century and seeks to understand how such transformation laid the foundation for the wide use of the term 'resilience'. It will then reflect on the criticisms of the use of the term by authors in the fields of development and critical security studies and suggest the potentials and difficulties of using the concept in the current critical juncture.

While acknowledging the general tendency of resilience thinking to ignore structural injustices and material causes of suffering and to legitimise measures to maintain a state of stability at lower material levels in the global South whose populations are deemed vulnerable and potentially dangerous, this chapter also reckons the possibility of using the term 'resilience' to critically analyse the situation in which structural injustices and causes of suffering are persistently reinstated and revived. Concurrently, the chapter provides a cautious view as to the use of the term considering the difficulties in avoiding the trap of legitimizing and normalizing Northern-centric, Western-centric or racist structures and discourses in the current humanitarian and development sector in which structural racism and power imbalances are deeply and resiliently entrenched and reminds of the need to be conscious of the goals and strategies if we are to use the term 'resilience'.

Changes in discourse in the development and humanitarian sector

In 1946, roughly half a century before the concept of resilience took centre stage, the World Health Organisation (WHO) Constitution stated that the 'enjoyment of the highest attainable standard of health' was 'one of the fundamental rights of every human being without distinction of race, religion, political belief, economic or social condition' (WHO 1946). Underlying this assertion was the widely held view that science and modern medicine would ultimately modernise Southern states and populations (Pupavac 2008). The premise of modernisation was evident in the mainstream Western development theory in the 1950s and 1960s. Development was equated with economic development, urbanisation, and industrialisation, while the spread of education and the penetration of liberal democracy were expected to transform pre-modern societies into modern ones. The lack of capital accumulation and technological innovation was seen as the main constraint to economic development, and state-led economic policies were devised to achieve import substitution industrialisation (Aghion and Bolton 1997; Pieterse 2009: 45–46). During the same period, Eastern countries also supported industrialisation under a state-led planned economy (Gilman 2003), and both the Eastern and Western blocs promoted state-led aid for economic development, especially selective aid to regimes close to their own blocs to expand their influence (Gilman 2003).

However, by the 1970s, development theories began to be criticised for not matching reality. The Declaration on the Establishment of a New International Economic Order (NIEO), adopted by the United Nations Special Session on Natural Resources in 1974, included the need for a fundamental reorganisation of the world order in such areas as international trade, international finance and technology transfer.[1]

Meanwhile, Western states and the World Bank argued that investments in human capital were important and that development assistance should meet 'basic human needs' (BHN), such as health, education, nutrition, safe water and housing (Kapoor 2008: 156–158). At the same time, criticism of modern society, industrialisation and its negative impact on ecosystems was growing and structuralism and post-structuralism took root in various fields of study, prompting scholars to critically analyse modern ideas. Against this backdrop of scepticism about modernity and attention to environmental issues, it was widely asserted that development projects should not try to instil Western materialistic values in the global South nor take the form of providing aid to Southern governments for material development but should focus instead on meeting BHNs by utilising local knowledge, resources, and capacities (Delgado 1995: 4–5). It was now deemed a priority to use local resources to ensure primary health care (PHC)[2] rather

---

1 UN Document. A/RES/S-6/3201. Declaration on the Establishment of a New International Economic Order.

2 At a joint meeting of the WHO and the United Nations Children's Fund (UNICEF) in 1978, PHC was defined as follows: 'Primary health care is essential health care based on practical, scientifically sound and socially acceptable methods and technology made universally accessible to individuals and families in the community through their full participation and at a cost that the community and country can afford to maintain at every stage of their development in the spirit of self-reliance and self-determination' (WHO and UNICEF 1978: 3).

than providing materials and technology to construct advanced medical facilities. Moreover, the oil crises of 1973 and 1979 increased awareness of the possibility that economic development in the global South would lead to competition for and depletion of natural resources and generated the idea that the entire world should not necessarily emulate the model of industrialised countries.

Later, in the 1980s, the idea of BHN was side-lined, as the accumulated debt problem was increasingly seen as an urgent and paramount development issue. In response to this situation, the structural adjustment policies promoted by the International Monetary Fund (IMF) and the World Bank began to dominate development assistance (Paloni and Zanardi 2006: 2–3). As Southern states lost influence in international politics, the establishment of the NIEO was increasingly overshadowed.

In the late 1980s, however, structural adjustment policies came under criticism for not having triggered the recovery of economic growth in the global South and for negatively affecting the lives of the poor. In 1987–1988, the report of the United Nations Children's Fund (UNICEF), 'Human-Faced Structural Adjustment' (Cornia et al. 1987, 1988), criticised structural adjustment for not paying enough attention to its negative impact on vulnerable people. In promoting 'structural adjustment with a human face', UNICEF advocated the idea of 'selective primary health care', which focused on increasing the survival of people (especially children) in crisis situations that resulted from structural adjustment policies (Pupavac 2008). This approach, however, focused primarily on mere survival, which was a lower-level goal than the PHC of the 1970s and far lower than 'the highest attainable standard of health' enshrined in the aforementioned 1946 WHO Constitution (Pupavac 2008).

Criticism against economically defined development and concerns over increased poverty under structural adjustment were further intertwined with environmental concerns. The World Commission on Environment and Development (commonly known as the Brundtland Commission)[3]

---

3 The commission was established in accordance with a UN General Assembly resolution in 1983 and was chaired by Norwegian politician Gro Harlem Brundtland.

launched its report in 1987 and advocated the concept of sustainable development as development that meets the needs of the present without compromising the ability of future generations to meet their own needs (World Commission on Environment and Development 1987). The report argued that economic and social development goals should be defined in light of their sustainability and that values should be promoted that keep consumption levels within an environmentally compatible range. There was now widespread recognition that development should be promoted in a long-term sustainable manner while taking the poor into account rather than simply pursuing economic growth and promoting economic policies without regard to their impact on the poor.

Criticism of development policies that centred on material development, industrialisation and economic growth continued and were further reinforced after the end of the Cold War. In the early 1990s, under the influence of neo-institutionalist economics,[4] aid agencies began to share the view that the market could not function by merely correcting government economic policies and that government should not be simply minimised but rather should play its role properly. Addressing institutional shortcomings through judicial and legislative reforms, democratisation, capacity building of those working in public institutions and anti-corruption measures were deemed necessary for the market to properly function and reduce poverty.

Part of the reason for this attention to the institutions of aid-recipient countries was that the collapse of the Soviet Union and the end of the Cold War made it possible for Western countries to advocate democracy as a universal value. Western countries claimed that democratisation of

---

4 New institutionalist economics generally believes that human rationality is incomplete (limited rationality) and that human beings subjectively recognise the real world and act on it, while their ability to collect, process and communicate information is limited; it focuses on institutional aspects, such as social norms that support this recognition and action. For instance, Douglas North et al. (North, Wallis and Weingast 2009) argue that social norms and social institutional frameworks play an important role in shaping individual behaviour and thinking, that economic development is correlated with the kind of social norms and social institutional frameworks which make democracy work, and that this correlation needs to be examined regarding the question of why poor countries remain poor.

aid-recipient countries would be essential for effective development aid and tried to promote democratisation by making it a condition of bilateral aid, including democratic elections and multi-party systems (Singer 1994).

The World Bank also became actively involved in legal and administrative reform to bring about transformation within aid-recipient states, despite having a mandate[5] prohibiting its intervention in political affairs (Inoue 2006: 69–70). The 'Governance and Development' report published by the World Bank in 1992 defined governance as 'the manner in which power is exercised in the management of a country's economic and social resources for development' (World Bank 1992: 1). The report argued that governance in this sense was not 'political' as long as it was recognised in the context of resource allocation and poverty reduction and set out its policy of involvement in the field of governance. Following this report, the World Bank produced a strategy for combating corruption in 1997 (World Bank 1997a) and the World Development Report 1997 on the role of the state in a changing world (World Bank 1997b) as well as a strategy for public sector reform and strengthening governance in 2000 (World Bank 2000a) and the World Development Report 2002 on building institutions for markets (World Bank 2002). The World Bank recognised the role of the state in addressing market failures and improving equity and advocated ensuring aid effectiveness and reducing poverty by addressing legal, judicial, and anti-corruption measures.

In this context, it was deemed necessary for Southern states to have the will and ability to voluntarily take ownership of the formation and implementation of poverty reduction policies in a manner appropriate to their national contexts. For example, 'Towards the 21st Century: Contributions through Development Cooperation' (OECD DAC 1996), adopted by the OECD DAC in 1996, set a development goal of halving absolute poverty by 2015 and emphasized the need for developing states to take ownership to achieve this goal. Based on the agreement reached at the IMF–World Bank Joint Development Committee and the Interim Committee in September 1999, the IMF and the World Bank also decided

---

5 IBRD Articles of Agreement, Article IV, Section 10 (Political Activity Prohibited).

that aid-recipient states themselves should prepare three-year Poverty Reduction Strategy Papers (PRSPs) that would identify the causes of poverty and strategies for poverty reduction in their own states; donors, international organisations, civil society and the private sector were also expected to be involved in the preparation and monitoring of PRSPs. Some donor countries and international organisations adopted a policy of 'selectivity' (Hout 2007), by which they would selectively provide aid to countries that exhibited good governance, good PRSP documents and their proper implementation.

However, as Graham Harrison (2004) argued, the focus on institutions and the emphasis on 'ownership' and 'partnership' does not necessarily mean that aid-recipient states would be allowed to make independent decisions about the direction and content of their policies. Instead of implementing policies as instructed or ordered by external agencies, aid-recipient states were now supposed to formulate and implement 'appropriate' policies in partnership with domestic and foreign actors and to take ownership of them; aid was to be provided selectively and preferentially to countries that demonstrated the will and ability to do so. Aid-recipient states were expected to transform the values and behaviours of people within the government through the process, while aid agencies and external 'experts' were assumed to be deeply embedded in policy-making and monitoring processes.

Moreover, the attention to institutions for development has been extended not only to the judiciary, legislature, and executive powers in countries of the global South, but to the whole of society. In 1998, Joseph Stiglitz, Senior Vice President and Chief Economist of the World Bank, delivered a speech titled 'Towards a New Paradigm for Strategies, Policies and Processes of Development' in which he argued that past attempts to introduce certain economic policies into the global South had failed because they were based on too narrow a concept of development and that future new development paradigms should aim at catalysing change and transforming whole societies and should involve the transformation of institutions and the creation of new capacities (Stiglitz 1998). It was now argued that trying to force wholesale social transformation on

Southern populations would not be effective and that participation, consensus and ownership by the global South would be the key (Stiglitz 1998: 16). Empowering people at every stage and transforming the entire society from the level of individual minds and behaviours through their participation and ownership came to be seen as the central themes of development. At the same time, global, external, and structural factors, such as the unequal relationship between the global North and South, were increasingly overshadowed in international policy debate on the causes of underdevelopment.

The United Nations Development Programme (UNDP) also recognised the minds, attitudes, behaviours, and relationships of Southern individuals as the key to development. The annual Human Development Report (UNDP 1990), which UNDP began publishing in 1990, relied on Amartya Sen's concept of capability (Sen 1992)[6] to define human development as a process of expanding people's choices and argued that the goal of development should be the expansion of choices, not economic growth or mere satisfaction of basic needs. The report also argued that human development itself should be the goal of development rather than being positioned as complementary to economic development and proposed the Human Development Index (HDI) – a composite of statistics on factors, including life expectancy, years of schooling and national income per capita – as an indicator to measure the degree of human development.

The Millennium Development Goals (MDGs),[7] based on the Millennium Declaration adopted at the United Nations Millennium Summit in September 2000, called for the cooperation of diverse actors to achieve its eight goals (eradicate extreme poverty and hunger, achieve universal primary education, promote gender equality and empower women, reduce child mortality, improve maternal health, combat HIV/AIDS, malaria and other diseases, ensure environmental sustainability and develop a global part-

---

6 Sen considered capability a set of selective functions representing a variety of being and doing that a person can achieve under economic, social and personal qualities, arguing that poverty is a lack of capability, and that development is the expansion of individual capabilities.

7 UN Document. A/RES/55/2. United Nations Millennium Declaration.

nership for development) by 2015. The MDGs were generally supported by governments, international organisations and NGOs, and poverty reduction was made central to policy debate on development. As Vanessa Pupavac (2008: 183–184) points out, while the aforementioned 1946 WHO Constitution declared that 'the highest attainable standard of health' was a basic human right, the MDGs' approach was primarily based on the recognition that the aforementioned 'selective primary health care' – originally conceived as the minimum level of health care to ensure the survival of people (especially children) in crisis situations caused by structural adjustment policies – was a human right.

In addition, concepts such as capability and human development in the 1990s increasingly emphasised happiness and fulfilment based on individual values and subjectivities. In the 2000s, such subjective aspects were further emphasised, and attempts were made to formulate indicators that incorporated subjective aspects. For example, the World Bank's 2000 report, 'The Voices of the Poor', argued that for the world's poor, '[w]ealth and wellbeing are seen as different, and even contradictory... Wellbeing and illbeing are states of mind and being. Wellbeing has a psychological and spiritual dimension as a mental state of harmony, happiness and peace of mind' (Narayan et al. 2000: 21), underscoring the importance of well-being rather than wealth. The 2011 UN General Assembly resolution 'Happiness: Towards a Holistic Approach to Development'[8] also recognised that 'the gross domestic product indicator by nature was not designed to and does not adequately reflect the happiness and well-being of people in a country'. It also stated that 'unsustainable patterns of production and consumption can impede sustainable development' and recognised 'the need for a more inclusive, equitable and balanced approach to economic growth that promotes sustainable development, poverty eradication, happiness and well-being of all peoples.' The resolution also invited UN member states to 'pursue the elaboration of additional measures that better capture the importance of the pursuit of happiness and well-being in development with a view to guiding their public policies.'

---

8 UN Document. A/RES/65/309. Happiness: Towards a Holistic Approach to Development.

The 2012 report of the United Nations Secretary-General's High-Level Panel on Global Sustainability also included a recommendation that the international community develop new indicators for sustainable development beyond GDP (United Nations Secretary-General's High-Level Panel on Global Sustainability 2012: 7, 14, 63, 85). In response, the UNDP indicated its willingness to consider revising the HDI and to develop more comprehensive indicators (Clark 2012). These developments resonated with the debate on the 'post-2015' development agenda as the deadline for achieving the MDGs approached. Goals that would encompass a broader range of themes than the eight MDGs mentioned above were considered, and the resulting Sustainable Development Goals (SDGs) were adopted at the UN Sustainable Development Summit in September 2015.

As we have seen, by the end of the 1990s, the dominant discourse in the development circle was no longer about GDP growth, material modernisation, macroeconomic stability, or a fundamental reorganisation of the world order. A widely shared view was that for development, governments should develop and maintain the will and capacity to formulate and implement 'appropriate' policies to reduce poverty and social and economic disparities while exhibiting ownership of such policies and demonstrating partnership with internal and external actors. At the same time, psychological, behavioural, and relational transformation at the individual and community level was deemed necessary to reduce poverty, and individual households were now expected to maintain their livelihoods in a sustainable manner by managing resources sufficient to meet their food security and immediate social welfare needs. While this transformation does not mean that the GDP concept entirely lost its significance in the policy debate on development or that material improvements were now completely neglected, the dominant discourse of the policy debate had cast attention on non-material elements, such as improved governance, empowerment, capacity building and the transformation of values and behaviours, and the level of material expectations for Southern populations was significantly lowered, as was seen in the change in recognition of a human right to health care.

Moreover, since the 1990s, poverty, underdevelopment and governance deficiencies have been problematised as the key factors that would increase the risk or likelihood of violence and armed conflict (Hillier and Wood 2003: 34–35; OI 2005: 14; UN Millennium Project 2005: 183).[9] Development, conflict resolution and peace building came to be seen as interdependent, as exemplified by the statement in a 2005 report published by the Commission for Africa that '[c]onflict is one of Africa's classic vicious circles. There can be no development without peace, but there can be no peace without development' (Commission for Africa 2005: 25).[10] It was also widely argued that poverty reduction and transformation of Southern states and populations into partners in creating peaceful social change was essential to address the root causes of violence and conflicts in the global South.

The World Development Report 2000/2001 (World Bank 2000b) also linked development and peace issues and supported the idea of development oriented towards the transformation of the values and capacities of Southern people and governments. The report made combating poverty a major theme and set out a policy to fight poverty in three dimensions: opportunity, empowerment, and security. Here, too, the logic to put empowerment and the expansion of opportunity, rather than materialistic development, at the centre of the fight against poverty and to link it to security is explicitly expressed. A similar logic is evident in a 2005 report by the United Kingdom's Department for International Development (DFID) (2005). The report stated that '[t]he most important functions of the state for poverty reduction are territorial control, safety and security, capacity to manage public resources, delivery of basic services, and the ability to protect and support the ways in which the poorest people sustain themselves' (DFID 2005: 7). It further argued that fragile states which could not or would not deliver core functions to the majority of their people were

---

9 UN Document. A/59/565. A More Secure World: Our Shared Responsibility. Report of the High-level Panel on Threats, Challenges and Change, paras. 22–23; UN Document. A/59/2005. In Larger Freedom: Towards Development, Security and Human Rights for All. Report of the Secretary-General, para. 16.

10 Similar arguments can be found, for example, in Collier et al. (2003: 1).

more likely to descend into instability and perpetuate regional conflicts (DFID 2005: 7, 10).

As Vanessa Pupavac pointed out, understanding war and social conflict as arising from internal problems in the global South had not prevailed internationally in the Cold-War period (Pupavac 2004: 153). Rather than personal or cultural attitudes, it was the actions of the First World that had been blamed for conflict in UN debates, dominated by the views of the communist bloc and the newly independent countries (Pupavac 2004: 153). In stark contrast to much of the Cold War policy debate, the view that internal problems of the global South are the root causes of conflict was more strongly proposed by Western policy circles from the 1990s onwards. As the Soviet bloc dissolved and the Non-Aligned movement lost much of the leverage it was once able to exert, a tendency to seek the source of violence and armed conflict in the global South within the internal problems of states, communities and individuals became increasingly prominent in the international arena (Pupavac 2004: 154–156).

The lines between development aid, humanitarian aid and political policy became blurred as 'peacetime' and conflict and post-conflict policy issues came to be seen as overlapping and interrelated. It was argued in the 1990s that humanitarian assistance should be planned and implemented in coordination and collaboration with development assistance and other policies to contribute to desired consequences such as peace and sustainable development, and that humanitarian aid based on the existing humanitarian principles such as neutrality, impartiality and independence could have negative impacts on achieving such consequences (Curtis 2001). The 'new humanitarianism', which held that humanitarian assistance should be provided with the utmost calculation and care so as not to harm the achievement of desired consequences and should be withheld when it is likely to harm such achievement started to gain support among major aid organisations and donors (Curtis 2001). Humanitarian aid was now expected to play a role in empowering and transforming potentially dangerous Southern individuals, communities, and states into partners in creating peaceful social change. And it is precisely when the level of material expectations for Southern populations was significantly lowered, when transformation of society as a whole on Southern populations was deemed necessary not only to reduce poverty but also to prevent violence and conflict in the global South, when the boundaries between development aid, humanitarian aid and political policy became ambiguous, that the concept of resilience started to gain traction in the development and humanitarian sector.

Wide use and criticisms of the resilience concept

The year 1990 also heralded the beginning of the International Decade for Natural Disaster Reduction (IDNDR) which focused on the role of vulnerability in disaster reduction. Although the final report of the Scientific and Technical Committee of the IDNDR did not mention the term 'resilience',[11] its Programme Forum, held in 1999, adopted the International Strategy for Disaster Reduction (ISDR), whose Secretariat used the term 'resilience' extensively in its preliminary report released in 2002 (ISDR 2002) and its final version of the report published in 2004 (ISDR 2004a, 2004b). The preliminary version defined 'resilience/resilient' as:

> [t]he capacity of a system, community or society to resist or to change in order that it may obtain an acceptable level in functioning and structure. This is determined by the degree to which the social system is capable of organising itself, and the ability to increase its capacity for learning and adaptation, including the capacity to recover from a disaster. (ISDR 2002: 24)

However, the final version defined 'resilience/resilient' slightly differently, as:

---

11 UN Document. A/54/132/Add.1 – E/1999/80/Add.1. International Decade for Natural Disaster Reduction, Report of the Secretary-General, Addendum, Final report of the Scientific and Technical Committee of the International Decade for Natural Disaster Reduction.

[t]he capacity of a system, community or society potentially exposed to hazards to adapt, by resisting or changing in order to reach and maintain an acceptable level of functioning and structure. This is determined by the degree to which the social system is capable of organizing itself to increase its capacity for learning from past disasters for better future protection and to improve risk reduction measures. (ISDR 2004a: 16–17, 2004b: 6)

In 2005, the Hyogo Framework for Action 2005–2015: Building the Resilience of Nations and Communities to Disasters, agreed upon at the World Conference on Disaster Reduction, referred to the definition adopted in the ISDR's final report in 2004.[12]

These and similar definitions have been widely used in policy circles in the development and humanitarian sector ever since,[13] and a great number of policy and scholarly documents have promoted resilience as a key concept in addressing matters such as human development, poverty reduction, disaster risk reduction, peace building, and provision of humanitarian aid. For instance, UNDP's 'Human Development Report 2014: Sustaining Human Progress: Reducing Vulnerabilities and Building Resilience' stated that resilience underpins any approach to securing and sustaining human development (UNDP 2014: 5). The report problematised vulnerabilities of '[h]undreds of millions of poor, marginalized or otherwise disadvantaged people', argued that 'failures to protect people against vulnerability are mostly a consequence of inadequate policies and poor or dysfunctional

---

12 UN Document. A/CONF.206/6. Report of the World Conference on Disaster Reduction. Kobe, Hyogo, Japan, 18–22 January 2005, Resolution 2, Hyogo Framework for Action 2005–2015: Building the Resilience of Nations and Communities to Disasters, p. 9.

13 For instance, the Sendai Framework for Disaster Risk Reduction 2015–2030, the successor instrument to the Hyogo Framework for Action (HFA) 2005–2015: Building the Resilience of Nations and Communities to Disasters, uses a slightly different definition of resilience as '[t]he ability of a system, community or society exposed to hazards to resist, absorb, accommodate to and recover from the effects of a hazard in a timely and efficient manner, including through the preservation and restoration of its essential basic structures and functions'. See UN Document. A/CONF.224/L.2. Sendai Framework for Disaster Risk Reduction 2015–2030, p. 2.

social institutions' (UNDP 2014: 16), called for the need to reduce vulnerability through creating more-responsive states, better public policies and changes in social norms, and argued that '[r]esponsive institutions and effective policy interventions can create a sustainable dynamic to bolster individual capabilities and social conditions that strengthen human agency – making individuals and societies more resilient' (UNDP 2014: 10).

Since its major debut in international policy debate in the development and humanitarian sector in the 2000s, the concept of resilience has been widely criticised. Critics generally argue that the wide use of the concept of resilience since the 2000s was precipitated by the aforementioned shift in the dominant discourse in policy debate in the development and humanitarian sector from GDP growth, material modernisation, macroeconomic stability or a fundamental reorganisation of the world order to internal and non-material transformation, such as improved governance, empowerment, capacity building and the transformation of norms, values and behaviours of individuals, communities and governments in the global South. The term 'resilience' started to gain traction in the development and humanitarian sector, exactly when such a shift had already materialised and the idea that the transformation of the whole of societies of Southern states had come to be seen as necessary not only to reduce poverty but also to prevent violence and conflict in the global South. Thus, the criticism of this concept has been accompanied by a critical examination of the dominant discourse on which it supposedly rests.

For example, Mark Duffield (2005, 2007, 2009, 2010) points out that concepts such as sustainable development and resilience are based on fundamentally different thinking from the 1960s pursuit of material development and modernisation and that the focus in resilience thinking is on enabling people of the global South to maintain a state of stability at lower material levels through self-reliance and to acquire and maintain resilience that enables them to respond and adapt to such events as disasters and conflicts. The anti-modernist and anti-materialist ideas behind the resilience concept has led to an excessive focus on non-material aspects and has diverted attention away from material causes of suffering. Coupled with the tendency in international policy circles since the 1990s to seek root causes of underdevelopment and violence in internal dysfunctions, resilience thinking tends to focus and problematise the day-to-day of Southern individuals, communities and states and may lead to overly intrusive interventions into non-material elements, such as the hearts, minds, values, and cultures of Southern populations. The resilience concept ignores or downplays global, structural, political, and material causes of suffering, and its orientation is to shift the responsibility away from the state to the individuals and communities and thereby legitimise measures focused on pacification and containment without ensuring significant material development.[14]

Some critics also argue that although resilience thinking at first glance seems to focus on people's capability and power, it is in fact based on an assumption of general human vulnerability and tends to seek root causes and solutions in internal problems of states and populations. As mentioned in the Introduction of this chapter, some advocates of this concept claim that the concept of resilience may enable an approach that overcomes the limitations of imagining Southern populations as powerless and that respects their autonomy and potential (Almedom and Tumwine 2008; FAO 2016). In contrast, some critics point out that the view that sees humans as vulnerable and the idea to seek to strengthen resilience are two sides of the same coin, and they problematise the logic that runs through both. For example, Frank Furedi (2007: 180–183) and Vanessa Pupavac (2012) argue that the concept of resilience assumes general human vulnerability and legitimises interventions into people's values, culture, and social relationships. David Chandler (2014) and Brad Evans and Julian Reid (2014) also argue that the resilience concept reflects a transformation from the idea of the self as capable of judging one's own goals and interests and changing the environment to emancipate oneself to a vulnerable self whose capacity to do so is dubious.

According to Chandler (2014), the resilience concept is fundamentally based on the denial of human agency in changing the world we live in.

---

14 Some authors point to similarities with colonial discourses or with Nazi discourse to encourage adaptation and non-resistance in concentration camps (see Reid 2012).

While the concept ostensibly appears to pay attention to the power and capability of humans, it assumes the fundamental vulnerability of humans. The concept is not based on classical liberal subjects who emancipate themselves through growth and the transformation of their circumstances, and therefore there is no longer the starting assumption of a transformative subject who drives progress and emancipation. The ontology underlying the concept of resilience is the view that humans are inherently and fundamentally vulnerable and that in such an idea, the ability of humans to anticipate crises, ensure safety, and transform the external environment is doubted, which is why crises and dangers are assumed to be unavoidable and ever-present. Citizens are now imagined as unable to change the world; they are ever vulnerable to risk and dysfunction and are therefore only expected to learn to cope with crisis and suffering. In this thinking, the principle of non-interference in the internal affairs of states and the private lives of individual citizens is fundamentally questioned. Once the rational capacity to make choices is denied to the human subject, there is no basis for liberal representative political and legal theory, and concepts, such as citizenship, are hollowed out and lose their political meaning. This thinking is fundamentally anti-modern and is based on fundamental doubts about the bases of modern thoughts and institutions, such as the human ability to think, create and change the world we live in. Human agency is doubted; political possibilities are closed off, and (modern) aspirations for material development are denied and even stigmatised. Critics argue that such 'resilient subjects' are not imagined as political subjects capable of judging and acting to transform the environment and free themselves from dangers but rather as subjects who can only adapt to immutable situations and alleviate suffering. Such a view is fundamentally anti-modernist and nihilistic.[15]

---

15 Chandler also points out that the concept of resilience seems to legitimise intrusive interventions into the everyday lives of Southern populations while human rationality is universally degraded and doubted (Chandler 2014). As a result, no one, not even those who intervene, is assumed to have the moral capacity to conceptualise the good, and no one is expected to provide a universal solution or linear path towards resilience. Thus, attempts are made to shift responsibility to local populations under the name of participation and ownership, yet these same local populations are also assumed to lack human agency.

Recently, criticisms of resilience thinking have resurged as part of the move to challenge racism embedded in the development and humanitarian sector as the Black Lives Matter movement spread across continents and sectors following the killing of George Floyd by Minneapolis police officer Derek Chauvin in the United States on 25 May 2020. Although a number of development and humanitarian aid organisations initially issued statements and articles to support the movement and to educate the public about racism, this was soon followed by a cascade of criticisms of the racism prevalent in the sector itself (Bruce-Raeburn 2020; Peprah 2020; Slim 2020). As the Black Lives Matter movement brought to the fore the systemic, structural and widespread nature of racism, a logical conclusion was that the sector may not have been immune to it. While the criticism provoked resistance and discomfort in some quarters of the sector, others claim that continuing to ignore the ubiquity of race and racism is no longer an option (Nwajiaku-Dahou and Leon-Himmelstine 2020). Structural racism, racist concepts and racist behaviours and attitudes that have long been overlooked in the sector have now been discussed in social media using hashtags – such as #DecolonizeDevelopment, #AntiRacistInAid, #RethinkingHumanitarianism, #AidToo – and humanitarian aid and advocacy organisations have been openly called out (Adelman 2020; Majumdar 2020; Oppenheim 2020). Attitudes, values, and norms that are underpinned by racist and racializing ideas are also being critically examined, and resilience is no exception. Some criticise the tendency of resilience thinking to seek to pacify the populations in the global South at lower material levels without addressing material causes or wider structural and political causes of suffering as a part of racist thought in the sector. Such criticism is well summarised by Angela Bruce-Raeburn, who argued that we should 'stop enabling the use of words such as "resilient" and "resilience" when referring to people who live every day with uncertainty' (Bruce-Raeburn 2020). She points out that these words 'serve to perpetuate inequalities while telling vulnerable people how strong they are – and therefore changing absolutely nothing in their lives' (Bruce-Raeburn 2020).

Potentials and difficulties of using the resilience concept

As discussed above, there have been numerous criticisms of the concept of resilience in the development and humanitarian sector. What, then, are the implications for using the concept in the sector? This section analyses the critical literature introduced in the previous section and suggests the potentials and difficulties of the use of this concept in the sector.

The criticisms are convincing and seem to be based on detailed examinations of international policy debate. The tendency of resilience thinking to ignore structural injustices and material causes of suffering and to legitimise measures to maintain a state of stability at lower material levels in the global South is evident and cannot escape the criticism of its racist undertones. Other points – such as the concept of resilience being based on an assumption of general human vulnerability and justifying interventions into people's values, culture, and social relationships – seem to be exemplified by major documents, including the Human Development Report series. For example, the aforementioned report published by the UNDP in 2014, titled 'Human Development Report 2014: Sustaining Human Progress: Reducing Vulnerabilities and Building Resilience', argued that while everyone is in principle vulnerable to some extent, some people are more vulnerable than others (UNDP 2014: 18–20). It asserted that '[c]onceptions of security require a view of the human person that includes physical and psychological vulnerability, strengths and limitations, including limitations in the perception of risk' (UNDP 2014: 77). The report claimed that everyone is fundamentally vulnerable to varying degrees and cannot always be fully aware of risk and that '[p]rogress has to be about fostering resilient human development' (UNDP 2014: 1). Diverse issues – including provision of basic social services, full employment, social inclusion and early childhood development – were proposed to holistically build resilience to a wide range of risks.

At the same time, we may need to avoid the tendency that is seen in many of the critics' arguments – the tendency to view the concept of resilience as stable and static. Critics of this concept often interpret its use as accepting more or less the same meaning, the same framework of thought and the same underlying concept of the self as generally imparted by agencies in the global North. The diffusion of this concept is often perceived as a monolithic process, enabling the creation of resilient subjects, and, hence, the possibility of various actors interpreting or localising the concept – unconsciously transforming or consciously manipulating the semantic content of the concept – tends to be ignored. In fact, the concepts used by actors of the global North – such as civil society, empowerment, and trauma – have been diversely interpreted and utilised by local actors to justify and receive political and financial support for a wide range of policies and practices which might not have otherwise been supported (Enomoto 2011). Like these concepts, the concept of resilience also has the possibility of facing tensions between diverse actors over meaning in individual contexts (Wandji, Allouche and Marchais 2021). Different meanings may be assigned to the term in translation, and its meaning may be transformed by the conscious use of the term as a rhetorical tool.

In fact, there have been some scholarly efforts to utilise the term 'resilience' to critically analyse the situation in which structural injustices and causes of suffering, such as colonialism, apartheid and xenophobia are persistently reinstated and revived (Nyamnjoh, 2016; Umeya 2020). While the legitimacy of the term 'resilience' may be enhanced through the process, there is also the possibility that the semantic content of this concept may be diffused or transformed by the conscious (and sometimes unconscious) actions of a wide range of actors. Thus, it is hard to conclude that the use of this concept necessarily leads to legitimation of the discourse that expects Southern populations to respond to, adapt to, and recover from difficulties and suffering without significant material development and without addressing political and structural matters.

Conversely, the outcome of the battle over the semantic content of a particular concept depends on the power relations in the arena in which it takes place. As became clear through the debate after the Black Lives Matter movement, the development and humanitarian sector has long been deeply imbued with structural racism and power imbalances between the global North and South. Ideologies and practices of the powerful have been routinely normalised and legitimised, and the triumph of the term 'resilience' in the sector may be understood precisely as the resilience of structural racism and power imbalances.

Since 2020, terms such as decolonisation of aid and anti-racism have quickly become buzzwords and have already been co-opted and utilized by organizations situated in or originating from the global North. As such terms may be in danger of becoming catch-all terms for any change initiative, the sector is clearly at a crossroads as to which path to take. Concurrently, recognizing racism in the development and humanitarian sector overall sheds light on a tough question for practitioners and researchers – how should we address Northern-centric, Western-centric or racist structures, systems, practices, terminology and concepts including 'resilience'? At this critical juncture, we are currently more than ever confronted with the difficult choice as to whether and how to use this the term 'resilience' and may need to be conscious of our goals, the feasibility of achieving them and the strategies we must take in achieving them if we are to use the term.

Acknowledgements

This work was supported by the JSPS KAKENHI Grants 18H03606, 16H06318, 16H05664, 16K17075, and 21K13250.

References

Adelman, L. 2020. 'I've seen first-hand the toxic racism in international women's rights groups'. *The Guardian*, 17 August. https://www.theguardian.com/global-development/2020/aug/17/ive-seen-first-hand-the-toxic-racism-in-international-womens-rights-groups (accessed 30 April 2022).

Aghion, P. and P. Bolton. 1997. 'A theory of trickle-down growth and development'. *Review of Economic Studies* 64 (2): 151–172.

Almedom, A.M. and J.K. Tumwine. 2008. 'Resilience to disasters: A paradigm shift from vulnerability to strength'. *African Health Sciences* 8 (S): 1–4.

Bruce-Raeburn, A. 2020. 'Opinion: the hustle – white saviors and hashtag activism'. *Devex*, 12 June. https://www.devex.com/news/opinion-the-hustle-white-saviors-and-hashtag-activism-97463 (accessed 30 April 2022).

Chandler, D. 2014. *Resilience: The Governance of Complexity*. Abingdon and New York: Routledge.

Clark, H. 2012. *Beyond GDP: Measuring the Future We Want*. Opening statement at UNDP event on measurement of sustainable development at Rio+20, Rio Centro, Brazil, 20 June.

Collier, P., L.Elliot, H. Hegre, A, Hoeffler, M. Regnal-Querol and N. Sambanis. 2003. *Breaking the Conflict Trap: Civil War and Development Policy*. Washington DC and Oxford: World Bank and Oxford University Press.

Commission for Africa. 2005. *Our Common Interest: Report of the Commission for Africa*.

Cornia, G.A., R. Jolly and F. Stewart. (eds). 1987. *Adjustment with a Human Face, Volume 1: Protecting the Vulnerable and Promoting Growth*. Oxford: Oxford University Press.

————— 1988. *Adjustment with a Human Face, Volume II: Ten Country Case Studies*. Oxford: Oxford University Press.

Curtis, D. 2001. 'Politics and humanitarian aid: Debates, dilemmas and dissension'. *Humanitarian Policy Group (HPG) Report* 10.

Delgado, C.L. 1995. 'Africa's changing agricultural development strategies: Past and present paradigms as a guide to the future'. *Food, Agriculture, and the Environment Discussion Paper* 3.

DFID (Department for International Development). 2005. *Why We Need to Work More Effectively in Fragile States*. London: DFID.

Duffield, M. 2005. 'Getting savages to fight barbarians: Development, security and the colonial present'. *Conflict, Security and Development* 5 (2): 141–159.

————— 2007. *Development, Security and Unending War: Governing the World of Peoples*. Cambridge and Malden: Polity Press.

————— 2009. 'Liberal internationalism and the fragile state: Linked by design?' In M. Duffield and V. Hewitt (eds). *Development and Colonialism: The Past in the Present*. pp. 116–129. Woodbridge and Rochester: James Currey.

————— 2010. 'The liberal way of development and the development-security impasse: Exploring the global life-chance divide'. *Security Dialogue* 41 (1): 53–76.

Enomoto, T. 2011. 'Revival of tradition in the era of global therapeutic governance: The case of ICC intervention in the situation in Northern Uganda'. *African Study Monographs* 32 (3): 111–134.

European Commission. 2014. *Beyond GDP: Measuring Progress, True Wealth, and the Well-Being of Nations*. http://ec.europa.eu/environment/beyond_gdp/index_en.html (accessed 30 April 2022).

Evans, B. and J. Reid. 2014. *Resilient Life: The Art of Living Dangerously*. Cambridge and Malden: Polity Press.

FAO (Food and Agriculture Organization). 2016. *Evaluation of FAO Strategic Objective 5: Increase the Resilience of Livelihoods to Threats and Crises*. Rome: FAO.

Furedi, F. 2007. *Invitation to Terror: The Expanding Empire of the Unknown*. London and New York: Continuum.

Gilman, N. 2003. 'Modernization theory, the highest stage of American intellectual history'. In D.C. Engerman, N. Gilman, M. Haefele and M.E. Latham (eds). *Staging Growth: Modernization, Development, and the Global Cold War.* pp. 47–89. Amherst: University of Massachusetts Press.

Harrison, G. 2004. *The World Bank and Africa: The Construction of Governance States.* Abingdon and New York: Routledge.

Hillier, D. and B. Wood. 2003. *Shattered Lives: The Case for Tough International Arms Control.* Amnesty International and Oxfam International.

Hout, W. 2007. *The Politics of Aid Selectivity: Good Governance Criteria in World Bank, U.S. and Dutch Development Assistance.* Abingdon and New York: Routledge.

Inoue, J. 2006. 'Policies of the UN system and the EU promoting "good governance" and "anti- corruption" in developing countries: For fighting against poverty'. *Keio Law Journal* 4: 63–101. (in Japanese)

ISDR (International Strategy for Disaster Reduction). 2002. *Living with Risk: A Global Review of Disaster Reduction Initiatives, Preliminary Version.* Geneva: ISDR.

————— 2004a. *Living with Risk: A Global Review of Disaster Reduction Initiatives, 2004 Version.* Volume I. Geneva: ISDR.

————— 2004b. *Living with Risk: A Global Review of Disaster Reduction Initiatives, 2004 Version.* Volume II. Geneva: ISDR.

Kapoor, I. 2008. *The Postcolonial Politics of Development.* Abingdon and New York: Routledge.

Majumdar, A. 2020. 'Bearing witness inside MSF'. *The New International,* 18 August. https://www.thenewhumanitarian.org/opinion/first-person/2020/08/18/MSF-Amsterdam-aid-institutional-racism (accessed 30 April 2022).

Narayan, D., R. Chambers, M.K. Shah and P. Petesch. 2000. *Voices of the Poor: Crying out for Change.* New York: Oxford University Press.

North, D.C., J.J. Wallis, B.R. and Weingast. 2009. *Violence and Social Orders: A Conceptual Framework for Interpreting Recorded Human History.* Cambridge: Cambridge University Press.

Nyamnjoh, F.B. 2016. *#RhodesMustFall: Nibbling at Resilient Colonialism in South Africa.* Bamenda: Langaa.

Nwajiaku-Dahou, K. and C. Leon-Himmelstine. 2020. 'How to confront race and racism in international development'. *Overseas Development Institute,* 5 October. https://www.odi.org/blogs/17407 how to confront-race-and-racism-international-development (accessed 30 April 2022).

OECD DAC (Organisation for Economic Cooperation and Development, Development Assistance Committee). 1996. *Shaping the 21st Century: The Contribution of Development Co-operation.* Paris: OECD.

OI (Oxfam International). 2005. *Paying the Price: Why Rich Countries Must Invest Now in a War on Poverty.* Oxford: Oxfam International.

Oppenheim, M. 2020. 'Nobel Women's Initiative leaders "profoundly shaken" as seven staff step down'. *Independent,* 7 August. https://www.independent.co.uk/news/world/nobel-womens-initiative-staff-resign-laureates-diversity-a9659571.html (accessed 30 April 2020).

Paloni, A. and M. Zanardi. 2006. 'The IMF, World Bank and policy reform: Introduction and overview'. In A. Paloni and M. Zanardi (eds). *The IMF, the World Bank and Policy Reforms.* pp. 1–23. Abingdon and New York: Routledge.

Peprah, D. 2020. 'Opinion: We must translate anti-racism statements into action'. *Devex,* 30 June. https://www.devex.com/news/opinion-we-must-translate-anti-racism-statements-into-action-97591 (accessed 30 April 2020).

Pieterse, J.N. 2009. *Development Theory.* 3rd ed. London, Thousand Oaks, New Delhi and Far East Square: Sage.

Pupavac, V. 2004. 'War on the couch: The emotionology of the new international security paradigm'. *European Journal of Social Theory* 7 (2): 149–170.

———— 2008. 'Changing concepts of international health'. In D. Wainwright (ed). *A Sociology of Health.* pp. 173–190. London, Thousand Oaks, New Delhi and Far East Square: Sage.

———— 2012. 'Global disaster management and therapeutic governance of communities'. *Development Dialogue* 58: 81–97.

Reid, J. 2012. 'The disastrous and politically debased subject of resilience'. *Development Dialogue* 58: 67–79.

Sen, A. 1992. *Inequality Re-examined.* Oxford: Clarendon Press.

Singer, H.W. 1994. 'Aid conditionality'. *IDS (Institute of Development Studies) Discussion Paper* 346.

Slim, H. 2020. 'Is racism part of our reluctance to localise humanitarian action?' *Humanitarian Practice Network,* 5 June. https://odihpn.org/blog/is-racism-part-of-our-reluctance-to-localise-humanitarian-action/ (accessed 30 April 2022).

Stiglitz, J.E. 1998. 'Towards a new paradigm for development strategies, policies, and processes'. Paper given at the 1998 Prebisch Lecture at United Nations Conference on Trade and Development (UNCTAD), Geneva, Switzerland, 19 October.

Umeya, K. 2020. 'The resilience of apartheid and xenophobia: from a glimpse on the ethnicity and citizenship in "Uber" business in South Africa'. *Tabunka Shakai Kenkyu* 6: 317–338. (in Japanese)

UNDP (United Nations Development Programme). 1990. *Human Development Report 1990.* New York and Oxford: Oxford University Press.

———— 2014. *Human Development Report 2014: Sustaining Human Progress: Reducing Vulnerabilities and Building Resilience.* New York: UNDP.

United Nations Secretary-General's High-Level Panel on Global Sustainability. 2012. *Resilient People, Resilient Planet: A Future Worth Choosing.* New York: UN.

UN Millennium Project (United Nations Millennium Project). 2005. *Investing in Development: A Practical Plan to Achieve the Millennium Development Goals*.

Wandji, D., J. Allouche and G. Marchais. 2021. 'Vernacular resilience: An approach to studying long-term social practices and cultural repertoires of resilience in Côte d'Ivoire and the Democratic Republic of Congo'. *STEPS Working Paper* 116.

WHO (World Health Organization). 1946. *Constitution of the World Health Organization*. http://apps.who.int/gb/bd/PDF/bd47/EN/constitution-en. pdf (accessed 30 April 2022).

WHO and UNICEF (World Health Organization and United Nations Children's Fund). 1978. *Primary Health Care: Report of the International Conference on Primary Health Care, Alma-Ata, USSR, 6–12 September 1978*. Geneva: World Health Organization. http://whqlibdoc.who.int/publications/ 9241800011.pdf (accessed 30 April 2022).

World Bank. 1992. *Governance and Development*. Washington DC: World Bank.

———— 1997a. *Helping Countries Combat Corruption: The Role of the World Bank*. Washington DC: World Bank.

———— 1997b. *World Development Report 1997: The State in a Changing World*. New York: Oxford University Press.

———— 2000a. *Reforming Public Institutions and Strengthening Governance: A World Bank Strategy*. Washington DC: World Bank.

———— 2000b. *World Development Report 2000/2001: Attacking Poverty*. New York: Oxford University Press.

———— 2002. *World Development Report 2002: Building Institutions for Markets*. New York: Oxford University Press.

World Commission on Environment and Development. 1987. *Our Common Future*. Oxford: Oxford University Press.

Resilience

through

Livelihood

Diversification

03

Resilience and the political economy of diversification: The case of Il Chamus, Baringo County, Kenya, 1980–2018

Peter D. Little

Introduction

Uses and abuses of the concept resilience are widespread both in the applied and basic social sciences. Some individuals – especially development practitioners – support its importance while others argue that it de-politicizes disasters (shocks) and avoids underlying political economic causes (Özden-Schilling 2022). The latter position is not a new concern in development studies and anthropology (Ferguson 1994; Li 2014), but instead it is resurrected through research on resilience and its applications (see Barrios 2016). This chapter builds on an earlier contribution of mine (2021) that looked at long-term patterns of pastoralist diversification in north-central Kenya and whether or not they contributed to resilience. Although it includes much of the materials from my earlier article, it adds a section on the political economic context to help explain why widespread pastoralist diversification and its different forms have taken place at all. It also adds a discussion about who has benefited – become more "resilient" – from political economic changes and diversification, and who has not. Regardless of the different contexts, what is clear from recent anthropological and ecological research is that considerable social and economic changes have taken place in East African pastoralism during the past 30 years or so, which have introduced both challenges and opportunities for these populations (Anderson and Bollig 2017; Fratkin and Roth 2004). Most important among these changes has been a general decline in per capita livestock holdings and a corresponding need to diversify, both in diet and economic pursuits, in order to survive and, in some cases, prosper.

The chapter uses longitudinal data from the Il Chamus community of Baringo County, Kenya, who I argue are an extreme case of a pastoral community with extraordinary challenges as well as a range of non-pastoral opportunities. It also is a community that has been strongly marginalized, even when compared to other pastoralists and agro-pastoralists, and has been subjected to considerable violence and losses of their rangelands. The response by many has been to seek alternatives to livestock production. Some of these non-pastoral opportunities, such as charcoal production, actually jeopardise pastoralism but help to cope with short-term difficulties. The chapter highlights *temporality* as a key feature of understanding resilience in terms of maintaining a livelihood and set of relationships that sustain individuals and communities over the long-term, and assist them in the short-term to deal with periodic shocks (drought, human and animal diseases, political violence, and other disasters). As used here, the concept of resilience encompasses a political economic dimension that helps to explain how pastoral livelihoods evolve not just in response to natural disasters, such as drought, but also to the challenges and opportunities that widespread political and economic changes present. In this sense, the article's treatment of resilience is limited to a specific ethnographic context and set of factors, and it does not attempt a comprehensive review of its varied applications, interpretations, and theories (see Adger 2000; Folke 2006; Holling 1973).

The article also uses the lens of class and social differentiation (especially gender) to suggest that increased patterns of diversification, including remittances, occur among both better-off and poor households headed either by females or males. It asks these additional questions: (1) resilience for whom and (2) livelihood diversification for whom? Lucrative diversification options, especially in towns, are found only among better-off households (top two to three per cent), including those who once had significant numbers of livestock; while for poor families, economic diversification is a survival strategy. For better-off households, diversification is an investment strategy, especially in urban rental properties and education, but for the large majority diversification is about survival (Little et al. 2001 and 2009). It will be shown that, as per capita livestock numbers declined and livelihood insecurity grew, the dependence on cash incomes and expenditures for most households increased, as did wealth inequalities.

A short note on methods is required before continuing. The paper draws on field research conducted intermittently during 1980–81, 1984; 1999–2005; and 2016–18 for a total of about 2.75 years. The most recent research (2016–18) includes two periods of detailed interviews with heads and/or principal members of 100 households in two Il Chamus locations, Salabani and Ngambo; structured and unstructured interviews with more than sixty non-rural individuals residing in or near Nairobi, Nakuru and Marigat towns; discussions with sixteen small groups of community members (five to eight individuals) differentiated by gender and age; and participant observation. The field research during 1980–1981, included an eighteen-month study of regional trade and household economy that covered a sample of 56 households in three locations of Il Chamus and interviews with more than fifteen traders and numerous government and development officials in Baringo; and the 1999–2005 sample included a study of livelihood diversification and risk ranking among 30 households in Baringo, as well as follow-up interviews with more than 40 family members from the 1980–1981 field research. The analyses in the following sections should be treated as preliminary. Further analyses, especially of ethnographic materials, are required to provide better context to the mainly descriptive statistical data presented in the article. While recognising the many shortcomings of relying on reported incomes and expenditures, in-person observations of housing, livestock holdings and other key assets, as well as knowledge of education costs and salary ranges for different occupations, served as important cross-checks on these data.

Context

The Il Chamus are a Maa-speaking group of the central Rift Valley who are related to Maasai and Samburu, and who reside in Marigat and Mukutani Divisions in Baringo County, Kenya. The Il Chamus' semi-arid environment of approximately 600 mm of annual rainfall is drought prone with climate shocks occurring on average about every five to six years (Kenya 1994). Because of an extremely high evapotranspiration rate, it supports arid vegetation normally found in much drier parts of Kenya (Little 1992). Major drought emergencies where rainfall over successive seasons is fifty per cent or more below averages occur about every ten to eleven years and can eliminate up to fifty per cent or more of local cattle (McPeak et al. 2012; Little 1992). The most serious droughts since 1970 occurred in 1973–74, 1979–80, 1984, 1991–92, 1999–2000, and 2011. For most of the past 100 years, Il Chamus have pursued mainly pastoralist and agro-pastoralist livelihoods and a ritual and social system that centres on livestock, especially cattle and their exchange (Little 1992 and 2016). In the mid-to-late nineteenth century, they were an important grain producer for the long-distance caravan trade as well as a refuge from drought and warfare for Maa-speaking pastoralists, such as Samburu and Maasai (Anderson 2003). Thus, they have engaged with markets for more than 100 years, but as will be shown latter in the chapter the intensity of market dependence for food purchases and sales of livestock, charcoal, and labour has intensified since 1980.

Similar to most ethnic groups in colonial Kenya, the Il Chamus were restricted to a native "reserve" called Njemps (a colonial mispronunciation of Chamus) that was territorially undersized for a group where pastoralism was practiced. On the reserve's borders to the north, south, west were two much larger pastoralist (Pokot) and agro-pastoralist (Tugen) groups and to the east was large-scale European settlement focused on cattle ranching. With these borders imposed by the colonial state, the growing pastoral sector of the Il Chamus had little possibility of territorial expansion beyond the borders of the state-demarcated Njemps Reserve (for the history of the area, see Anderson 2002; Little 1992).

With constraints on herd mobility during long dry seasons and droughts, Il Chamus were forced to practice a form of semi-sedentary pastoralism that overused certain rangelands and resulted in massive livestock losses in poor rainfall years. They also had begun to practice seasonal labour migration by the 1940s to work on European-owned ranches in neighbouring Laikipia County. The work was on the same lands that the group had depended on for seasonal grazing before they were expropriated for European use. Although small-scale irrigation had always played a role in Il Chamus livelihoods, the community also began to pursue risky rainfed farming during the 1950s and 1960s. In the 1950s the colonial state also expropriated a large amount of excellent seasonal pastures (in excess of 4,000 hectares) and perennial water sources to promote large-scale irrigation and to resettle populations from outside Il Chamus. This intervention further constrained livestock production and stimulated the growth of local towns, such as Marigat. The push on settlement and irrigation continued in the post-colonial era. In the 1980s diversification was not just a pursuit of poor but also wealthier Il Chamus households who through grain production were able to hold on to their livestock by avoiding sales to purchase food (see Little 1983 and 1985). In short, diversification into farming at the time helped to sustain pastoralism in the area, a finding that Lesorogol (2008) and McCabe et al. (2010) also found to be true for Samburu of Kenya and Maasai of Tanzania, respectively. In the past ten years, the culture of sending remittances back to the community has played an equally important role in sustaining pastoralism and a pastoralist identity.

Other political economic processes further constrained pastoralism in the area. From 2000 to 2018 the livestock-based economy declined significantly and non-pastoralist activities grew due to at least three important factors. First was the further loss of pastures as a result of significant *Prosopis juliflora* invasion (henceforth referred to as *prosopis*). The introduction of *prosopis*, referred to locally as the "devil plant," in the 1980s was a disastrous measure to combat "desertification," an unproven concept that had its origins in the colonial period. By the early 2000s much of the key pasture areas in the area, especially important dry season pastures around Lake Baringo, had been taken over by *prosopis* and were unusable to livestock. To quote a recent publication: "What had once been an open landscape with patches of acacia trees was now a mass of gnarly, bushy trees subject to damaging floods, of little utility for a livestock-based economy" (Little 2019: 139). Second, large-scale violent raids by neighbouring groups, especially Pokot, led to further losses of livestock with some families divesting completely out of cattle for fear of attracting armed raiders. As a small ethnic group of approximately 35,000, Il Chamus had little chance against the much larger Pokot and Tugen, who as members of the larger Kalenjin community could also draw on considerable political support (Little 2016). Finally, land encroachment on Il Chamus territory by these groups, often backed up by the potential for violence, further constrained pastoralism in the area and pushed more families to diversify into other activities. In all of these cases, the Kenyan state has done little to resolve the political issues negatively impacting Il Chamus and their livelihoods.

Trends in pastoralism and livelihoods, 1980-2018

Although the Il Chamus economy already experienced diversification prior to the 1980s, especially with regard to cultivation and labour migration, the scale of recent trends is unprecedented (see Little 1985 and 1992; Little et al. 2009). The political-economic processes described above that strongly disrupted pastoralism facilitated diversification into non-pastoralist activities. Table 3.1, for example, compares the average livestock holdings per household and per capita as well as the holdings of different animal species (cattle, goats and sheep) at two different points in time: 1980–81 and 2017–18. As the data show, per capita and household (HH) livestock holdings as measured by Tropical Livestock Units (TLUs)[1] declined more than 75 per cent since 1980–81, while those poor households with less than three TLUs increased from ten to 38 per cent of total

---

1 1 TLU =1 head of cattle or 10 head of goats or sheep.

households. Other data from 2001–02 show that by this time, household livestock holdings in the area were down considerably compared to levels in 1981. Those households with more than 50 TLU (the equivalent of fifty cattle or the equivalent number in goats and sheep) are now only three per cent of households, whereas they represented more than 25 per cent of the community in 1980–81. Even accounting for inter-annual changes in livestock herds based on occurrences of droughts, animal disease outbreaks and livestock raids, the decline in livestock holdings over an approximate 36-year span has been dramatic.

Not surprisingly, the percentage of income derived from livestock has declined while other forms of revenues have grown. New forms of cash, such as pensions/government socialsecurity payments, also occur in recent years, while others, such as fishing, have declined. Irrigated farming, in turn, has grown considerably since 2000 and is being pursued wherever feasible, including along the shores of Lake Baringo, the banks of local rivers (Pekerra and Molo Rivers), and on or near government-supported irrigation schemes, such as the Pekerra Irrigation Scheme. Although mean annual precipitation of 600–650 mm would seem sufficient for rainfed farming, it is poorly distributed during the year with long periods of very little rain early in the main growing season (long rains) (Little 1992: 24). The growth in irrigation accounts for the relatively large percentage of income now derived from agriculture (Table 3.2).

The drop in the importance of livestock production and sales in household incomes was more significant between 2000–01 and 2018 than between 1981 and 2000–01. Part of the reason for this is the decline in livestock since 2005 due to large-scale raids by Pokot and the cumulative

Table 3.1 Changes in livestock herds and species composition, Il Chamus

Year	Cattle Per HH/ Per Cap	Goat Per HH/ Per Cap	Sheep Per HH/ Per Cap	TLUs Per HH/ Per Cap	% < 3 TLUs
1980–81a	16/2.38	29/4.3	73/10.77	26.2/3.86	10
2017–18b	4.6/0.77	9/1.36	16/2.42	6.1/0.92	38

Notes: (a) Based on sample of 56 households. (b) Based on sample of 50 households.
   (c) HH=household.

impacts of *prosopis* expansion. A large percentage of households now purchase fresh milk from shops and traders who transport it from neighbouring Nakuru County and south Baringo (Eldama Ravine), a stark indication of the steep reduction in subsistence herds. Many herds in Nakuru and Eldama Ravine are comprised of high-yielding milk producers, such as exotic Friesian and Aisha breeds or mixed breeds of European and local cattle, and some of these are owned by Il Chamus who have settled outside Baringo. The irony of pastoralists now purchasing fresh milk attests to the scale of change that has occurred in the local economy, as well as the capacity of some individuals to respond to novel market opportunities.

In the majority of interviews conducted during 2017–18, respondents associated the better-off members of the community with salaried employment, ownership of businesses, large numbers of livestock, a large farm and, in some cases, a 'modern' house and rental properties. The poorer households, in turn, were identified with ownership of few livestock, casual waged labour, small rainfed farms and charcoal making. Until the late 1990s, the wealthiest Il Chamus households were those with large numbers of livestock, especially cattle, and perhaps an irrigated farm. They also frequently headed polygynous households with the wealthiest

Table 3.2 Composition of household income, 1980–2018

Year	Livestock[a]	Farming	Fishing	Non-pastoralist/ non-farming	Pension/govt payments
1980–81	64	14	3	17	0
2000–01[b]	42	13	0	45	NA
2017–18	20	35	0	40[c]	6

Notes:
(a) Approximately 80 per cent of income value of livestock was derived from livestock sales.
(b) These data are from the Pastoral Risk Management (PARIMA) Project where the author served as a Co-Principal Investigator (see Little et al 2008, McPeak et al 2012).
(c) This includes salaried and casual waged income, remittances, and self-employed business incomes from livestock trading, charcoal making and selling, and retail shop keeping. For 2017 the calculation excludes from the sample one petrol station owner with extraordinarily high income that would have greatly skewed the relative importance of business income.

males having four or five different wives, which is a very rare social phenomenon today.

Further evidence that benefits from diversification have been unequally distributed relate to waged employment. For a salaried government job or a management position in the private sector, a post-secondary diploma or a university degree usually is a prerequisite. Those families who could afford to send their children to good secondary and post-secondary schools often were the better-off pastoralist households who could sell livestock to pay education fees. Several individuals indicated that they had reduced their livestock holdings because of the need to sell animals to pay for school fees, with the result that many of today's educated young elite are from families who recently owned relatively large numbers of livestock (in some cases, more than 100 cattle and several hundred goats and sheep). For those families who are unable or choose not to educate their children, waged employment usually means working as a casual (daily or temporary) labourer on an irrigated farm or in town, or as a paid herder for another family.

Earlier studies showed the important role that remittances and access to formal education play in building resilience to droughts and food insecurity emergencies (Little et al. 2009). Although the amounts of remittances are not large in many cases, they still are an important buffer (insurance) against drought and hunger since their flows are only minimally impacted by weather. Salaried employment affects a relatively small percentage of households, but it makes an important contribution for those who are fortunate enough to have access to it. It also serves as a future aspiration for many youth today. This is an important reason why households make incredible sacrifices to pay for the education of their children and other relatives. For example, education expenditures currently account for about thirty per cent of total household cash incomes, although, as noted above, some of these expenses are paid directly by remittances from family members employed in towns and do not show up in household transaction data. Remittances to drought-impacted families actually increase during emergencies and symbolise a continued commitment to the Il Chamus community even when amounts may be very small (< US \$20 monthly).

In addition to changes in income patterns over time, available data allow comparisons of changes in self-identification of livelihoods or occupations by household heads and spouses at two different time periods, 2000 and 2017–18. Table 3.3 does not include all the different occupations but only the main ones that respondents identified. While Table 3.2 reveals how incomes actually are earned, Table 3.3 reflects what household heads and spouses identify as their main livelihood even if it did not contribute as much to income as another activity. For example, association with non-farm/non- pastoral occupation – e.g., a businessperson or casual labourer – is not as significant a form of identity as one might assume, given the important contribution they make to household income. This may reflect a bias towards livelihoods that are perceived as more culturally acceptable, such as herding and farming or a combination of the two, rather than working as a charcoal maker or casual labourer on another person's farm.

Non-pastoral activities and diversification strategies

This section of the paper looks at current patterns of asset diversification and business ownership, including some of their gender and class (wealth) dynamics.

Table 3.3 Principle livelihood identified by household head and main spouse, 2000 and 2017

	2000[a]	2017–18[b]
Farming	28%	40%
Herding	39%	31%
Charcoal	0%	14%
Wage employee	4%	6%
Food/cash-for work	0%	4%
Trade (incl firewood sales)	17%	11%
Casual labor	11%	2%
Shopkeeper	1%	7%

Notes: (a) Sample size is 30 households. (b) Sample size is 100 households.

Asset diversification

As was indicated earlier, other forms of wealth (assets) have emerged as important for enhancing stable livelihoods over time. Land, in particular, has emerged as an important asset for which better-off households are investing. The wealthiest households have been investing in land outside Il Chamus, particularly in neighbouring Nakuru County and Mochongoi Division, south Baringo. Even within Il Chamus territory, some individuals have begun to buy and sell plots on the Pekerra Irrigation Scheme – even though government ultimately owns this land – and around Lake Baringo and Eldume where irrigation is possible. Note that in 1980 there were only a small minority of Il Chamus who had interest in buying/leasing land in the Pekerra Scheme. While some individuals who own lands in Nakuru have private land titles, buying and selling land is occurring in parts of Il Chamus where land is held communally and where property titles are not distributed.

Several other noteworthy findings about assets are revealed in the 2017–18 study. First, although different types of business activities have become very important income-earning strategies, they do not represent significant amounts of durable assets. Local business activities include trading, such as livestock, grain and charcoal trading, or retail shop keeping, but they do not involve valuable equipment (assets). Second, investments in improved housing, especially since virtually no households are mobile today, has also grown in importance, including the purchase of concrete and stone material for walls, and of metal sheets for roofing and walls. This type of asset indicates the extent to which many households are sedentary today, even though livestock still must move to take advantage of dispersed grazing and water resources. Finally, although other forms of assets have emerged, livestock still is valuable as a store of wealth. At an average price of about KSH 30,000 ($300) per head of cattle or per ten small stock even poorer households may have much of their wealth stored in livestock. After land livestock is the most valuable asset, accounting for 37 per cent of average asset value per household. It also is a more liquid asset than land, which is not frequently sold or bought. If one has an immediate need for cash, livestock is considerably more likely to be sold than land.

New patterns of business diversification

Faced with a series of political, weather, and environmental (*prosopis*) shocks during the past 35 years and with opportunities created by increased commodification and small town growth, small-scale business activities have become critical for sustaining local welfare among many Il Chamus. In fact, more than 60 per cent of households have at least one or more members who engage in some form of business. The commercial activity that has grown the most in recent years is charcoal making and trading. However, it is seen as a culturally inferior form of work by many in the community. Yet, because of the *prosopis* invasion in the area – with its negative impacts on pastoralism – charcoal making and trading of wood-based products has become an important form of business. Local sentiments about charcoal making as a livelihood are mixed with some people strongly opposing it and others realising they have little choice but to be involved with charcoal making. Different opinions by Il Chamus are expressed as follows:

> … proposis has really ruined pastoralism and we will look into how we can get rid of it … Il Chamus used to depend on cattle but now they depend on charcoal. They have become charcoal producers (field notes, June, 2007).

> …the *ilmanie* [key wetland grazing zone] is no more. The prosopis and the Pokot have made us very poor now. About 75 % of Chamus are making charcoal, but it involves lots of work and it is dirty work. [Interviewee looks in disgust]… charcoal burning is not a living … I won't do it (field notes, May 2017).

> Yes, the Maasai [who are related to Il Chamus] do not like charcoal making. They say it is not a good living. The money you get from charcoal is not good (field notes, May 2017).

As the above quotes demonstrate, there are strong sentiments against charcoal making/selling, but some individuals now feel that, because of its potential for earning income to survive, they have to engage in it. There are differences by gender, class and age in sentiments towards charcoal making. Most wealthy individuals with livestock and other investments are against it, especially male household heads. Young men have become involved as charcoal traders and poorer individuals (male and female) are now involved in the dirty work of producing charcoal, although they may not like the fact that they have to do it.

Table 3.4 shows the different business activities with which individuals are now involved. The gender of household head is noted to show aspects of gender-based diversification. Data were collected both on the business activities of the household head and the principal spouse. Most households headed by males had only one or at most two wives, a major change since the 1980s when polygamy was widespread (Little 1992). With this caveat in mind, several important observations can be inferred from the findings. First, four types of businesses – Shop Keeper and worker, Local Drink [liquor] selling, Transport (motorbike operator) and Charcoal/Firewood – account for 76 per cent of total business activities, but there are significant differences based on the gender of the household head (Table 3.4). All shop owners and shop workers were from male-headed households (M-HH) and there was a more balanced distribution of different activities for them. For female-headed households (F-HH), two activities – Charcoal/Firewood and Local Drink selling (local beer or *busaa*) – accounted for 77 per cent of their total businesses. Thus, with minor exception, those F-HH involved in businesses are very likely to pursue these two activities while male-headed households pursue a greater range of activities. Secondly, there also is a significant difference between different business activities in the incomes earned. As has been mentioned previously, one male household head in our sample owned a petrol station which has very high revenues. However, there are other activities that earn significant incomes, including grain and livestock trade and shop keeping (especially shop ownership). In fact, there is a marked bifurcation in mean incomes for different businesses, with many of them having relatively low monthly incomes of 4,000 KSH (US $40) or less but a few, such as grain and livestock trade, with very high incomes.

Thirdly, the two activities – charcoal making/selling and local drinks selling – most closely associated with F-HHs earn relatively low incomes. There is, however, one female grain trader and one female livestock trader in the study who earn relatively high incomes. Male-dominated activities like shop keeping and livestock trading earn relatively high incomes. Males and male-headed households are associated with more lucrative businesses than female-headed units. Finally, there is a significant range in incomes for some activities that confirms a significant amount of economic differentiation based on different businesses. There are especially wide differences in incomes from livestock and grain trade, shop keeping and charcoal trade, in some cases differences up to 800 per cent. For example, 45 per cent of households who indicate shop keeping as an activity report

Table 3.4 Self-employed business activities, Il Chamus

ACTIVITY	% of TOTAL	% male-headed HHs activity	% female-headed HHs activity	Estimated monthly mean income (range) (KSH)[a]
Petty trade	2	2	0	NA
Grain Trade	5	6	3	30,600 (15,000–80,000)
Livestock Trade	8	9	3	29,000 (18,000–40,000)
Shop Keeper and worker[b]	17	24	0	10,214 (2,000–16,000)
Transport (motorbike)	15	17	7	3, 682 (2,000–5,000)
Charcoal/Firewood	27	21	46	3,192 (1,000–10,000)
Local Drink selling	17	12	31	3,307 (900–7,000)
Local food making	4	3	7	3,000 (1,000–5,000)
Weaving	3	5	0	1,875 (1,500–2,000)
Selling clothes	1	0	3	5,000
Petrol station owner	1	1	0	45,000
TOTAL	100	100	100	

*Sample size = 100 households

Notes:

(a) Only Ngambo households (n=50) are used for income calculation because of reliability, although the incidence of different businesses derives from the larger sample (n=100). Figures were calculated per household members present, which is 76 per cent of listed household members, and did not include those who were living and working outside Il Chamus.

(b) Only about 30 per cent of individuals were owners of shops where they worked. Most were wage workers but still considered themselves as shop keepers.

incomes of less than KSH 10,000 monthly, but 25 per cent of households indicate earnings of more than KSH 40,000 monthly. Thus, the majority of individuals in this category are likely to be shop workers and not owners, and those 25 per cent who earn more than KSH 40,000 monthly are the shop owners. In sum, diversification into business activities has clear benefits, but it also has aggravated household wealth differences in Il Chamus.

Cash remittances

As was discussed earlier, cash remittances are received by a very large percentage of households but the amounts are relatively small compared to other livelihood sources. Unlike in the past, most cash remittances are not invested in livestock but are for maintaining household expenses. On average eighty per cent of households which report receiving remittances receive less than KSH 20,000 (US $200) annually. This finding implies that: (1) the majority of those working outside Il Chamus are employed as drivers, security guards, farm labourers or other occupations with relatively low salaries; (2) costs of living in towns outside Il Chamus are relatively high, which leaves the majority of migrants with minimal surplus to remit to rural family members; (3) the relatively low value of remittances means that only a small percentage of Il Chamus are highly dependent on them for their survival; and (4) it is culturally important to send some level of remittances to family members no matter how small, especially since those back home may have incurred considerable costs in educating and supporting the migrant who is employed. This latter point (4) supports the findings of Lisa Cliggett (2003) in her study of migration and remittances in Zambia. In this work, she shows that among the Gwembe Tonga of southern Zambia the cultural and symbolic aspects of remittances, often represented as a 'gift', outweigh their economic significance.

If we consider, however, that remittances may be in the form of direct payment for school fees or for health costs rather than cash transfers, the value of remittances is likely to be much higher than reported here. Earlier discussion reported that school fees use about thirty per cent of total household income. Health costs, in turn, are equivalent to seven percent of total household income. These proportions, especially those for education payments, are unrealistically high for many Il Chamus households unless part of these costs are covered by external transfers from those working outside the area. Thus, to understand the important roles and full impacts of remittances, including payments for education and school fees, it is important to assess information both from senders and receivers of remittances. At this point I have only examined the recipients of remittances in the rural Il Chamus homeland, and not those who send remittances but reside outside Il Chamus. In the rest of this section, I focus on Il Chamus who work and live outside Baringo, most of whom send cash or in-kind payments to rural family members. Importantly, they continue to identify as part of the larger Il Chamus community by investing in and participating in significant rituals, such as initiation and age-set rituals, and ceremonies, such as weddings and funerals, and by maintaining relationships with clan and age-mates back in the rural areas.

In the 2017–2018 study we interviewed 33 Il Chamus individuals who work and live outside Baringo, mainly in Nairobi city and Nakuru County. We were able to collect remittance data for 27 of them. These individuals represent a new wave of migration that differs greatly from earlier patterns. In the early 1980s, when the Il Chamus had very little formal education, it was mainly the very poorest individuals, many of whom lost livestock due to drought, who migrated to work outside the community. They often returned to Baringo after a few years once they had accumulated enough money to buy animals or to farm. These early migrants largely moved out of dire necessity to work as labourers on large farms and ranches in Laikipia County and the Naivasha area, both locations referred as *mashambani* ('farm areas'). These enterprises were mainly European-owned or formerly owned by Europeans who sold them to wealthy Kenyans, often to elite families with national political connections. Very minimal remittances came back to the community from these migrants, since they earned miserably low incomes (< $1/day). By contrast to the earlier pattern, many migrants who have received education mainly go to Nakuru County or to large urban centres, such as Nairobi, rather than to *mashambani*. Moreover, some of them include the wealthiest and best educated Il Chamus rather than the poorest seeking unskilled agricultural work ('contract labour'). Virtually no migrants today, even casual labourers, move to Laikipia or Naivasha to work. Those who move out to seek jobs but have few skills and education often end up working on a farm, in a household or in a business owned by an Il Chamus in Nakuru County, Nairobi or another location.

The sample of individuals working outside Baringo County differs from the rural household sample in two important aspects. First, rural households involve remittances from any non-resident family member outside Il Chamus and include individuals employed in towns in Baringo, such as Marigat or Kabarnet (the administrative capital of Baringo County). All of the migrant sample, however, were individuals residing and working outside Baringo, where incomes are much higher than in Baringo towns. Second, the non-rural sample had considerably better education (33 per cent were university graduates) and held better positions (approximately 50 per cent held salaried jobs) than members of rural households. In the latter case, only about six to eight per cent of households had a family member with a university degree, while only 20–21 per cent of them had members working in salaried positions outside Il Chamus. Not surprisingly, the amount and range of remittances is considerably higher for the urban than for the rural sample, with many urban-based individuals employed in lucrative salaried positions. Here the average level of remittances sent is more than seven times higher than remittances received in our rural sample, with a significant amount of the remittances in the form of payments for school and health costs. Additionally, almost the entire sample of non-rural migrants remit money back home, except a few individuals who are engaged in low-paying casual labour. Another important finding is the relatively high number of livestock (TLUs) owned by migrants and the fact that most urban Il Chamus still invest in livestock (84 per cent own livestock) but rely on family members or hired labourers to herd their animals. In fact, the numbers of livestock owned by migrants living outside Baringo are more than double those reported for households residing in Il Chamus territory. Since livestock of Il Chamus residing outside Baringo often are kept by family members back in Il Chamus, they provide an important external resource in support of pastoralism. By keeping their livestock with families back in Il Chamus, urban migrants are assisting them with maintaining herds and a pastoral livelihood even when their own herds have decreased in recent years. This finding confirms what many respondents indicate: livestock remain an important asset and form of cultural identity for Il Chamus even for those employed in towns.

Gender-based diversification and inequalities

There are important gender-based differences in livelihood diversification strategies and income levels among Il Chamus. The earlier discussion of non-pastoral activities showed that women tend to be associated with less remunerative businesses than men. Before turning to issues of gender-based differentiation and diversification, it is important to discuss general patterns of inequality among all households, both male-headed (M-HH) and female-headed households (F-HH).

As noted earlier, increased income diversification and investment is associated with growing inequality, not just between F-HH and M-HH but between the wealthiest and poorest households. For example, when you disaggregate households into different wealth quartiles based either on annual income or total asset value, patterns of inequality become very apparent. Prior to the 2000s it was possible to look at inequality mainly in terms of just one variable, livestock ownership. With increased diversification and the growing importance of farming, waged employment, business activities and new forms of wealth (assets), such as land and improved housing, it is unrealistic to use only livestock as an indicator of inequality. Those in the poorest quartile of households have annual incomes and assets that are estimated to be less than one per cent of those in the wealthiest quartile. Because of the increased market for land in Il Chamus, especially irrigable farmland, asset values are less unequal than incomes, and represent an important resource for the poor. However, it would greatly reduce their welfare and resilience to withstand future shocks if they ever did sell their farmland.

In considering the role of livestock among different households it is the wealthiest and most diversified households with significant farming, wage and business incomes who control almost double the average numbers of livestock that other households do. Similar to what was revealed for Il Chamus migrants and their ownership of livestock, these data also demonstrate that the highest income earners with significant non-pastoral income sources and high asset values, still continue to sustain and invest in livestock. This pattern suggests that a new kind of pastoralism has emerged with increased absentee herd ownership along with more customary forms of pastoralism for a smaller percentage of the population.

One customary institution that remains prevalent is the loaning of livestock. For example, owners frequently lend livestock to relatives and others in different grazing areas both to reduce labour requirements and to spread risk against losses in a specific location. As one well-educated Il Chamus who works outside Baringo noted, 'I have lent my livestock to households in three different locations in case there is a problem in one place. It is foolish to leave them all in one location' (field notes, January 2019). Loaning out livestock is also a strategy to reduce labour needs, especially among those who work outside Il Chamus, and to assist those families who have few animals. Some poor households may have few animals but adequate labour. Additional analysis is required to understand the different reasons for current patterns of livestock exchange, but initial assessment is that it still is widely practised despite widespread changes in ecology and economy. In fact, more than eighty per cent of households indicate that they had either loaned or borrowed livestock during 2017, a percentage that is actually as high as recorded in the 1980s. What could be the reason for this? First, with the invasion of *prosopis* in most parts of western Il Chamus, households, including those working outside Baringo, are intentionally loaning animals to families in areas that have not been badly affected by the plant. Second, with the high number of youth attending school and/or engaging in different income-earning strategies, households may be lending livestock to reduce labour needs. Finally, it is likely that those who receive livestock through borrowing or other types of exchange are taking care of the animals of individuals who work outside the area. In short, despite insecurity, *prosopis* and other challenges, Il Chamus have been able to maintain customary livestock exchange strategies which are important to their identity as pastoralists.

Turning now to gender-based differentiation, M-HH earn more revenues than F-HH in total and by different income sources, with the exception of remittances where F-HH receive about 60 per cent more revenues than M-HH. F-HH also control about 33 per cent fewer animals than M-HH. Most significant differences between the two groups were in wage income where F-HH actually had no annual wage income during the research period (2017–2018); business income where M-HH had 230 per cent higher income than F-HH; and total household income where F-HH had only 64 per cent of the income that M-HH had. However, in self-employed business F-HH are more involved in the less remunerative activities, such as charcoal making and sales of local alcohol drinks, than are M-HH (see Table 3.4). With minimal livestock holdings, F-HHs are more dependent on remittances than M-HH and are particularly vulnerable to food insecurity and shock-induced poverty. Because women rarely remarry after divorce or the death of their husbands, there are a relatively high percentage of female-headed households in the area.

As Table 3.5, shows there is a higher percentage of F-HH than M-HH in the poorest quartiles of both income and total assets. For example, only three per cent of M-HH are in the poorest income quartile, but 31 per cent of F-HH are in that category. In terms of assets almost 38 per cent of F-HH are in the poorest asset quartile, with only twelve per cent of them in the richest asset quartile. As indicated earlier, the fact that irrigable and other lands near Marigat town now have a market value means poor households

Table 3.5 Wealth quartiles and gender-based inequality, Ngambo

Indicator	% Male Heads	% Female Heads
Poorest Quartile (4) TLUs	21 %	31 %
Richest Quartile (1) TLUs	24 %	19 %
Poorest Quartile (4) Income	3 %	31 %
Richest Quartile (1) Income	32 %	0 %
Poorest Quartile (4) Assets	18 %	38 %
Richest Quartile (1) Assets	26 %	12 %

Note: One TLU (Tropical Livestock Unit) = 1 cattle or 10 small stock (goats and sheep).

who may have possessed land for several years have experienced an increase in the value of their assets but not their incomes.

Conclusions

The chapter has used the lens of political economy to examine how external forces dating back to the colonial period have affected livelihood and asset diversification among Il Chamus. It has also argued that the temporal dimension of the resilience concept is useful for understanding how opportunities to diversify over time allow some pastoralist families to cope with and even prosper in a shock-prone environment while also maintaining an identity as pastoralists. By contrast, it has also helped to explain how other families are unable or unwilling to diversify despite the stark reality of diminishing livestock holdings, thereby making temporality (time) a distraction rather than a benefit. Moreover, the chapter argues that pastoralism by itself can no longer sustain most households, but must be combined with other strategies, especially town-based, to adjust to a range of short- and long-term demands and contingencies. A longitudinal perspective allows one to observe both the 'booms (recoveries)' and 'busts (shocks)' that families and individuals (better-off/poor, male/female) encounter and the positives and negatives associated with widespread political and economic changes that increase reliance on markets and towns.

Livestock, especially cattle, remain an important part of Il Chamus cultural identity and it is frequently noted 'without livestock one cannot be an Il Chamus'. However, the capacity or willingness to be flexible in pursuing different assets and livelihood strategies, many of which provide minimal economic returns, does not mean the majority of families are materially better off than they were in the 1980s. In fact, as the chapter has suggested, the opposite is true: only a privileged minority are materially better off today than they were 35+ years ago. Political economic forces have worked against the community and starting in the colonial period these pressures have constrained pastoralism and sowed the seeds of class differentiation which are increasingly apparent today. What one might label 'customary pastoralism' with mobility and a heavy dependence on livestock continues, but for a considerably smaller percentage of households than in the past. As other studies have shown, diversification can complement pastoralism instead of replace it, and in this case one could even argue that diversification allows pastoralism to continue for some families (see Homewood et al. 2009; McCabe et al. 2010; Mc Peak et al. 2012). Right now, those who work in salaried jobs outside Baringo are actually able to invest more in livestock and keep larger herds, than those who remain in rural areas. Understanding the dynamics of rural-urban linkages and pastoralist diversification is complex, but an important finding of this chapter confirms that diversification mainly benefits the better off with large herds and high incomes. Pastoralism will continue in the area, but in a different form with relatively high levels of absentee herd ownership, waged herding and inequality.

Acknowledgements

Some of the research reported here was supported by the USAID-funded Pastoral Risk Management in East Africa (PARIMA) programme (1999–2005) and the John Templeton Foundation under a project titled 'Cross-Cultural Insights into Well-being among Vulnerable Populations in Eastern Africa (2016–2018)'. I wish to acknowledge the warm and welcoming support of the Il Chamus community, especially Lempaysan Keis, Eunice Lepariyo, Clement Lenachuru, Nickson Lolgisoi, Calvin Longeshele, and Samuel Sekeu. I also wish to thank the following individuals for their assistance: John McPeak, Abdillahi Aboud, Mark Risjord and Danielle Veal. A follow-up visit to Kenya in 2019, as well as write-up support was provided by the 'The Humanitarian-Development Gap and the "Resilience" of African Nomadic Peoples' project (JSPS-KAKENHI Grant 18H03606) based at the University of Shizuoka. I especially am grateful to the assistance from the head of the project, Shinya Konaka. Two anonymous reviewers provided especially helpful comments on an earlier version of the manuscript. It goes without saying, however, that the contents of this paper are the sole responsibility of the author and should not be attributed to any of the above individuals or institutions.

References

Adger, W.N. 2000. 'Social and ecological resilience: are they related?' *Progress in Human Geography* 24 (3): 347–364.
https://doi.org/10.1191/030913200701540465

Anderson, D.M. 2003. *Eroding the Commons: The Politics of Ecology in Baringo, Kenya, 1890s-1963.* Oxford, UK: James Currey.

Anderson, D.M., and M. Bollig (eds). 2017. *Resilience and Collapse in African Savannahs.* Abingdon, UK: Routledge.

Barrios, R. 2016. 'Resilience: A Commentary from the Vantage Point of Anthropology.' *Annals of Anthropological Practice* 40 (1): 28–38.
https://doi.org/10.4324/9781315267647

Cliggett, L. 2003. 'Gift remitting and alliance building in Zambian modernity: Old answers to modern problems. *American Anthropologist* 105 (3): 543–552.
https://doi.org/10.1525/aa.2003.105.3.543

Debsu, D.N., P.D. Little, W. Tiki, U. Kitron and S. Gargliardi. 2016. 'Mobile phones for mobile people: The role of Information Communication Technology (ICT) among livestock traders and Borana pastoralists of southern Ethiopia.' *Nomadic Peoples* 20 (1): 35–61. https://doi.org/10.3197/np.2016.200104

Ferguson, J. 1994. *The Anti-Politics Machine: Development, Depoliticization, and Bureaucratic Power in Lesotho.* Minneapolis: University of Minnesota Press.

Folke, C. 2006. 'Resilience: The emergence of a perspective for social–ecological systems analyses'. *Global Environmental Change* 16: 253–267. https://doi.org/10.1016/j.gloenvcha.2006.04.002

Fratkin, E., and E. Roth (eds). 2004. *As Nomads Settle: Social, Health, and Ecological Consequences of Pastoral Sedentarization in Northern Kenya.* New York, NY: Kluwer Academic/Plenum Publishers.

Galaty, J.G. 2016. 'Reasserting the commons: Pastoral contestations of private and state lands in East Africa'. *International Journal of the Commons* 10 (2): 709–727. https://doi.org/10.18352/ijc.720

Holling, C.S. 1973. 'Resilience and stability of ecological systems'. *Annual Review of Ecology and Systematics* 4: 1–23. https://doi.org/10.1146/annurev.es.04.110173.000245

Homewood, K., P. Kristjanson, and P. Trench. (eds). 2009. *Staying Maasai?: Livelihoods, Conservation and Development in East African Rangelands.* New York: Springer.

Kenya, Government of. 1994. *Range Management Handbook of Kenya, Vol. II, 6: Baringo District.* Nairobi: Ministry of Agriculture, Livestock Development, and Marketing.

Lesorogol, C.K., 2008. 'Land privatization and pastoralist well-being in Kenya'. *Development and Change* 39 (2): 309–331. https://doi.org/10.1111/j.1467-7660.2007.00481.x

Li, T.M. 2014. *Land's End: Capitalist Relations on an Indigenous Frontier.* Durham: Duke University Press.

Little, P.D. 1983. 'The livestock-grain connection in Northern Kenya: An analysis of pastoral economics and semi-arid land development'. *Rural Africana* 16: 91–108.

———— 1985. 'Social differentiation and pastoralist sedentarization in Northern Kenya'. *Africa* 55 (3): 243–261. https://doi.org/10.2307/1160579

———— 1992. *The Elusive Granary: Herder, Farmer, and State in Northern Kenya.* Cambridge, UK: Cambridge University Press. https://doi.org/10.1017/CBO9780511753077

———— 2016. 'A victory in theory, loss in practice: Struggles for political representation in the Lake Baringo-Bogoria Basin, Kenya'. *Journal of Eastern African Studies* 10 (1): 189–207. https://doi.org/10.1080/17531055.2016.11 38665

———— 2019. 'When green equals thorny and mean: The politics and costs of an environmental experiment in East Africa'. *African Studies Review* 62 (3): 132–163. https://doi.org/10.1017/asr.2019.41

Little, P.D., K. Smith, B.A. Cellarius, D.L. Coppock, and C.B. Barrett. 2001. 'Avoiding disaster: Diversification and risk management among East African herders'. *Development and Change* 32 (3): 401–433. https://doi.org/10.1111/1467-7660.00211

Little, P.D., J. McPeak, C. Barrett, and P. Kristjanson. 2008. 'Challenging orthodoxies: Understanding pastoral poverty in East Africa'. *Development and Change* 39 (4): 585–609. https://doi.org/10.1111/j.1467-7660.2008.00497.x

Little, P.D., A. Aboud, and C. Lenachuru. 2009. 'Can formal education reduce risks for drought-prone pastoralists? A case study from Baringo District, Kenya'. *Human Organization* 68 (2): 154–165. https://doi.org/10.17730/humo.68.2.n70t617197x4w778

Little, P.D., W. Tiki, and D.N. Debsu. 2014. 'How pastoralists perceive and respond to market opportunities: The case of the Horn of Africa'. *Food Policy* 49 (2014): 389–397. https://doi.org/10.1016/j.foodpol.2014.10.004

McCabe, J.T., P. Leslie, and L. DeLuca. 2010. 'Adopting cultivation to remain pastoralists: The diversification of Maasai livelihoods in Northern Tanzania'. *Human Ecology* 38 (3): 321–334. https://doi.org/10.1007/s10745-010-9312-8

McPeak, J., P.D. Little, and C. Doss. 2012. *Risk and Social Change in an African Rural Economy: Livelihoods in Pastoralist Communities.* London and New York: Routledge. https://doi.org/10.4324/9780203805824

Özden-Schilling, T. 2022. Promising resilience: Systems and survival after forestry's ends. *American Anthropologist* 124: 64–76.

Livelihood Diversification and Resilience among the East Africa Pastoralists

Toru Sagawa

Introduction

'Seeing like a pastoralist'

Many development projects targeted to pastoral societies in Africa have failed (Sandford 1983; Anderson 1999; Catley, Lind and Scoones 2013). Researchers pointed out that one of the main reasons of failure is the interveners' assumption that pastoralism is a backward and outdated activity and a desirable alternative livelihood for pastoralists is a sedentary (irrigated-) agriculture. State and development agencies tend to see the world with this sedentarist-centred perspective. In contrast, the pastoral scholarship has insisted that it is necessary for more suitable development to take a perspective of 'seeing like a pastoralist' (Catley, Lind and Scoones 2013) or 'seeing like a herder' (Semplici 2019) and to learn from pastoralists' creative practices in uncertain ecological and social environments.

What does 'seeing like a pastoralist' mean concretely? In reading ethnographies focused on pastoral societies in East Africa, their common perspective would emerge. Pastoralists locate pastoral activities and livestock at the centre of their livelihood and value system. This does not mean that pastoralists think only about livestock. It must be emphasized repeatedly, because such a stereotype as 'irrational pastoralists who cling to only livestock' has functioned as an ideology which legitimates sedentarist-centred development policies.

The characteristics of pastoralists' perspective is that when they evaluate a specific activity and thing, there is a tendency that they consider its relation to pastoral activities and livestock. I call it the relational perspective of the world centred on the pastoral activities and livestock. Pastoralism, cultivation, gathering, fishing, and other activities are not evaluated independently but relationally by pastoralists. When the natural and social contexts surrounding pastoralism and livestock have changed, evaluations of and attitudes towards other activities and things have also changed. While pastoralists surely consider that livestock is important, their perspective is not closed to everything else.

This perspective resonates with the resilience concept under inquiry in this book. Resilience is often defined as a capacity to respond to and to bounce back from problems and disturbances (e.g., Walker and Salt 2012). However, in the Introduction of this book, Konaka, Little and Semplici locate resilience in the dynamic context of pastoralism. They identify the 'capacity for change' as an important aspect of resilience and outline the contextual, ontological, and relational perspectives of resilience. This corresponds with Chandler and Coaffee's (2017) observation that resilience refers not only to 'bouncing back' but also to 'bouncing forward'. In this chapter, I attempt to show how the Daasanach, pastoral people in East Africa, deal with dynamic social change by 'bouncing forward' and focusing on diversifying their livelihood.

Livelihood diversification[1]

Many studies have focused on livelihood diversification by pastoral peoples in East Africa since the 1980s.[2] Ellis (2000: 14) defines livelihood

---

1 This chapter is based on my previous article (Sagawa 2021a).

2 See: Chamus (Little 1992), Borana (Desta and Coppock 2004; Boku and Oba 2010), Rendille (Fratkin and Roth 2005; Sun 2005), Maasai (Homewood et al. 2009; McCabe et al. 2010, 2014), Samburu (Lesorogol 2008), Turkana (Opiyo et al. 2015; Watete et al. 2016), Pokot (Bollig 2016), Karimojong (Caravani 2019), and comparative studies of 11 communities (Little et al. 2001).

diversification as "the processes by which households construct an increasingly diverse portfolio of activities and assets in order to survive and to improve their standard of living." Pastoral peoples have attempted to diversify and rearrange their livelihood not only for survival or 'bouncing back' but also to improve their lives or 'bounce forward' under uncertain natural and social environmental conditions.

Pastoral scholars point out that general factors driving diversification are drought and famine, population growth, loss of common property resources, penetration of the market economy, and urbanization (Fratkin 2013). The common options for livelihood diversification are selling milk and other livestock products, adopting cultivation, gathering wild products, engaging in petty trade and wage labour. Sun (2017) showed that the Rendille diversify their livelihood by appropriating emergency food aid as well as other goods and money supplied by the government and development agencies.

The central theme of research on livelihood diversification is to clarify the economic background, contribution, and consequences of non-pastoral activities for each household. Comparative studies in 11 communities of northern Kenya and southern Ethiopia showed that the degree and purpose of livelihood diversification differ in terms of the households' economic status. To the poor, diversification is a means for survival, and to the rich, a means for accumulation and investment (Little et al. 2001; McPeak and Little 2017). Among the Karimojong of Uganda, livelihood diversification promotes social differentiation in the community (Caravani 2019). Others have focused on the gendered aspect of diversification and found that women play important economic roles in new activities (Fratkin and Smith 1995; Smith 1998).

Based on these studies, the driving factors or motives for diversification have been classified. Katherine Homewood identified three motives of diversification: 1) poverty strategies driven by necessity, 2) risk management strategies to make the best of changing ecologies and economies, and 3) strategies of wealth investment and accumulation (Homewood 2008: 238). However, this classification and previous research are generally inclined towards economic explanations. I argue instead that the diversification strategy is not a sole 'adaptation' under changing ecological and economic environments. When pastoralists decide whether to be involved in new activities or to use and consume new products, they consider how the new activity and product relate with pastoral activities and livestock.

McCabe et al. (2014: 390) showed that social and cultural issues also have important influence on pastoral diversification processes. It found that, among the Maasai of Tanzania in Ngorongoro, the main reason for adopting cultivation was changing tastes in food and the influence of neighbouring peoples such as the agro-pastoral Iraqw (McCabe et al. 2010). Although most Maasai engaged in cultivation, "young men viewed cultivation as un-appealing and a non-Maasai way of life" centred on pastoralism (McCabe et al. 2014: 392). They also investigated why young Maasai men migrated to towns searching for work. Young men desire to acquire bridewealth and to be independent of their fathers. As many youths migrated to urban areas, working outside the traditional livestock sector to accumulate resources became a new cultural norm among them (McCabe et al. 2014).

These studies clearly showed that pastoralists diversify their livelihood driven not only by economic necessity but also by considering how the new activities and products relate with pastoral activities and livestock, and the diversification process itself can change the values and relations. In this chapter, I examine how the pastoralists started to engage in fishing activities which have been negatively evaluated by many pastoral peoples in East Africa.

Negative value on fish and fishing

Previous studies on East African pastoral societies have focused on pastoralists, placing higher cultural value on pastoral activities and livestock. In the process, researchers observed that pastoralists looked down on or despised non-pastoral subsistence activities, the people who engage in such activities, and their products.

Pastoral people in East Africa have rarely been "pure pastoralists." They have often engaged in other subsistence activities and acquired crops and other food items from neighbouring peoples and merchants. However, for pastoralists, people with many livestock are 'wealthy' and people who constantly depend on other activities such as agriculture, hunting, gathering, and fishing are 'poor.' For example, among the Turkana, "a man poor in livestock, even though a wage earner, is thought of as a poor man" (Hogg 1982: 165). In the Borana society. "(f)amilies that do not own cattle have no position…, even when their granaries are full" (Boku and Sjaastad 2010: 1171). Although pastoral society seems to be an 'egalitarian society', this egalitarianism has been achieved by excluding the poor who lose their livestock from 'we pastoralists' (Broch-Due and Anderson 1999). The excluded poor are compelled to depend on non-pastoral activities to a great degree, if not in total.

Among non-pastoral activities, fishing is especially negatively evaluated by some pastoralists. The ideal meal for the Samburu of Kenya consists of meat, milk, and blood of livestock, but they depend on other foods as well. Although they eat crops, honey, and the meat of some wild animals, they do not recognize reptiles, birds, or fish as food (Holtzman 2009: 113). The Turkana, whom the Samburu regard negatively as a people who "eat almost anything" (Holtzman 2009: 95), also evaluate food hierarchically, with livestock products at the top, crops in the middle, and fish at the bottom (Broch-Due 1999: 63). Although fishing has recently become an important source of livelihood, those who have livestock among the Turkana refuse to eat fish because of "its strong and offensive smell" (Lokuruka 2006: 208).[3]

While pastoralists look down on people without livestock, they themselves are at risk of becoming despised because of the uncertain natural and social environment. Any pastoralist can lose most or all his livestock at once. For example, many livestock died in East Africa at the end of the nineteenth century when bovine pleuropneumonia, rinderpest,

---

3 The degree of hatred for catching and eating fish varies among East African pastoral societies. For example, while the Nuer despised the neighbouring Shilluk whose livelihood mainly depended on fishing, they liked to engage in fishing and enjoyed eating fish (Evans-Pritchard 1940). There is also variety in the same community. The Chamus consider fish is a low-status food. While many members of rich and very rich homesteads have never eaten it, fish is the second most important food and often the most important source of protein for members of poor and very-poor homesteads (Little 1992: 121).

and smallpox were prevalent. Some of the Samburu found shelter with the Elmolo, a fishing community on the south-eastern shore of Lake Turkana, and survived the crisis by catching and eating fish. However, they were not satisfied with their lives as fishing people. They obtained livestock by exchanging ivory with merchants and/or raiding neighbouring groups and returned to their livestock-centred livelihood (Sobania 1988). Fishing was a temporary activity during the emergency period for pastoral people.

Since the 1980s, some people started engaging in fishing more regularly. In 2007, 78 per cent of the Turkana people who engaged in fishing because of losing livestock identified themselves as "fishing" when they were questioned regarding their main subsistence activity (Yongo et al. 2010 cf. Derbyshire 2021). However, the authors did not pose the questions of how people entered the negatively evaluated activity and whether the cultural value of fishing had changed or not in the process of many people engaging in fishing.

In this chapter, I examine how the Daasanach youth, who are neighbouring people of the Turkana, started to engage in fishing activities that they had previously evaluated negatively. In the next section, I will outline the Daasanach's customary evaluation of fishing activities and fishing. Then, I analyse the specific context in which some Daasanach youth started fishing activities, how they legitimized their choice to take-up fishing activities, and how other Daasanachs viewed their choice. Last, I discuss how the relational perspective of the world reflected their diversification process.

Fish and fishing people among the Daasanach

I have conducted field research on the Daasanach of southwestern Ethiopia since 2001. The data in this chapter were mainly collected via field research within a few weeks during each of the following years: 2009, 2011, 2012, 2013, and 2015.

The Daasanach live around the border area of Ethiopia, Kenya, and South Sudan. The majority of them live in the Daasanach Woreda (District) in the South Omo Zone of Ethiopia and some in northwestern Kenya. The population on the Ethiopian side was approximately 65,000 in 2016. Their main subsistence activities were pastoralism and flood-retreat cultivation. The Omo River flooded approximately every July. The floodwater brought fertile soils from the Ethiopian highlands and improved land productivity. When the water began to retreat, people herded their livestock and sowed sorghum in the rich soil of the floodplain. Although the annual rainfall is approximately 350–400 mm, the flooding had enabled the Daasanach to produce abundant and stable pasture and crops (Carr 1977; Almagor 1978; Sagawa 2010a).

Until recently, they maintained their livelihood with food items either produced themselves or acquired from neighbouring groups. Although the town of Omorate was built in the mid-1980s, employment opportunities were few, and the majority of the people depended on activities in the village. However, since the late 2000s, their living space has become the target of various large-scale development projects, such as commercial farming. Today, their way of life is drastically changing (Sagawa 2021b).

The Daasanach land is located in the lower reaches of the Omo River, where the river flows into the northern shore of Lake Turkana. Although there is plenty of fish in the river and lake, the majority of the Daasanach hated and avoided catching and eating fish. I observed many situations in which the Daasanach negatively evaluated fishing and eating fish. The pastoral Daasanach told me that fish smelled too bad to eat. The people who engage in fishing are those with no livestock, and they reluctantly eat fish only when they have no other food.

Around 2010, some of them started to engage in fishing, which seemed sudden from my perspective. When I visited P village for the first time in two years in August 2011, some villagers were not there. I asked the remaining villagers about their whereabouts, and the villagers only answered, "they moved to the south." In the south, which means the northern shore of Lake Turkana in this context, merchants from the Kenyan side came and bought fish from the Daasanach. In March 2009, a domestic investor opened a commercial farm near P village. There is likely some connection between starting to fish and opening commercial farms; however, I could not understand why they engaged in fishing of all activities. In this chapter,

I examine the reasons they have started fishing and how this has changed the cultural value of subsistence activities.

Uri Almagor, who conducted field research among the Daasanach in 1968 and 1969, analysed the relationship between fishing people, the Dies, and pastoral people, the other Daasanach who had livestock (Almagor 1987; see also Carr 1977: 159, 170, 198). The Dies are fishing experts and sometimes engage in hippopotamus hunting. Their settlement consists of only Dies members and is located on the northern shore of Lake Turkana. The Daasanach consists of eight territorial groups (*en*) that roughly share living space and pasture land.[4] However, the Dies are not a territorial group but a social category in which membership is designated by one's subsistence activity.[5]

Almagor focused on the evaluation of smell by Daasanach. The pastoral people regarded the smell of livestock products and pastoral people as good. In contrast, the smell of fish and fishing people were considered "bad almost to the point of revulsion" (Almagor 1987: 469).[6] There was a case of a marriage between a pastoralist's son and a fisherman's daughter.[7] In that case, the fisherman's daughter needed to stay at a pastoralist's house, such as that of the groom's father for many years to get rid of the fish smell from her body and become a "pastoral girl." This prompts the question of why pastoral people are sensitive to the smell of fish.

---

4 I have conducted research mainly among the Randal and Inkabelo (or Shiir) territorial groups. The latter identify themselves as 'the genuine Daasanach.' They tend to look down on members of other territorial groups, especially people without livestock. Although I have also conducted research among other territorial groups, the following description would strongly reflect the Inkabelo's viewpoint.

5 The members of the Riele territorial section also frequently engage in fishing. Riele also refers to a half-man and half-fish monster. There is a rumour among people of other territorial sections that Riele people are closeted with a Riele monster in the night.

6 Almagor wrote that, for outside observers, the smell of pastoralists who smear ghee and chyme on their body in many rituals is "much stronger and more evocative than that of fishermen" (Almagor 1987: 473).

7 In contrast, there were no cases in which a son of a fisherman married a daughter of a pastoralist (Almagor 1987).

The Daasanach recognize fish as an infertile creature which do not engage in sexual intercourse. If the bad smell of fish and fishing people infects livestock, pastoral people believe that the livestock would become infertile; thus, ideally, fish and fishing people must be isolated from livestock and pastoralists. Pastoral people ate fish during food shortage periods. However, they did not catch the fish themselves but received it from their Dies friends. When Almagor asked pastoralists about their reason for eating fish that might harm their livestock, the people reasoned that "since the food was proffered as a gift, they might insult their hosts by refusing it" (Almagor 1987: 470). Pastoralists insisted that they eat fish not because they want to eat it, but because they need to consider their social relations.

Yvans Houtteman, who conducted long-term field research among the Daasanach in the mid-1990s, also described the negative value of fish and fishing by the Daasanach (Houtteman 2011: 33). My observations from field research from 2001 to 2006 were consistent with those of previous studies. The act of catching and eating fish was not only despised but also regarded as harmful to livestock and pastoralism.

Fishing pastoralists

Almagor wrote that there were a few hundred Dies on the shore of Lake Turkana in the late 1960s (Almagor 1978: 52, 1987: 461). Since around 2010, the number of people engaging in fishing has been increasing. When I visited H village located about 20 km north of the Kenyan border and near the northern shore of Lake Turkana in March 2015, about a hundred households had built a fishing camp and were engaged in fishing. There were other camp sites in the district where a similar number of people began fishing.

Most people engaged in fishing were young men in their 20s and 30s. They moved to a fishing camp with their wives and small children. As two or three boats were needed to set and draw fishing nets, they called unmarried brothers or other kinsmen to work together. When they could not find such help, they cooperated with members of other households.

Two points should be emphasized here. First, few elderly people fished. Second, many fishing people have a moderate number of livestock. Their livestock was herded away from the fishing camps by their sons and other kinsmen. In the past, only impoverished people who lost their livestock or people with livestock during food shortages engaged in fishing. Today, people holding livestock have also begun to fish.

Taking up fishing is not so difficult in economic and technical terms. The necessary tools for fishing are wooden canoes and nets. A cow is exchanged for a canoe made by the Elele (one of the territorial groups in the Daasanach). The Elele live around the town of Omorate, so people acquire a canoe near town and take it to the fishing campsite on a car by paying the car owner. They exchange one small livestock for a net from a Somali merchant in the border area. The way to row a canoe, to use a net, and to dry fish can be learned easily from colleagues. They sell dried Nile perch and Tilapia. Approximately 15 cm long fish were sold for four birrs in 2015. Cow's milk in a long calabash container was sold for approximately eight birrs in the town. In short, two dried fish paid for half a day's cow milk. People recognized the price of fish to be considerably high.

In a fishing camp, people ate sorghum that they had harvested and/or maize flour that they bought from a Kenyan Somali merchant. When their livestock was herded on pasture land near a fishing camp, cow's milk was brought from the livestock camp. Some people ate fish daily, those they could not sell to the merchants because of their small size, while others only ate them sometimes.

Building commercial farms

One of the triggers that made the Daasanach start fishing was the building of commercial farms. In the Daasanach district, some foreign and domestic investors started to operate commercial irrigated farms along the Omo River in the late 2000s. These plantations were built on lands that the Daasanach had originally used for residences, cultivation, and pastures. There was no compensation payment, and many people were displaced. One of the plantations was built near P village on the eastern side of the

Omo River by a domestic investor in March 2009. I have conducted research in the village since 2001.

Considering the time frame, it seems probable that impoverished people who lost their land to commercial farms started fishing. However, in P village, although the total leased land was 1000 hectares, only 30 hectares were cultivated by 2010. These lands had rarely flooded, so they were not used for cultivation. Thus, it was possible to continue their lives, remaining around P village. In addition, people who lived in other villages where no plantations were opened had also started fishing. In short, they started fishing not solely because they lost their previous base of subsistence activities, but because they had more active intentions.

To examine this intention, it is important to consider that people must have perceived symptoms of drastic social change and held a sense of uneasiness regarding the future before the opening of the farm. In August 2009, when I expressed my surprise at the significant change in the landscape by the opening of a farm, people explained to me that many other unidentified small buildings were built before the farm opened. When I walked with the villagers to the town for two hours, people directed my attention to these foreign structures. As one of the villages discovered newly dug shrubs, he spat out, "Highlanders (*ushumba*) built their houses everywhere! *Garlam* (damn it)!"

"Highlanders" refer to the people who have come to the Daasanach land from the north such as the Amhara. The Daasanach first encountered highlanders at the end of the nineteenth century when the imperial Ethiopian Army conquered them. The imperial army raided livestock, women, and children. Thus, the Daasanach view highlanders as people who represent a suppressive Ethiopian state that attempted to deceive and rob them.

During my research in 2009, people also told me about the "weaknesses" of livestock, such as being prone to disease, vulnerable to drought, and easily raided by an enemy. I rarely heard such statements before. When I asked them why they said this, the people replied that at the local government meeting, government officials listed negative elements of livestock and pastoralism. The officials persuaded them to leave pastoralism and instead, engage in irrigated farming and plantation work.

Right before the plantation was built, the Daasanach became concerned that their living space and pastoralism were being encroached by highlanders. Their concerns became a reality when a commercial farm was opened. Not only was the land expropriated, but seasonal labourers from the highlands stayed near the village, and the sound of the morning radio and the drinkers' voices until midnight echoed loud in the village. Highlanders in the labourers' camps sometimes hit their livestock with sticks. As they recognized the encroachment in their everyday living space, they felt a need to search for new means of livelihood.

Trading partners

However, there was a time interval between the farm opening and people starting to fish. A farm near D village opened in the first half of 2009, and many pastoralists only started fishing in the first half of 2010. Interestingly, in 2009, an external actor who purchased fish had already existed. A fishing company from the Ethiopian highland had operated on the northern shore of Lake Turkana since the mid-2000s. This company brought frozen fish to northern towns, such as Arba Minch and Addis Ababa. However, only a few Daasanachs continually traded fish with them. One of the reasons was that the company brought labourers from the highlands. Another reason was the Daasanach's negative perception of the highlanders. Some of them traded with the highlander's company but soon stopped because of the latter's disagreeable attitude. Such experiences reproduced and reinforced the negative images of the highlanders.

The primary trading partners from 2010 were the Kenyan Somali merchants who came to the Daasanach land across the border. As the demand for fish increased in the town, merchants advanced to the northern shore of Lake Turkana, where fish had not been fully commercialized. According to Somali merchants, the quality of fish in Lake Turkana is higher than that in other, polluted, lakes in Kenya. They take fish from the Daasanach land to the town of Irelet, Kenya, by car, to the Karakole crossing of Lake Turkana by ferry, and fish-consuming towns in western Kenya, Uganda, and the Democratic Republic of Congo by car.

The Daasanach classify the Somali as "our people" (*gaal kinnyo*). The Daasanach land was under Italian control from 1937 to 1941. The Italian Army organized the Daasanach youth as a defending army and gave them guns and other goods. The Daasanach elders called the Italians "our elder brother or father." As the Italian Army came and acted in concert with the Somali people, the Daasanach regarded the Somali as intimate colleagues under Italian leadership. The other reason the Daasanach prefer the Somali as a trading partner is due to their superior trading position. As Kenyan citizens, Somali merchants cannot stay on the Daasanach side indefinitely and are obliged to depend on the Daasanach for their fish supply. I have mentioned that the Daasanach recognized the price of dried fish as relatively high. As the Daasanach has an advantageous position, there is a certain level of initiative to negotiate the price. In addition, Somali merchants supplied beads and clothes from the Kenyan side. The Daasanach bought them for their own purposes and sold them to the highlanders in the town. Somali merchants are partners with whom the Daasanach have historically built good relations and can expect better economic benefits than the highlanders.

Dynamics of cultural value hierarchy

As desirable trading partners appeared in front of the people who were searching for new means of livelihood, the conditions were ripe to start fishing. However, considering the low cultural value of fishing activities and eating fish, it is too simple to assert that economic motives destroyed the cultural barrier within a short period. Rather, how pastoralists chose to enter fishing under the tension between the conventional cultural value and the new opportunity for a livelihood should be examined. Interestingly, many youths did not start fishing immediately after the opening of commercial farms, but only after observing the farm's management and operations, evaluating farming, and legitimizing their engagement in fishing.

Some pastoralists who started fishing contrasted fishing activities with working on a commercial farm. According to their narratives, while a farm opposes pastoralism or destroys the base of pastoral activities, fishing can coexist with pastoralism and even intensify the base of pastoral activities. It is easily understandable that people evaluate a farm negatively because the farm's land was partly appropriated from their pastureland. However, people talked about additional things, such as the farm being consistently anti-pastoral after its opening.

This evaluation was done by contrasting a commercial farm with conventional flood-retreat cultivation on two points. In flood-retreat cultivation, a livestock owner does not need to pay compensation if his livestock intrudes on another's cultivated land and damages crops. However, an owner must pay compensation when his livestock intrudes on a commercial farm. The intruded livestock is captured and taken to the labourers' camps. If no compensation is paid, livestock is not returned. Second, people make the livestock eat leaves and stems in the field after harvest in flood-retreat cultivation. The cultivator has the right to herd after harvest. However, if none of his livestock is around, anyone can herd. In August 2009, the Daasanach, who thought this rule also applied to a commercial farm, started to herd livestock on a farm after the maize harvest. However, the highlanders who managed the farm ran to them and shouted, "Drive livestock out! We sell leaves and stems in the town." The Daasanach were disgusted with the stinginess of the highlanders and left. In flood-retreat cultivation, livestock is superior to crops, and cultivation contributes to the development of pastoralism by supplying livestock with leaves and stems. On a farm, crops are superior to livestock, and a farm makes no contribution to developing pastoralism.

In addition to the negative value of farm management, working on a farm has a declining effect on pastoralism. On a farm near P village, there were approximately 25 Daasanach workers in August 2009. Their motive for working was to buy small livestock with their salaries. Their main work was to watch the farm and drive people and livestock away with a rifle. However, they regarded this work as "bad work" (*ujichi dedewa*) for three reasons.

First, in this work, they had to point a gun at their friends and their livestock. Second, far from saving money, this work indirectly decreased the number of livestock they owned. In 2009, small adult livestock was sold for about 400 birrs, and their salaries were about 200 to 450 birrs per month. Thus, they could buy small livestock with one- or two-months' salaries. However, as workers must watch the farm all night remaining alone, they bought alcohol to combat cold and loneliness. As a result, some of them spent much of their salary on alcohol. One man even needed to sell his small livestock to pay back the debt accrued from buying alcohol. Third, they needed to work continuously for at least one month to earn a salary. If they took off three or four days in a month, they were fired. If something happened to their livestock herded by others in the livestock camp, they could not swiftly move to the herd. Working on a commercial farm thus decreased the mobility necessary for pastoralism.

In contrast, the land use pattern of fishing does not compete with pastoralism. While it seems to be a simple fact, it had been difficult for the Daasanach to imagine an activity and land use pattern that was superior to herding and pastoralism. As the commercial farm that excluded livestock appeared, fishing in the river and lake became recognized positively because it did not disturb pastoral activity. The context in which pastoral and fishing activities had been located was changed by the introduction of a new development project and the relationality of each activity was rearranged.

In addition, pastoralists who started fishing began to observe how fishing contributed to the maintenance and development of pastoralism. They used the money acquired from selling fish to buy small livestock and medicine for livestock. As I mentioned, in 2009, people frequently talked about livestock being prone to disease. They later came to believe that new livestock and medicines bought from fishing revenues compensated for such weakness. Furthermore, because they could buy the necessities for livelihood with this money, they did not need to sell their milk, which meant the children could drink it. Finally, from fishing, they could make money within a few days – setting nets in the morning, collecting fish in the evening, drying them the next day, and selling them to merchants the day after that. They could continue or break this cycle at their own convenience. Fishing did not reduce the mobility necessary for pastoralism.

Then, did the negative cultural value of the smell of fish change? When people engaged in fishing for subsistence, cultural values about catching fish, the smell of fish, and eating fish were generally negative. However, after their fishing activity was commercialized, fishing pastoralists came to evaluate catching and selling fish positively. At the same time, some in H camp started to eat small fish and catfish daily, and at least, they did not tell me they were unhappy to do that.[8] In 2015, a young man in his 20s who had recently begun to eat fish told me, "Fish have a bad smell, but the taste of fish meat is the same as that of livestock meat. Good." He evaluated the smell and taste separately and accepted eating fish. In contrast, others, particularly women who did not engage in fishing, were reluctant to eat fish. The wife of the abovementioned young man continued to refuse to eat fish because of its smell, although her household economically benefited from selling fish. Her husband opined, "She is going to eat it soon." As the cultural value of fishing increased, evaluation of the smell of fish and eating them also changed and became more diverse among the Daasanach.

Social attitudes towards fishing pastoralists

The final question concerns how people who have not engaged in fishing regard young pastoralists who have started fishing. It might be assumed that they would negatively evaluate the youth because the smell of fish and fishing people might infect their livestock and make them infertile (Almagor 1987). However, I did not observe this attitude regarding pastoralists who had started fishing. This lack of criticism is consistent with a general attitude of respecting individuals' decisions (Sagawa 2010b). However, considering the strong negative cultural value of fishing, the specific context of why the other Daasanach do not criticize the youth for their decision should be analysed.

---

8 The Daasanach eat only boiled fish. Although they ordinarily eat fish without seasoning, if there is hot pepper (*barabare*) in the house, they flavour fish with it.

I examined this lack of negative evaluation by focusing on the comment "they moved to the south" mentioned above. When I asked about the absentees' whereabouts in 2011, villagers just answered, "They moved to the south." In my understanding, this expression meant that people moved to the floodplain to herd livestock. I asked them, "Do they herd small livestock or cattle?" They eventually answered, "No, fishing." It seemed that they did not dare to refer to fishing directly and instead emphasized the youth's physical movement. While "south" (*uro*) geographically refers to the floodplain of the Omo River and the northern shore of Lake Turkana, there are polysemous implications in their expression "moved to the south."

One implication is the relationship between the Daasanach and the Ethiopian state. In D village, not only commercial farms but also other development policies, including sedentarization, were planned in the late 2000s. This space was rapidly changed to a "legible" space by the state (Scott 1998). In contrast, as "the south" was far from town and located in a floodplain where it is difficult to build commercial farms, people imagined it as an ideal space where they could live an independent life away from the highlanders' influences. Although other villagers could not move there for various reasons, the youth's movement to the south can be positively interpreted as the movement to search for freedom and autonomy (cf. Scott 2009), even though they engaged in fishing.

Another implication is the relationship between the youth and elders in the Daasanach. A generation set is a group that comprises multiple genealogical generations: grandfather-father-son. All Daasanach men belong to a specific generation set (*haari*). In the P village near a commercial farm, the eldest generation set with strong ritual power was *Nimor*. The generation set of their sons was *Nigolomogin*, which holds a village-level government office and is connected with local governments. The generation set of *Nigolomogin*'s sons was *Hele Nigolomogin*, which had little power in both the generation set system and the government.

It is rumoured that when the local government decided to build a commercial farm in the P village, the government gave "gifts" to a few *Nimor* elders in exchange for their support, which should have ensured that other Daasanach accepted the decision. However, others, especially the young *Hele Nigolomogin*, reacted against the *Nimor* elders who supported the government's plan. Nevertheless, despite their opposition to the plan in the local government's public meeting, it was implemented, and a farm was opened. As the relation between the *Nimor* and *Hele Nigolomogin* worsened, some members of the latter "moved to the south" and some of those who moved started fishing.

Considering the timing and social context, it is clear that the reasons for their movement were not only economic incentives but also a desire to at least temporarily escape the worsening relationship with their elders. As many studies have indicated, avoidance is a simple yet important way to prevent or mitigate conflicts (Baxter 1972). Moving to a different location and maintaining physical distance contributes to easing the heightened emotions of opposing parties. As people remaining in the village recognized these processes and motives, the youth's decision to "move to the south" was accepted as a reasonable action to maintain the community's integrity.

Concluding remarks

This chapter clarifies the motives and processes by which young pastoralists have started fishing. I found that they changed and restructured their livelihoods resiliently. They came to recognize that the highlanders were encroaching on their living space and tried to maintain a livestock-centred livelihood by starting to fish. As a result, fishing has become an acceptable activity that complements pastoralism. In addition, people who did not engage in fishing accepted their choice as a desirable reaction to external pressures.

Peter Little distinguished non-pastoral activities that are competitive with or harmful to pastoralism from those that support or strengthen the pastoral sector. The former activities, such as cultivation in key grazing zones, negatively affect the ability of pastoralism to access important resources for livestock and cause resource-based conflicts. The latter activities add value to the pastoral sector and/or provide pastoralists with new sources of income and value that complement pastoralism (Little 2009, 2013).

In this chapter, I have shown that pastoralists also make such a distinction, and the standard for this distinction can change over time. Krätli (2017) classified two perspectives for analysing the behaviour and perceptions of pastoral people: the substantive and relational perspectives. The substantive perspective assumes stability and focuses on states rather than processes. From this perspective, the negative value of a certain activity is its inherent attribute. In contrast, the relational perspective regards changes and instability as normal and focuses on processes rather than states. The latter is a suitable perspective for examining the behaviour and perceptions of pastoral people.

This perspective is identified by Konaka, Little and Semplici in the Introduction, emphasizing the "capacity for change" as an important aspect of resilience and outlining the contextual, ontological, and relational perspectives of resilience. From this point of view, the evaluation and attitude towards an activity is shaped by the relationships between it and pastoralism. As the social context surrounding pastoralism has changed, the evaluation and attitude towards other activities have also been rearranged.

Until recently, fishing had been regarded as an activity that was contrary or harmful to livestock and pastoralism for the Daasanach. However, as commercial farms appeared and excluded pastoral activities and livestock, the social context in which fishing had been devalued changed and people came to recognize it as better than working in commercial farms. Furthermore, as young pastoralists engaged in fishing, they came to see it as a desirable activity that maintained and intensified pastoralism because fishing did not compete with pastoralism in terms of land use and it secured mobility. Moreover, it did not disturb pastoral activities, and the income from selling fish increased the number of livestock that could be purchased while also improving livestock health. Activities such as pastoralism, irrigated-agriculture, and fishing are not evaluated independently but relationally and contextually by pastoralists. As a result, people have resiliently adapted new activities and have thus addressed livelihood crises brought by development projects. This process highlights the 'capacity for change' based on the relational perspective of the world centred on pastoral activities and livestock.

Finally, I will briefly mention the latest situation in Daasanach. Since 2015, when the Gibe III Dam was completed on the upper side of the Omo River, flooding has rarely occurred. The livelihood strategy, depending on pastoralism and cultivation in the floodplain, has mostly failed. As the water flow of the river and the water level of Lake Turkana are expected to decrease further because of the dam, it is estimated that the number of fish will also decrease (Gownaris et al. 2016). Many of the fishing pastoralists abandoned fishing and converted to making a living strongly dependent on maize and wheat supplied through the food security policy of the Ethiopian government (Sagawa 2021b). The resilience of the pastoral society in this new context remains very uncertain.

References

Almagor, U. 1978. *Pastoral Partners: Affinity and bond partnership among the Dassanetch of south-west Ethiopia*. Manchester University Press.

———— 1987. 'The cycle and stagnation of smells: Pastoralists-fisherman relationship in an East African society'. *Res: Anthropology and Aesthetics* 13: 107–122.

Anderson, D.M. 1999. 'Rehabilitation, resettlement and restocking: Ideology and practice in pastoralist development'. In D.M. Anderson & V. Broch-Due. (eds). *The Poor Are Not Us: Poverty and pastoralism*. Oxford: James Currey. pp. 240–256.

Baxter, P.T.W. 1972. 'Absence makes the heart grow fonder: Some suggestions why witchcraft accusations are rare among East African pastoralists'. In M. Gluckman. (ed.). *The Allocation of Responsibility*. Manchester: Manchester University Press. pp. 163–191.

Boku, T. and G. Oba. 2010. 'Is poverty driving Borana herders in southern Ethiopia to crop cultivation?' *Human Ecology* 38 (5): 639–649.

Boku, T. and E. Sjaastad. 2010. 'Pastoralists' conceptions of poverty: An analysis of traditional and conventional indicators from Borana, Ethiopia'. *World Development* 38 (8): 1168–1178.

Bollig, M. 2016. 'Adaptive cycles in the savannah: Pastoral specialization and diversification in northern Kenya'. *Journal of Eastern African Studies* 10 (1): 21–44.

Broch-Due, V. 1999. 'Remembered cattle, forgotten people: The morality of exchange and the exclusion of the Turkana poor'. In D.M. Anderson and V.

Broch-Due. (eds). *The Poor Are Not Us: Poverty and pastoralism*. Oxford: James Currey. pp. 50–88.

Broch-Due, V. and D.M. Anderson. 1999. 'Poverty and the pastoralist: Deconstructing myths, reconstructing realities'. In D.M. Anderson and V. Broch-Due. (eds). *The Poor Are Not Us: Poverty and pastoralism*. Oxford: James Currey. pp. 3–19.

Caravani, M. 2019. 'De-pastoralisation' in Uganda's Northeast: From livelihood diversification to social differentiation'. *Journal of Peasant Studies* 46 (7): 1323–1346.

Carr, J.C. 1977. *Pastoralism in crisis: The Dasanetch and their Ethiopian lands*. The University of Chicago.

Catley, Andy, Jeremy Lind and Ian Scoones. (eds). 2013. *Pastoralism and development in Africa: Dynamic changes at the margins*. New York: Routledge.

Catley, Andy, Jeremy Lind and Ian Scoones. 2013. Development at the Margins: Pastoralism in the Horn of Africa. In Andy Catley et al., (eds.) *Pastoralism and Development in Africa: Dynamic Changes at the Margins*. New York: Routledge. pp. 1–26.

Chandler, D. and J. Coaffee. 2017. Introduction: Contested paradigms of international resilience. In D. Chandler and J. Coaffee (eds). *The Routledge Handbook of International Resilience*. New York: Routledge. pp. 20–28.

Derbyshire, Samuel F. 2021. *Remembering Turkana: Material histories and contemporary livelihoods in north-western Kenya*. New York: Routledge.

Desta, S. and D. Layne Coppock. 2004. 'Pastoralism under pressure: Tracking system change in southern Ethiopia'. *Human Ecology* 32–4: 465–486.

Ellis, F. 2000. *Rural Livelihoods and Diversity in Developing Countries*. Oxford: Oxford University Press.

Evans-Pritchard, E.E. 1940. *The Nuer: A Description of the Modes of Livelihood and Political Institutions of a Nilotic People*. Oxford: Clarendon Press.

Fratkin, E. 2001. 'East African pastoralism in transition: Maasai, Boran, and Rendille'. *African Studies Review* 44 (3): 1–25.

————— 2013. 'Seeking alternative livelihoods in pastoral areas'. In A. Catley, J. Lind and I. Scoones. (eds). *Pastoralism and Development in Africa: Dynamic change at the margins*. Oxford: Routledge. pp. 197–205.

Fratkin, E. and E.A. Roth. 2005. *As Pastoralists Settle: Social, health, and economic consequences of the pastoral sedentarization in Marsabit District, Kenya*. Berlin: Springer.

Fratkin, E., and K. Smith. 1995. 'Women's changing economic roles with pastoral sedentarization: Varying strategies in alternative Rendille communities'. *Human Ecology* 23 (4): 433–454.

Gownaris, N.J., E.K. Pikitch, J.Y. Aller, L.S. Kaufman, J. Kolding, K.M.M. Lwiza, K.O. Obiero, W.O. Ojwang et al. 2016. Fisheries and Water Level Fluctuations in the World's Largest Desert Lake. *Echohydrology*. 10 (1): e1769.

Holtzman, J. 2009. *Uncertain Tastes: Memory, ambivalence and the politics of eating in Samburu, northern Kenya*. Berkley: University of California Press.

Homewood, K. 2008. *Ecology of African Pastoral Societies*. Oxford: James Currey.

Homewood, K., P. Kristjanson and P. Chevenix Trench. (eds). 2009. *Staying Maasai?: Livelihoods, Conservation and Development in East African Rangelands*. Berlin: Springer.

Houtteman, Y. 2011. *Living in the Navel of Waag: Ritual Traditions among the Daasanech of South West Ethiopia*. Ph.D Thesis, Universiteit Gent.

Krätli, S. 2017. 'Pastoral Localization of Humanitarian Aid: The Need to Re-qualify the Pastoral Context'. *African Study Monographs, Supplementary Issues* 53: 141–146.

Lesorogol, C.K. 2008. 'Land privatization and pastoralist well-being in Kenya'. *Development and Change* 39 (2): 309–331.

Little, P.D. 1992. *The Elusive Granary: Herder, Farmer, and State in Northern Kenya*. Cambridge: Cambridge University Press.

———— 2009. *Income diversification among pastoralists: Lessons for policy makers*. COMESA (Common Market for Eastern and Southern African) Comprehensive African Agriculture Development Programme Policy Belief 3.

———— 2013. 'Reflections on the Future of Pastoralism in the Horn of Africa'. In A. Catley, J. Lind and I. Scoones (eds). *Pastoralism and Development in Africa: Dynamic change at the margins*. New York: Routledge. pp. 243–249.

Little, P.D., K. Smith, B.A. Cellarius, D.L. Coppock, and C.B. Barrett. 2001. 'Avoiding Disaster: Diversification and risk management among East African herders'. *Development and Change* 32: 401–433.

Lokuruka, M.N.I. 2006. 'Meat is the meal and status is by meat: Recognition of rank, wealth, and respect through meat in Turkana culture'. *Food & Foodways* 14 (3/4): 201–229.

McCabe, J.T., P.W. Leslie and L. Deluca. 2010. 'Adopting cultivation to remain pastoralists: The diversification of Maasai livelihoods in northern Tanzania'. *Human Ecology* 38 (3): 321–334.

McCabe, J.T., N.M. Smith, P.W. Leslie and A.L. Telligman. 2014. 'Livelihood diversification through migration among a pastoral people: Contrasting case studies of Maasai in northern Tanzania'. *Human Organization* 73 (4): 389–400.

McPeak, J.G. and P.D. Little. 2017. 'Applying the concept of resilience to pastoralist household data'. *Pastoralism: Research, Policy and Practice* 7 (14): 1–18.

Opiyo, F, O. Wasonga, M. Nyangito, J. Schilling and R. Munang. 2015. 'Drought adaptation and coping strategies among the Turkana pastoralists of northern Kenya'. *International Journal of Disaster Risk Science* 6 (3): 295–309.

Sagawa, T. 2010a. 'Local potential for peace: Trans-ethnic cross-cutting ties among the Daasanech and their neighbors'. In C. Echi-Gabbert and S. Thubauville. (eds). *To Live with Others: Essays on cultural neighborhood in southern Ethiopia*, Köln: Rüdiger Köppe Verlag. pp. 99–127.

————— 2010b. 'War experiences and self-determination of the Daasanach in the conflict-ridden area of northeastern Africa'. *Nilo-Ethiopian Studies* 14: 19–37.

————— 2021a. 'Dynamics of cultural value on non-pastoral activities among the Daasanach in East Africa'. *Nomadic Peoples* 25 (2): 206–225.

————— 2021b 'Large-scale development projects, food security policy, and livelihood of agro-pastoralists in southwestern Ethiopia'. In M. Takahashi, S. Oyama and H. A. & Ramiarison. (eds). *Development and Subsistence in Globalising Africa: Beyond the dichotomy*. Bamenda: Langaa RPCIG. pp. 19–41.

Sandford, S. 1983. *Management of Pastoral Development in the Third World*. Chichester: Wiley.

Scott, J.C. 1998. *Seeing like a State: How certain schemes to improve the human condition have failed*. New Haven: Yale University Press.

————— 2009. *The Art of Not Being Governed: An anarchist history of upland southeast Asia*. New Haven: Yale University Press.

Semplici, G. 2019. 'Seeing like the herder: Climate change and pastoralists' knowledge: Insights from Turkana herders in northern Kenya. In International Labour Organization. (ed). *Indigenous Peoples and Climate Change: Emerging Research on Traditional Knowledge and Livelihoods*. Geneva: International Labour Organization.

Smith, K. 1998. 'Sedentarization and market integration: New opportunities for Rendille and Ariaal women of northern Kenya'. *Human Organization* 57 (4): 459–468.

Sobania, N. 1988. 'Fishermen herders: Subsistence, survival and cultural change in northern Kenya'. *Journal of African History* 29: 41–56.

Sun, X. 2005. 'Dynamics of continuity and changes of pastoral subsistence among the Rendille in northern Kenya: With special reference to livestock management and response to socio-economic changes'. *African Study Monographs, Supplementary Issue* 31: 1–94.

————— 2017. 'Strengthening local safety nets as a key to enhancing the food security of pastoralists in East Africa: A case study of the Rendille of northern Kenya'. *African Study Monographs, Supplementary Issue* 53: 19–33.

Watete, P.W., W-K. Makau, J.T. Njoka, L. AderoMacOpiyo and S.M. Mureithi. 2016. 'Are there options outside livestock economy? Diversification among households of northern Kenya'. *Pastoralism: Research, Policy and Practice* 6 (3).

Walker, B. and D. Salt. 2012 *Resilience Practice: Building Capacity to Absorb Disturbance and Maintain Function.* Washington, DC: Island Press.

Yongo, E.O., R.O. Abila and C. Lwenyga. 2010. 'Emerging resource use conflicts between Kenyan fisherman, pastoralists and tribesmen of Lake Turkana'. *Aquatic Ecosystem Health and Management* 13 (1): 28–34.

Resilience
and Identity

05

Mobile Identities: Resilience, belonging, and change among Turkana herders in northern Kenya

Greta Semplici

Loporucho welcomes us with an apology. 'Nowadays, the sun does not let us treat visitors properly'. He means he cannot slaughter one of his goats for us. Not that he would really do so, as he is unsure of the purpose of our visit. However, the sun has little to do with the impossibility of him offering us a goat. He is old and has now parted from his livestock. He has moved walking distance away from a growing village and hopes to gain something from the road and shops. We learn he has multiple large stocks, 'some of which must be already in Uganda by now', in the dry season. Nonetheless, making such a non-offer was enough to act, to perform, as this complies with the practice of wealthy families when receiving visitors.

A single string necklace with big red beads matches the colour of Loporucho's hat. A pink *echuka* (blanket) around his hips, folded on his legs as he is sitting on his *ekicholong* (head stool). Around his wrist he shows an *abarait*, a circular knife protected by a piece of stretched goat skin, used to cut meat during *akiriket* (meat ceremonies). An *aburo* (herder's staff) and *esebo* (hunting club), of which I have noted an increased display in times of drought, both lie on the ground next to him. And finally, a ram's horn, where he stocks his tobacco, hangs from his left shoulder. As we talk, he is firm in asserting that he does not know the 'things of town' such as schools and 'those modern ways to earn shillings.' 'For us *raiya*[1] is all about animals and making *ng'akibuk* (sour milk) in *etwo* (gourd).'

Introduction

This chapter is about resilience and identity, an overlooked dimension in the resilience scholarship, which nonetheless is a fundamental part of everyday life: conferring meaning, support and purpose. It draws on fourteen months of ethnographic fieldwork in Turkana County, in the northern Kenyan arid regions, and it is part of a wider study on resilience, pastoralism and drylands (Semplici 2020). In this chapter I show, through a range of mixed sources of data (including field notes, field diaries, interviews and focus groups), the critical importance of the 'feeling of belonging' to explain forms of everyday resilience. While in some cases certainly important, the predominant focus on shocks and disasters in the resilience scholarship has led to neglecting an understanding of resilience beyond the 'shadows of the crisis' (Bakewell and Bonfiglio 2013), with the risk of ignoring the history and social context of the crisis itself (Little and Leslie 1999). Restoring an experiential and ordinary approach to resilience can help instead unravel its long lineage and history of accumulated experiences, situatedness and relationships. From this perspective, the role of identity – or better, a sense of solidarity built around the feeling of belonging – appears as a strong element of resilience. It certainly is a key signifier for life in the Turkana drylands. Key traits and features around a collective social identity were strongly remarked by my hosts and interlocutors. They all expressed a clear idea of what living their lives entails, and how they go about it: because they are trained and accustomed to the place, because they are strong and well equipped with knowledge, customs and social norms adapted to a life in the desert, because they are *raiya*, as they mean it: a person who lives in the reserve, a herder, an original Turkana.

---

1 Local term which holds a variety of meanings indicating a livelihood (pastoralism), a place (rural areas), a way of being (dietary, aesthetics, and other convivial habits): see Rodgers 2020.

Based on the conceptual underpinnings of 'cultural resilience', various recent ethnographies have sought to pursue analyses of social dimensions of resilience by looking at the ways in which cultural traits survive despite a rapidly changing world (Fortier 2009; Galaty 2013). However, these studies maintain a restrictive approach to resilience by lingering on a tired dichotomy, tradition–modernity, that does not account for the changes that occur within culture (Derbyshire 2020). With this chapter I intend to contribute to this literature by showing that resilience rests in both: in the local identity used as a social construct of belonging and solidarity (thus similarly to other ethnographies), but also in its responsiveness to change and distinctive malleability, or, in other words, in the mobility of identities. This chapter shows how, far from being statically anchored in tradition, identities are flexible, disrupting dichotomies built up along 'symbolic boundaries' (Lesorogol 2008), and are accommodating change.

The need for a group identity as well as an underlying flexibility of this identity is particularly important in contexts of rapid socio-economic, political and cultural changes, such as those occurring in Turkana. From merely being known as one of the world's richest archaeological sites and cradle of human beings, the arid and most north-west county of Kenya, is now subject to great international attention. In this context, power is shifting away from the ruling authority of elders and local seers, who used to manage and control the region, to progressively (and seemingly irreversibly) go into the hands of urban elites. This is a long process, which indeed started during British colonisation (Awuondo 1990; Broch-Due and Sanders 1999) but which has apparently been accelerating in recent years. Hatcher reports a growing feeling among Nairobi citizens that Turkana 'will be one of the best counties in terms of investment and development' (Hatcher 2014: 37). This sense of economic growth is associated with devolution plans emerging from the 2010 new Constitution, with the nascent oil industry in the south of Turkana (Okenwa 2020) and with expanding refugee operations in Kakuma and Kalobeyei camps, referred to as models for the international community (Betts et al. 2019). By the end of my fieldwork, there were numerous low-cost flights connecting Nairobi and Lodwar, the Turkana capital town, several times a day, landing in the newly-built small airport. Lodwar town was growing fast, with increased sections of tarmac roads and streetlights, and an increasing number of hotels, restaurants, and tourism companies. Streets were full of *boda-boda* drivers (motorbike riders) coming from other parts of Kenya, because 'in Lodwar there are more opportunities', as one driver explained when, out of sheer curiosity, I asked why he came from Kitale to Lodwar. When I left, I had the feeling of leaving a town in ferment.

In response to changes of such magnitude, as well as to finding position in *relation to* them, perhaps similarly to Bourdieu's theory of 'distinction' (1984), there tends to emerge a need for a feeling of belonging and solidarity based on a shared identity. Such identity is built *relationally*, with-in group, but also with-out group, namely in relation to what is perceived as *other* by means of 'symbolic boundaries' (Lesorogol 2008). Such a sense of identity is however never static as it undergoes perpetual re-articulations. In other words, precisely because of the incessant expansion of towns, 'development' and 'modernity', a strong sense of localised identity is necessary, but also the skills to confront, to respond, to adapt. In this way, not only livelihoods diversify as largely documented in studies about resilience and pastoralism (Little et al. 2001; McCabe et al. 2014; McCabe, Leslie and DeLuca 2010), but also identity, material culture and ritual life undergo processes of continual adjustment. Resilience, then, rests also on such skills of re-adjustment, which I will refer to here as 'capacity for change'.

Changes taking place in Turkana, while promising to many, are also accompanied by an 'impenetrable fog of misinformation and distrust' (Derbyshire 2020: xiii). Indeed, those who promote development interventions are rarely familiar with Turkana people, their livelihoods and societal dynamics or the socio-ecological context. From the perspective of town dwellers and other promoters of change, the life of Turkana herders remains tied to a sense of tradition which inhibits growth and development. However, tradition has little to do with their lifeways. As I show by the end of this chapter, tradition proves malleable and fluid, and individual identities too. What emerges more strongly is the reproduction of a collective social imaginary through everyday performances of being *raiya*. I argue that it is by understanding how people construct their collective identities, in this case the *raiya* category, and how these flexibly respond to changes in the surroundings, that one can understand resilience.

In the next section, I introduce the *raiya* identity, which I claim contributes to the resilience of Turkana herders by marking belonging and giving meaning, purpose and peer-support. I continue by explaining how this identity is built relationally, with respect to town counterparts. The associated practices, adornments and morality of both *raiya* and urbanity act as 'symbolic boundaries' which signify a sense of distinction and in turn reinforce the feeling of belonging and solidarity, through a process of *othering* that is also used to make sense of observed changes in society. Such distinction between *raiya* and their town counterparts is of course blurry. I conclude the chapter by unpacking contradictions and ambivalence between the self-presentation of the *raiya* identity and the everyday life of my hosts, through an analysis of memories, narratives and daily practices. Bush and towns are interconnected in ways that also contribute to a perpetual evolution of the *raiya* identity. In such ambivalence, I argue, resilience nests. These contradictions show a certain degree of responsiveness to changes, including peoples' skills to craft, navigate and bridge different (imagined) social words. These contradictions also show the coexistence of fluid identities among Turkana herders – *raiya*.

Raiya, a Turkana collective identity

> '*Aponopono: you could have told her that there is something known as thirst*
>
> *you could have told her that there is something known as sun*
>
> *you could have told her that we have learnt to stay-without*'.
>
> (Field note, young male herder)

We were preparing my visit to the sons of one of my host families during fieldwork, who were migrating with family shoats.[2] We met at their family

---

2 Mixed flock of sheep and goats.

wells on a day we knew they would be taking shoats to drink. Aponopono, the elder brother, was worried: if I joined them, I would lack the comforts he assumed (rightly) I am used to, like cool water and shade from the sun. At the same time, he was praising his own ability to survive despite undesirable hardship. In Turkana, herders walk long distances with the danger of scorpion bites and fighting with sharp thorns while herding livestock, fetching water or firewood or gathering wild fruit. While carrying out these tasks, they have adapted to a modest diet largely composed of fresh or fermented milk, sorghum or millet when available, berries and fruits, occasionally meat, and blood at cattle camps during dry seasons. They have adjusted to hunger and thirst and describe with pride the harshness they have grown accustomed to. Indeed, while trying to understand what a local term for the word resilience could possibly be, a recurring expression was presented to me: *anaikis nghichan*, to go-through, to overcome problems, to be accustomed to place. Despite having stopped asking this question quite rapidly, because I felt it was not providing me with novel answers nor was it an innovative approach to the study of resilience, it eventually

Photo 5.1 Being *raiya*. Photo by Greta Semplici, Turkana County. Kenya

occurred to me what my informants and hosts were trying to say. They were presenting me with some of the key features of their shared identity: endurance, strength, ecological-situatedness. They were delineating their 'identity of survivor' around which Turkana herders have built a collective persona: *raiya*.

Raiya, as a social construct, is a crucial element for the definition of resilience among Turkana herders. It in fact gives strength, value and meaning to one's own life in the everyday. It is a polythetic category, one which holds several meanings, including a livelihood (pastoralism), a place ('the bush'), a behaviour (those who uphold custom).[3] In the Turkana drylands, a part of being *raiya* implies being tough, while lamenting difficulties would be a sign of weakness[4] or would expose one to the risk of being accused of witchcraft.[5] Such an image of strength and endurance was presented to me during fieldwork through attitudes towards everyday matters, such as the ability to abstain from food, sleeping hungry and yet never failing to meet daily responsibilities. It was also reproduced in the reconstruction of event calendars,[6] such as the improvised focus group described below.

> It was *ekaru emuudu* (the year when everything finished). It was a very
> dangerous period, a year when most animals were attacked by diseases
> within a scorching drought. Before, people were rich and healthy, grass

---

3 See Rodgers 2020 for a detailed account of the meanings and evolution of the term *raiya*.

4 This behaviour is different towards outsiders or in relation to 'things' of foreigners (i.e. aid food). In these cases, weeping, begging, lamenting, asking for more are all very common and locally understood as rightful behaviour, part of an optimal foraging strategy (not a sign of weakness) and subject to different rules and norms than prevail among fellow insiders. That is, begging is external not internal.

5 To show weakness by moaning or constantly lamenting difficulties can be seen also as jealousy. Jealous people are feared and considered dangerous because they could curse people and livestock and cause bad events to others. Therefore, they are also among the first to be accused of witchcraft when such bad events do occur (Semplici 2020).

6 Events are commonly used to mark calendars in oral histories and are passed through generations (Derbyshire 2020; Little and Leslie 1999).

and trees used to reproduce in large numbers. Conditions are worse now. Rains disappeared. Grass withered. Wild fruits were absent. Disease became fierce and dangerous, and killed people and animals.' (Male elder, Key Informant Interview)

I was asking John Lopua, an elder from one of the small settlements born in the surroundings of a dirt road, to share with me one of his worst memories.

'People started eating dogs because they lacked food to eat' [someone else added]

'Dogs were not the only things we ate. We also ate dried animal hides (*ejomu*). We used to soak it in water to make it soft and later it was cooked in a pot'.

'Others roasted it'.

By then, Lopua's story had become a collective recollection of tragic details to which a group of people, gathered under the shade of a big *ewoi* (*Acacia tortilis*) tree, contributed. Multiple voices echoed:

'People sold their children [hired them for employment] to the farms in Kitale'

'And then rain came. But the grass that followed was not good grass as the surviving animals were getting sick with diarrhoea'

'...Many houses were destroyed. Animals died both in the field and in kraals. Most trees fell, floods carried away many properties. That rain was not a blessing. It was accompanied by thunderstorms. Both cold and thunder killed our animals.'

From this account, one of many I collected, there emerges an underlying story, one that goes beyond the struggle and the suffering. In some ways, and despite the tragedies narrated, it was not a story of vulnerability, at least not in the ways it was recalled to me. From the choice of words, from the tones of voice, the collective laughter and the gestural behaviour

of the people gathered under the acacia tree, there was something else –
something initially difficult to grasp. There emerged a certain 'pride' in
'having gone through', having survived and persevered until the 'season
of plenty'.

Drawing from Brewer and Gardner's idea of the social self as a more
inclusive self-representation in which relations and similarities to others
become central (1996), *raiya* becomes a strong marker of belonging, one in
which people find solidarity with others and a reference to endure. Among
the characteristics of being *raiya*, beside toughness, there are elements
of diet, fashion and other convivial habits. How to behave, what to wear,
what to eat and what not, are elements that were almost stereotypically
described as essential for 'being *raiya*'. I show below two drawings we
made with the participants in focus groups held in one of the villages
sampled for the study (Photo 5.2).

They represent a Turkana woman and a Turkana man. In these drawings
are all the elements which point to distinctive *raiya* identities. Garments
and accessories, including artisan jewellery, earrings and other ornaments

Photo 5.2 Representation of *raiya*. Source: drawings from FGD in Lorengelup,
Turkana Central

(some described in Table 5.1 below); objects, properties and articles (some reported in Table 5.2 below); number of children, livestock, wives (or co-wives), as well moral qualities such as being hard-working, trustworthy and generous were echoed in the discussions about the representation of *raiya* people.

These drawings represent the collective imaginary of wealthy *raiya*. Discussions about poverty were more complex and go beyond the primary objective of this chapter. As argued by Rodgers (2020), the *raiya* category acts as an element of social stratification within Turkana people as a whole, especially since the expansion of urbanity, when the need for distinction started growing.

Table 5.1 Female attire and adornments

Item	Translation	Example of usage
Abwo	Back skirt	Half skirt made of animal skin in a specific pattern, with hanging extensions on the sides. It is worn by women to cover the back, stretching from the waist to the feet.
Adwel	Front short skirt	It is worn at the front by women, stretching from the waist to just above the knee.
Akoloch	Calf teeth	Calf teeth used as ornaments worn by women around their head
Aruak	Animal fat	Solid fats, most fats come from animals' stomach or other parts such as around the kidney, heart, intestines. The fat is cooked and stored for future use. Turkana women and men normally mix Aruak with red clay (called ogre) and apply to the hair.
Aruba	Belt	It is worn around the waist by women, it is like a big belt, and is commonly used during dances. It is made from skin and patterned with small beads.
Egolos	Long front dress	It is made of animal skin and covers the entire front part of a woman, from just below the beads and stops just below the knees. It is normally decorated with beads or multiple patches of skin.
Ekude	Headband	It hangs on the head with small, decorated beads up-to the face.
Ngakoromwa	Necklace	Beads arranged in a thread and worn around the neck
Alagama	Wedding ring	It is a large wedding ring made from copper or aluminium wire, worn around the neck on top of the beads as a symbol of a married woman. It can be silver or golden depending on the age group.

Turkana language does not have a specific word to say poor – as reported by Müller-Dempf 2014: *ekechodon* (temporary lack of animals), *elongait* (no relatives), *ekebotoonit* (no animals). And indeed, scholars such as Anderson and Broch-Due, in the well-known text *The Poor Are Not Us*, have argued that the destitute are not Turkana herders as government narratives portray (Anderson and Broch-Due 1999). This is surely so, at least for the wealthy herders. However poor *raiya* do exist. Their poverty was described to me by subtraction of what it means to be wealthy, and entails rotten morals (laziness, untrustworthiness, transgression of local norms), reduced social capital (size of extended family and relationships) and compromised assets (livestock and material wealth).

To conclude, everyday life in the Turkana drylands is negotiated by means of a collective constructed social identity, *raiya*, which serves as threshold of reference. It allows finding solidarity with others through shared performances of being *raiya*: being tough and enduring, adhering to social norms including clothing and attire, nutriments and moral behaviour. It gives meaning and strength. It marks belonging, especially when used to contrast *raiya's* 'sense of place' with the 'feeling out of place' experienced by outsiders (non-*raiya*) who venture in the bush (Rodgers 2018). Indeed, a great part of being *raiya* entails a distinction from urban life, as discussed in the next section. Such a distinction, while in reality blurred, holds great significance for *raiya*, reinforcing solidarity through 'symbolic boundaries' (Leseregol 2008) performed to interpret broader processes of change.

Distinct social worlds

Everybody is laughing loudly. They are laughing at Ejomu when he exaggeratedly jokes, 'you would die of starvation if you travel with her!' Everything started when people were amused to see me and my assistant, Ejomu, drinking a whole mug of tea with salty water and fresh goat's milk, in one gulp. It was mid-day, and this cup of sugared tea contained our first calories of the day.

People hiding with us from the midday sun inside someone's hut

Table 5.2 Home utensils and articles.

Item	Translation	Example of usage
Agulu	Clay pot	Pot made from clay, used for boiling cereals like maize, wild fruits, etc.
Akaloboch	Spoon	Serving spoon; it is carved from wood into various sizes and shapes.
Akurum (and *akurum na ibole*)	Gourd	Elongated round shaped container used to store fresh milk, and to start fermentation before being transferred into *etwo*. It is also used for carrying milk while moving around to sell in the villages. The long lid (called *ibole*) is used to serve milk.
Akutwam	Calabash	Round shaped container made from animal hides into a dried ball. It is used for storing liquid fats from animals, for example ghee (*akuring*)
Atubwa	Bowl	It is a home plate/bowl, or animal drinking trough, can be carved in many shapes. Small ones serve as plates and the largest can serve as a trough for the animals to drink water. It is carved from woody trees.
Ebur	Container	Used for storing/reserving fatty pieces of meat, to be eaten in the future.
Egech	Stirring stick	It is made from fixing one end of a wooden stick to goat or sheep spine bones. Used for stirring soup inside a large pot so that the oil does not float, or to stir porridge or blood meals during cooking until fully cooked.
Ejomu	Sleeping mat	It is made from skin and hides. It occurs in many sizes but hides of big animals (cow, camel, donkeys) are preferred due to their large size and long lasting.
Elado	Whisk	Made from several tails of cows into a big thick tail fixed on a short stick, the handle is made from skin, it is used during traditional dances, it is wagged or swayed from side to side, or in all directions, during dancing to move the dances in unison, to give order or harmony to the dancers, to control the pattern of dancers. Mostly used in celebrations, weddings, village dances, welcoming prominent leaders, etc.
Elepit	Milking can/ jar	It can be carved in different shapes, but most often it looks like a tall wooden glass. It is used when milking livestock, where initially milk is gathered, before being transferred to *akurum*.
Etwo	Calabash	Round shaped and used to shake milk every dawn, after it has undergone fermentation, to allow milk, its fats, and coagulated particles to break, and mix properly. Clean pieces of rocks are sometimes added to allow the break-down of the fatty layer or coagulated layers

could not believe a town girl (and *muzungu*[7] too) would drink goat's milk with such ease. Beyond towns and road settlements, only the strongest can possibly make it, while 'town people fear sand' and 'eat children's food'. *Raiya* believe townspeople would not cope with hunger and thirst and would let themselves die craving for food if they lived as *raiya* do. Sentences like 'they went to sit and earn'; 'they are lazy and do not walk'; 'they only worry about food'; framed the conversation around our sugared tea, broken by sonorous laughs.

Raiya markedly distinguish themselves from their urban counterparts. Diet, fashion and convivial habits become expressions of different social worlds, which *raiya* use to position themselves with respect to the broader society that is rapidly changing. Indeed, in a world which is practically fully urbanites, one that sees pastoral livelihoods turning into a 'marginalised majority against expanding town elites' (Rodgers 2018), the need for self-identification becomes even stronger in order to reinforce belonging and find justification for observed changes. In this context, a process of othering is put in place by means of 'symbolic boundaries' (Lesorogol 2008) between *raiya* and town people.

Part of this distinction is built around food. Indeed, the *raiya* identity is tied to the identification of certain food, which is considered appropriate, satisfying and delicious (*ebob*), and most suited for pastoral living. There is food, in other words, which was presented to me as the 'food for *raiya*'. It satisfies hunger, allows undisturbed nights of rest and days of walking under the hot sun. This food is mainly represented by the pastoralist triad (milk, meat, blood), but not only, as it also includes a large variety of berries, sorghum, millet, wild game and other supplements. On the contrary, town people are seen by *raiya* as only eating children's food, such as rice and spaghetti, and as vulnerable to the lack of food, while *raiya* can stay-without (Semplici 2020).[8]

---

7 White person, Kiswahili.

8 One of the most predominant outcomes of being *raiya* is to learn to 'stay-without' since hunger is a salient part of life in the drylands. The capacity to stay-without thus results from the relationship with the environment, one where food is often a bet, and from a livelihood which takes individuals on long marches needing to travel light.

Despite such strong statements about the *raiya's* diet, in reality they consume a greater variety of items, including the so-called new foods, namely food brought by outsiders including beans, maize, tea and sugar, rice, oil, *Royco* (stock cube) and others. The *raiya* identity is thus apparently compromised, but it has reorganised in the polarisation of memories skewed towards a plentiful past compared to a shameful present of decay and scarcity, in which food becomes a 'locus of historically constructed identity' (Holtzman 2009). In such memories, despite strong elements of 'cultural nostalgia', taste serves as an assessment of the social world which manifests through food (Montanari 2006) and as a marker of identity.

Memories of taste

Memory is the 'process of making sense of experience, of constructing and navigating complex temporal narratives and structures and ascribing meaning not only to the past but to the present and future also' (Keightley 2010: 56). Some scholars have shown food to be a valued artefact for inner sensory memory (Counihan and Van Esterik 2013; Monsutti 2010; Sutton 2001). And indeed, with a good dose of nostalgia, through continuous swings between an imagined or real past and present, Turkana herders have reconstructed their history around the distinction between a past of plenty and a shameful present of poor food and dependency. Through such narratives, my interlocutors were not making historical claims; indeed, they also frequently recalled their past experiences of famine and other

Photo 5.3 Blood and maize. Photo by Greta Semplici, Turkana County, Kenya

shocks as reported earlier, for example about *ekaru emuudu*. Rather they were sharing a view of the world.

Years of plenty

'I was born in *ekaru amuje kimet*, the year to eat fat', remembers Alaar, one of the elders in one village, while walking through the surrounding plains.

> It was a time of plenty, animals were big and healthy and produced a lot of fat. That's why we called it the year to eat fat. All this land you see here around us, this was all grazing land. Those years we were eating a lot of meat and there also were many *alogita* (age group dances) during which we were given livestock to slaughter and celebrate. I met my first wife in one of those dances. She was beautiful, with red ochre on her body and beautiful beads. During the dry seasons *ewoi* (acacia) would still be full of *ngitit* (acacia pods), out of which our animals survived and never suffered. And we were drinking their blood without fear of making them weak. When it rained, we planted millet and sorghum along Turkwel River which we would stock for the dry season and eat together with animal fat and blood. All food was so delicious and filling. And it was so much that people did not fight. Many were crossing these lands when it rained, and we stayed together in peace because we were all Turkana and there was enough for everybody. I remember eating milk every day. We did not know about relief food. We were depending on our own. We did not know about money. Our animals were enough. Women were wearing only animal skin dresses and men were fierce. That time was a time of happiness. Children were happy and listening to the elders. Our food was so satisfying that we could stay the whole day looking at our animals without being disturbed by hunger.

From this account, years of plenty are not only valued for their nutritional intake (plenty of meat, meat soup, fat, blood, milk, and wild fruit), but also, and mostly, for the emotive and social effects. Years of plenty are described as years of happiness when people could dance without worrying; when people were full of energy; they were proud, as they did not need to be helped but 'depended on their own'. Social friction is also recalled as being minimal. Plenty of food was in fact seemingly fostering a moral personhood, leaving also time to rest from stress.

A past of plenty is marked in contrast with an impoverished present when food is lacking, and behaviour has degenerated. Such strong affection towards the past may be felt more strongly among elders, such as Alaar, but is also shared by younger herders in their attempt to comply to the image of being *raiya*. They might not recall the same images of the past, and they might not talk with such negativity about contemporary societal developments, but they do hold on to the perceived differences with respect to town people also through attitudes towards food. Young herders reaffirm the distinctive *raiya* capacity to abstain from food, compared to their urban counterparts' dependence on food, and maintain the disdain for the food that comes from towns.

Years of shame

'And now we live in these years of shame', Alaar continues after my exhortation to tell how those years of plenty compare to today.

> Almost everything is now lost. There is no grass, no livestock, no *ngitit*, no rain. Now we are eating the government [what the government brings]. We are eating ground maize and salt, and our body grows thin. We have stomach-aches, and adding salt is burning our body. When we drank blood our body became strong, we ate milk and got fat; when we ate meat there was nothing else to add to your meal. The oil we are using now leaves a glue at the bottom of the bottle. The same happens in our stomachs! *Mandeleo* (development) brought us shame. Children have back pains like old people and will remain with appetite in the heart.

A whole set of somatic meanings clustered around notions of health is attributed to 'new' foods (Barthes 1961), which spread through relief operations and *mandeleo*.

'Current food is full of diseases, like *edeke moruariiwan* [stomach pain].' (Field note, middle aged woman)

'All body parts are in pain. When we eat animal products our body becomes light and one can even decide to run. But nowadays after that we have eaten these new foods, we have known their bad effects and we refuse to eat them.' (Field note, middle-aged woman)

Frequently in the rural lands of Turkana one hears people lamenting new foods. Their claims sound definitive. The new food, of which maize is the most significant, is not liked because it is associated with diseases, which people say were absent in old foods, and because it is tasteless, not like eating meat when 'there was nothing else to add to your meal'. It is also extremely difficult to cook as it takes many hours and yet remains 'hard as stones' (and indeed it also comes with stones in the relief bags, carefully removed one by one, introducing a new time-consuming livelihood practice).

Akamaise, the youngest wife of Alaar: 'Someone puts *githeri* (beans and maize) on both sides of the mouth, tries to chew but spits maize out'.

Nawoi, one of the young sons: 'That's why the goat is following her like a dog!'

Lopwenya, the eldest brother herding family shoats: 'It has become the goats that are feeding on that white maize, just like a bird'. (Field note)

Maize compromises one's identity, even that of goats, now acting like dogs or birds. Indeed, new foods entail several changes of alimentary structure and family social organisation. Cooking for longer implies reducing time fetching water, wild fruit, firewood. Eating hot food implies reducing distances from the homestead to be served. And certain cooking utensils are also used at cattle camps where men are nowadays 'entering a woman pot'. With these words Lomekwi, one of my host fathers, meant that, nowadays, herding boys have learnt to cook like women; while in the years of plenty there was no need for those cooking practices because blood and milk were enough, and men could remain men, and not act like women.

In the social imaginary, new food has not only brought nutritionally less valued intake but also a whole set of moral degenerations that bring shame: women are promiscuous because of alcohol consumption, *ngimurok* (seers) reside in towns and become greedy, children lack respect for their elders, men are no longer fierce. These statements sounded definitive: a *raiya*, a real Turkana, repels new foods brought by outsiders, classifying the social world they have brought in with their foods as a world of general decay. And yet, these two worlds are certainly more intertwined and less polarised than they are typically portrayed.

Indeed, while new food was so disliked, there also was strong pressure on authorities such as local chiefs to demand more food aid, as I could observe in numerous *barasa* (village meetings). Also, there have always been some Turkana without enough livestock, who depend on alternative foods rather than the pastoralist triad (milk, meat, blood). Furthermore, the food triad has never been exclusive, as it has always included grains from farming communities through barter or other forms of exchange. Many young herders, while laughing at the town people's perceived weakness, also confided their wish to have permanent houses, a business and money to buy food, thus imaginatively stepping into the social world of their urban counterparts. Many admitted seeing benefits from processes of diversification that lead some members of the family to search for jobs in

Photo 5.4
Mixed social worlds. Photo by Greta Semplici, Turkana County, Kenya.

town, especially during dry seasons or droughts when they could support their families in the rural areas.

In other words, everyday life and needs blur the symbolic boundaries put in place to reinforce belonging and solidarity. Such contradictions are the result of a process of positioning in a world that is fast changing and emerge from the observation of everyday practices, beyond the social imaginary. During these episodes, I witnessed the formation of counter-narratives emerging from the implementation of new practices and traditions (Hobsbawm and Ranger 1992); narratives which hinge upon a representation of the self as pure and detached from *things of town and modernity*, but that are in reality embedded in daily practices which draw from both worlds – that of *raiya* and that of town dwellers – indicating a broader process of social, economic and political transformation and the coexistence of fluid identities.

Mobile identities

Anthropologists have presented evidence that discontinuity between self-description and social behaviour is the norm rather than the exception (Santopietro 2015), an incompatibility that Turner calls 'social drama' (Turner 1987). Mainstream narratives, such as the individual sense of solidarity with a certain social identity, do not necessarily correspond to an individual's everyday practices (difference between ideal and actual behaviour). Greater variation and heterogeneity emerge within everyday performances of being *raiya* than is typically imagined through its dichotomisation against urbanity. By 'stylistic playing', as reported in Rodgers' study of Turkana herders (2018: 149), or by reproducing social change, as reported by Holtzman about the Samburu adaptations to modernity (2009), *raiya* are also incorporating foods and traits representative of their urban counterparts.

However, such contradictions between narratives and practices do not imply that these accounts should be discarded. Rather, they show how reformulations of the self and collective identity undergo multiple, at times contradictory, processes (Gonzales 2015). Indeed, the *raiya* identity, as contradictory as it can be, still plays a crucial role for belonging and solidarity, and in turn people's resilience. Additionally, contradictions between principles and practices are not mutually exclusive. Rather, they show the coexistence of multiple identities and the fluidity of social worlds, which are not organised along dichotomies but navigate across symbolic boundaries. This is indeed true for everyone (everybody has multiple identities), and certainly I am not making a critique of identity theory in anthropology. Rather I am challenging the use of cultural resilience for indigenous populations, through the case of a pastoralist population, as restrictively discussed through a modernisation lens, one which limits reflections on identity and culture to a debate on the relevance of tradition. Such fluidity of social worlds, I instead argue, is a salient dimension of one's own resilience, for the capacity to change, remain attentive and mobile.

In what follows, I describe some of the contradictory elements found in everyday practices and narratives about development and modernity. I explain these contradictions in terms of responsiveness to change and mobile identities, which I claim are important ingredients of resilience, especially in relation to fast-changing societies.

*

The pastoralist diet (the expression of physical potential and beauty, independence and integrity, as described above), centred around the sacred triad of milk, meat and blood, is now at odds with an 'identity built around maize' (Holtzman 2009: 1). Ground maize-flour (*epocho*) and beans have become very common staple foods and are found in most homesteads. Grains are not entirely new to the Turkana diet, having always played an important role in exchanges with agriculturalists, but the variety of white maize made available through the provision of relief food and the spread of small rural shops, has instead emerged in the last sixty years (Holtzman 2009) and has entered daily and ritual food practices. For example, *akuuta* (wedding ceremony), one of the most loved ceremonies in Turkana, concludes with a big pot of *epocho* (maize flour) mixed with wild fruit, signifying the incorporation of new foods within traditional practices.

It is not rare to find salt, vegetable oil and *Royco*, and, in some families, even rice and spaghetti, though these remain treated mainly as children's food. I have met mothers walking miles to reach the closest road settlement to buy rice, sugar and milk powder for their youngest children who 'are refusing our food and eat only the government food'. And all herders also depend on purchased food, regardless of how much they dismiss it. The expansion of urbanity and the associated social world is, in other words, forcing the rearticulation of certain traits of being *raiya*.

While I was migrating with the young brothers of one of my host families and their shoats in the Turkana central plains, we hid our bags of maize and beans in holes dug in the ground to prevent squirrels from stealing them from us. And we also diverted from our route to restock our provisions from a small rural shop. But, mainly, we diverted in order to charge their mobile phones through solar panels leaning over the shop's store front, a service which cost them 20 *shilling* (approx. 20 cents). We watched music videos and listened to the phone ringtones at night while lying over a carpet of shrubs and sand, next to an improvised *anook* (kraal) for the livestock.

The contradictions between declared hostility to what comes from towns and daily practices of imitation and incorporation of town habits and articles is not confined to food practices alone but affects all aspects of social life, as western items replace traditional ones. In the picture below I want to draw attention to the few objects the two shepherd brothers were travelling with (Photo 5.5). Among these objects are a pot and a discarded metal container with a wire as handle, repurposed as a cooking pot to cook maize (as well as a plastic bag which contained maize and beans). The young herders explained that they had bought the pot at a local market and were proud to have found the discarded metal container next to an abandoned building site. In this way they have incorporated into their daily practices articles from a repertoire belonging to their urban counterparts, and they are weaving together two counterposed lifestyles in their social imaginary. The same approach to recycling was explained to me about abandoned plastic pipes once meant for boreholes or irrigation schemes that were never brought to fruition. That plastic is re-used in multiple ways including blade covers, small tobacco containers, armbands, and other decorations.

Turkana herders have also incorporated items into their wardrobe and accessories. It is not rare to see skirts, trousers and shorts, even in remote bushlands, as the catalogues of products sold in rural shops display. A striking example of outfits which do not suit a *raiya's* closet was the adornment shown by Lokir, one of my host fathers, during the wedding of his son (*akuuta*). He wore his favourite collared dark green shirt and exhibited long and well-cared for white feathers on his head, a black *elado*, from the tail of an ox, on his hands to shake during *edonga*. Around his neck he wore a fuchsia garland, possibly found abandoned on the ground at the end of some other formal event (Photo 5.6). It made me smile, but I soon realised how resourceful such a gesture appeared to other attendees at the wedding. The disdain for urban life was overcome with the appropriation of attire and articles felt to be glamorous, like the improvised scarf worn by Lokir, and practical, like the discarded metal used by the shepherd brothers. The capacity to select specific traits from the culture of *raiya's* urban counterparts 'removing the meaning conveyed and replacing with their own' shows that the constructed 'social imaginary' remains fundamentally open, polythetic and with 'fuzzy borders' (Piasere 2009: 91). All this shows features of constant reformulations of the self

Photo 5.5
The objects of the herders. Photo by Greta Semplici, Turkana County, Kenya.

and the malleability of traditions and social positionalities, challenging the strong dichotomisation presented above.

In addition to contradictions between daily practices and self-representations, I could also participate in the production of counter-narratives. On the one hand, the emphasis on a *raiya* diet and the omission of other types of food not perceived as strictly *raiya* reinforces the idea of a cultural identity, a sense of solidarity around one's own imagined and performative living: a nomadic lifestyle where the food available originates only from cattle, camels and flocks of goats and sheep. On the other hand, changes brought by the arrival of new food are also seen as signs of progress. Some tired elders trusted that, thanks to new foods, people would no longer die from hunger. Some young wives confided that new food is easier to obtain without endless struggles of following livestock. Some other young herders hoped that *mandeleo* (development) would help them build their own permanent house with a water tap directly into the homestead.

I explain the existence of these counter-narratives and ambivalences in daily practices to highlight the inherently mutable character of tradition and skills in navigating the complex social imaginary. These skills are perhaps the greatest contribution to people's resilience at the socio-cultural level, as they make it possible to weave dichotomies together and avoid fragmentation. Indeed, amidst broad social, economic and

Photo 5.6
Feathers, elado and garland. Photo by Greta Semplici, Turkana County, Kenya

political changes, individuals find themselves grappling between an ideal behaviour and a changing landscape that imposes new practices. With this perspective, I confirm Turner et al.'s idea that self-categorisation is inherently variable (1994), and dependent on contextual shifts (Brewer and Gardner 1996). The *raiya* identity is adaptive to changing conditions as their way of interpreting 'modernity'.

Fluid and malleable social identities coexist and can be activated at different times. The ability to navigate the social imaginary corresponds to Brewer and Gardner's theory of relational identity, (1996) by which the 'we' category, *raiya* in this case, coexists with one or several self-categories, all of which can be used at different times, creating flexible self-representations. The *raiya* category is thus in reality kaleidoscopic as it includes a set of cultural traits that, on a case-by-case basis, come together differently, creatively borrowing features from their urban counterparts when these are perceived to be in their own interest.

Through ambivalences and contradictions, Turkana herders have shown both a great responsiveness to changes, including the growth of non-pastoral populations (Rodgers 2018), and the fluid nature of their being. Admittedly, this is easier said than done. One of my host mothers, Akamaise, once shared her feelings about towns:

> '… the person who has not gone to school, the survivors of the new changes of life, nothing can suit them in the new life in town. Those who are not learned [those who did not go to school] grow thin in town [they would struggle to survive in town].' (Field notes)

Indeed, as argued by Ferguson, not everyone will be able to adapt, to change identity, to shift between different styles, as this will also depend on the ability to do so competently (Ferguson 1999). However, regardless of how difficult it is, flexibility of identity proves useful in breaking up dichotomies in a changing and variable social landscape. The ability to shape one's own identity gives Turkana herders greater access in a highly variable lifeworld. As argued by Rodgers (2018), the resilience of pastoralism does not involve the persistence of its forms, but the ways it re-articulates itself in the face of contemporary variabilities (cf. also Chatty 2013; Derbyshire 2020). Such rearticulation occurs at the livelihood level, as discussed in the literature about resilience and diversification (Little et al. 2001; McCabe et al. 2014, 2010), but it also manifests through crafting and re-crafting one's own identity, material culture and ritual life. Turkana herders' lifeway and livelihood are thus not disappearing, incorporated into the corrosive hands of modernity: rather, herders themselves are creatively reworking their values, self-representations and culture (also material culture – see Derbyshire 2020) to make a meaningful living. As argued by Chatty, there is little room for supporting theories of conservatism in the face of change (2013). Quite the opposite, Turkana herders show a flexible identity which engages with forms of development and 'modernity' while maintaining a sense of solidarity around certain distinctive traits of being *raiya*. That is, Turkana herders show intrinsic mutability but also a degree of resistance, never to be free from themselves – as posited in a song by Bob Dylan: 'Are birds free from the chains of the skyway?'

Conclusions

Resilience, as an analytical concept, has been strongly criticised and many have suggested it should be abandoned (Davoudi et al. 2012; Scott-Smith 2018). Nonetheless, it seems to me that the concept still has some utility. A 'post-resilience' era is seemingly far ahead of us, as resilience continues to dominate policy discourses. Thus, we, as researchers, are left with three options:

1. Accept resilience dogmatically and apply it to various cases;
2. Criticise resilience entirely in stark opposition to policymakers; or
3. Recreate and reconceptualise resilience, as something more significant and adaptive to the lives of the people we are working with.

With this chapter I embraced the third option. Resilience is inevitably fuzzy; and perhaps it should remain so, like the *raiya* identity. In this chapter, I have suggested that resilience is both personal and relational. It is about self-identification within and without a group of reference. It is about place and identity, and the relationship between place and identity (drylands as forging an 'identity of survivor'). It is about resistance (remembrance and memory) and about meaningful transformation. As argued by Chandler: 'Resilience is both about adapting to the external world and about being aware that in this process of adaptation the world is being reshaped' (Chandler 2014); and, I would add, reshaped in ways that make sense locally, like Lokir's scarf and the brothers' possessions. Thereby, resilience is strongly 'localised' (Konaka 2017), rooted in context, and it should not exist as a 'category for outsiders', as a view from above, but rather in the lived and symbolic experiences of people in their everyday lives, at eye level; it does not emerge from 'making a view of the world, but taking a view in it' (Ingold 2000: 42).

This chapter builds on recent trends in the resilience scholarship which increasingly support the investigation of socio-cultural dimensions of resilience (Adger 2000; Crane 2010; Folke 2006). However, I have tried to move beyond what I would call reductionist approaches to cultural resilience which hold onto static views of tradition as symbolic of the resilience of cultural identities. The *raiya* identity remains important, not necessarily in terms of tradition but in terms of belonging, and through the construction of symbolic boundaries which distinguish *raiya* from town people, in turn reinforcing within-group solidarity and positionality in a changing world. Nonetheless, I have shown that the distinction between *raiya* and town dwellers is far from sharp. As with other pastoralist groups, *raiya* 'continue to change, to adapt, and at times to spontaneously adopt techniques, technologies, and ideas they perceive to be in their own interests' (Chatty 1996: 190). The initial impression of a homogenous and united community of rural herders was contrasted by daily practices of imitation and appropriation of articles, attire, and lifestyles found in urban centres. I have explained this observation by resorting to the notion of 'ambivalence' proposed by Holtzman (2009) as a symptom of broader social, economic and political changes, and through the logic of 'fuzzy epistemology', which explains the fluidity of one's own 'being' (Piasere 2009). Resilience rests in the capacity for change, remaining open and as flexible as possible in decisions and practices. In other words, it rests in the mobility of being, employed as mechanism to navigate change and different cultural environments. In a world that is highly mobile and constantly changing, mobility is the strategy my hosts and interlocutors have employed. Such mobility is not only physical in terms of movement, but it also entails a mobility in the practices (livelihood diversification), culture (identity and material culture) and relationships (within and without). It is important we recognise all these forms of mobility, and never close borders – symbolic, physical, perceived borders of identities – because to be mobile in its many manifestation, including as a quality of identity, is to be resilient.

References

Adger, W.N., 2000. 'Social and ecological resilience: Are they related?' *Progress in Human Geography* 24 (3): 347–64. https://doi.org/10.1191/030913200701540465

Anderson, D., and V. Broch-Due. 1999. *The Poor Are Not Us: Poverty & pastoralism in eastern Africa*. Oxford: James Currey.

Awuondo, C.O. 1990. *Life in the Balance: Ecological sociology of Turkana nomads*. Nairobi: ACTS Press.

Bakewell, O., and A. Bonfiglio. 2013. 'Moving beyond conflict: Re-framing mobility in the African Great Lakes region'. *IMI Working Paper Series* (71): 34.

Barthes, R. 1961. 'Toward a psychosociology of contemporary food consumption'. In C. Counihan, P.V. Esterik and A. Julier. (eds). *Food and Culture: A Reader*. New York: Routledge. pp. 20–27.

Betts, A., N. Omata, C. Rodgers, O. Sterck, and M. Stierna. 2019. *The Kalobeyei Model: Towards self-reliance for refugees?* Oxford: Refugee Study Centre (RSC), University of Oxford.

Bourdieu P. 1984. *La Distinction: critique sociale du judgement*, trans. by Richard Nice. Cambridge, Mass.: Harvard University Press. Originally published 1979, Paris: Les Editions de Minuit.

Brewer, M.D., W. Gardner. 1996. 'Who is this "we"? Levels of collective identity and self representations'. *Journal of Personality and Social Psychology* 71 (1): 11. https://doi.org/10.1037/0022-3514.71.1.83

Broch-Due, V., and T. Sanders. 1999. 'Rich man, poor man, administrator, beast: The politics of impoverishment in Turkana, Kenya, 1890–1990'. *Nomadic Peoples* 3 (2): 35–55. https://doi.org/10.3167/082279499782409389

Chatty, D. 2013. *From Camel to Truck: The Bedouin in the modern world*. Revised second edition. Knapwell, Cambridge, UK: The White Horse Press.

Counihan, C., and P. Van Esterik. (eds). 2013. *Food and Culture: A Reader*. 3rd ed. New York: Routledge. https://doi.org/10.4324/9780203079751

Crane, T.A. 2010. 'Of models and meanings: Cultural resilience in social-ecological systems'. *Ecology and Society* 15 (4): art19. https://doi.org/10.5751/ES-03683-150419

Davoudi, S., K. Shaw, L.J. Haider, A.E. Quinlan, G.D. Peterson, C. Wilkinson, H. Fünfgeld, D. McEvoy, and L. Porter. 2012. 'Resilience: A bridging concept or a dead end?' *Planning Theory & Practice* 13 (2): 299–333. https://doi.org/10.1080/14649357.2012.677124

Derbyshire, S.F. 2020. *Remembering Turkana: Material histories and contemporary livelihoods in north-western Kenya*. London; New York: Routledge. https://doi.org/10.4324/9781003001263

Ferguson, J. 1999. *Expectations of Modernity: Myths and meanings of urban life on the Zambian copperbelt*. Berkeley; Los Angeles; London: University of California Press.

Folke, C. 2006. 'Resilience: The emergence of a perspective for social–ecological systems analyses'. *Global Environmental Change* 16 (3): 253–67. https://doi.org/10.1016/j.gloenvcha.2006.04.002

Fortier, J. 2009. *Kings of the Forest: The cultural resilience of Himalayan hunter-gatherers*. Honolulu: University of Hawaii Press. https://doi.org/10.21313/hawaii/9780824833220.001.0001

Galaty, J. 2013. 'The indigenisation of pastoral modernity: Territoriality, mobility and poverty in dryland Africa'. In M. Bollig, M. Schnegg, and H. Wotzka. (eds). *Pastoralism in Africa: Past present and future*. New York: Berghahn.

Gonzales, G. 2015. 'Individual and collective identities among Malian Kel Tamasheq refugees in Burkina Faso: Continuities and transformations'. Master's diss., Department of International Development, University of Oxford.

Hatcher, J. 2014. *Exploiting Turkana: Robbing the cradle of mankind*. Newsweek Insights.

Hobsbawm, E.J., and T.O. Ranger. 1992. *The Invention of Tradition*. Cambridge: Cambridge University Press.

Holtzman, J. 2009. *Uncertain Tastes: Memory, ambivalence, and the politics of eating in Samburu, northern Kenya*. Berkeley: University of California Press. https://doi.org/10.1525/9780520944824

Ingold, T. 2000. *The Perception of the Environment: Essays on livelihood, dwelling and skill*. New York: Routledge.

Keightley, E. 2010. 'Remembering research: Memory and methodology in the social sciences'. *International Journal of Social Research Methodology* 13 (1): 55–70. doi: 10.1080/13645570802605440.

Konaka, S. 2017. 'Introduction: The articulation-sphere approach to humanitarian assistance to East African pastoralists'. *African Studies Monographs* (53): 1–17. doi: 10.14989/218918.

Lesorogol, C.K. 2008. 'Setting themselves apart: Education, capabilities, and sexuality among Samburu women in Kenya'. *Anthropological Quarterly* 81 (3): 551–77. https://doi.org/10.1353/anq.0.0020

Little, M.A., and P. Leslie (eds). 1999. *Turkana Herders of the Dry Savanna: Ecology and biobehavioral response of nomads to an uncertain environment.* Oxford: Oxford University Press.

Little, P.D., K. Smith, B.A. Cellarius, D.L. Coppock, and C. Barrett. 2001. 'Avoiding disaster: Diversification and risk management among East African herders'. *Development and Change* 32 (3): 401–33. https://doi.org/10.1111/1467-7660.00211

McCabe, J.T., P.W. Leslie, and L. DeLuca 2010. 'Adopting cultivation to remain pastoralists: The diversification of Maasai livelihoods in Northern Tanzania'. *Human Ecology* 38 (3): 321–34. https://doi.org/10.1007/s10745-010-9312-8

McCabe, J.T., N.M. Smith, P.W. Leslie, and A.L. Telligman. 2014. 'Livelihood diversification through migration among a pastoral people: Contrasting case studies of Masaai in Northern Tanzania'. *Human Organization* 73 (4): 389–400. https://doi.org/10.17730/humo.73.4.vkr10nhr65g18400

Monsutti, A. 2010. 'Food and identity among Young Afghans'. In D. Chatty (ed.) *Deterritorialized Youth: Sahrawi and Afghan refugees at the margins of the Middle East.* New York: Berghahn Books.

Montanari, M. 2006. *Il cibo come cultura*. Bari: Laterza.

Müller-Dempf, H. 2014. 'Hybrid pastoralists – development interventions and new Turkana identities'. *Max Planck Institute for Social Anthropology Working Paper* 156: 31.

Neocleous, M. 2013. 'Resisting resilience'. *Radical Philosophy* 1 (178): 2–7.

Okenwa, D. 2020. 'Contentious benefits and subversive oil politics in Kenya'. In J. Lind, D. Okenwa and I. Scoones. (eds). *Land, Investment and Politics: Reconfiguring Eastern Africa's pastoral drylands.* New York: Boydell and Brewer. pp. 55–65. https://doi.org/10.2307/j.ctvxhrjct.10

Piasere, L. 2009. *I rom d'Europa. Una storia moderna.* 10 edizione. Roma: Laterza.

Rodgers, C. 2018. 'Rural, remote, raiya: Social differentiation on the pastoralist periphery in Turkana, Kenya'. Ph.D. diss, Anthropology, University of Oxford.

———— 2020. 'Identity as a lens on livelihoods: Insights from Turkana, Kenya'. *Nomadic Peoples* 24 (2): 241–54. https://doi.org/10.3197/np.2020.240205

Santopietro, G. 2015. 'Fish as food, fish for food. Produzione e Cultura Alimentare Di Una Piccola Comunità Di Pescatori Sul Lago Turkana, Kenya'. Master's diss., Antropologia culturale, Università di Torino.

Scott-Smith, T. 2018. 'Paradoxes of resilience: A review of the World Disasters Report 2016'. *Development and Change* 49 (2): 662–77. https://doi.org/10.1111/dech.12384

Semplici, G. 2020. 'Moving Deserts. Stories of mobilities and resilience from Turkana County, a Kenyan desertscape'. Ph.D. Thesis. Oxford Department of International Development, Oxford, UK.

Sutton, D.E. 2001. *Remembrance of Repasts: An Anthropology of Food and Memory*. Oxford: Berg.

Turner, V. 1987. 'The anthropology of performance'. *PAJ Publications*.

CHAPTER

06

Changing Land Laws and the Resilience of Samburu Pastoralist Women

Rahma Hassan

Introduction

This chapter builds on and expands the meaning of socio-ecological resilience to examine livelihood strategies of Samburu pastoralist women in the context of changing land laws and tenure. The social ecological concepts of resilience reflect the social processes of land access and the alternative strategies that Samburu pastoralist women apply to cope and even thrive despite the shocks in the environments they live in. Pastoralists in Kenya occupy vast rangelands with their livestock in mainly arid and semi-arid areas characterised by poor and unpredictable rainfall and dynamic environment (Behnke et al. 1999).

Mobility is a key feature that continues to define pastoralists' mode of production in these areas where movement is seen as the most effective strategy to sustain the pastoralists' mode of production in relation to the variability of the environment (Galaty 2013; Turner 2011). Indeed, pastoral mobility determines the capacity to reach dispersed and patched resources across vast land at the time when their nutritive value is best (Krätli and Schareika 2010). Pastoral mobility crucially relies on social pacts and agreements on access to land for grazing (Sullivan and Homewood 2003). These arrangements guide access to land and are critical for pastoralists given the variability that characterizes the rangelands in which they operate, which in turn affects pastoralists' mode of production (Scoones 1999). Under these circumstances, social arrangements and customary plans can be flexible and provide a buffer for the community. As such, they are an important source for pastoralists' resilience. For example, some of the unique pastoralist institutions of land management and customary land rights have been found to safeguard communities in times of drought, organising migration across distant territories as well as accessing resources (Bruce and Migot-Adholla, 1994). Access to land for pasture is thus a critical feature of pastoralists' livelihoods and their resilience.

On these premises, it appears important to discuss pastoralists' resilience in the context of pastoral tenure and land use regimes regulating access to and control over resources. In many instances, there are differing categories of rights in pastoralist areas. This situation mainly revolves around the need for coexistence between those who own land privately and the communal system. In most pastoral areas the categories of rights over resources range from communal rights that are private in nature, to those that are more communal in nature such as access to dry season forests or grazing around a water point (Ensminger 1996). While most of the land occupied by pastoralists remains communally held, recent changes in land laws in Kenya and the recognition of community land has presented new dynamics of access that, perhaps paradoxically, potentially disrupt how land is accessed.

In Kenya, following the 2010 Constitution, the government enacted the Community Land Act (2016) to recognize, protect and register community land rights, including land held by pastoralists. The Community Land Act, which Alden Wily (2018) calls the most supportive land legislation in Africa, provides for communities, including pastoralists, to utilize and manage their land in accordance with customary norms. Most significantly, this law also gives women, including pastoralist women, unprecedented rights to own and access community land, as well as the right of representation in local level land governance institutions (GOK 2016; Boone et al. 2019).

It occurs then, that land reforms have a strong impact on the ways pastoralists access their lived space, access resources, and make a living, including their ability to withstand shocks. The emergent need to adapt to changing land laws requires continuous negotiations for accessing resources and alters the ability of different groups to draw benefits from land resources (Ribot and Peluso 2013). The reforms, often accompanied by land law changes, take place in a context of continuously changing political, social, and economic processes and make it necessary to adapt to these dynamics in the society (Berry 1993). This is the context within which pastoralist livelihoods and resilience must be understood.

These dynamics compound the impacts of climate change and can make it very difficult for pastoralists to sustain their livelihood. Further, pastoralists seek to ensure access to different grazing areas at different times of the year, depending on the weather patterns (Niamir-Fuller, 1998). These different considerations and arrangements mean that pastoralists need to participate in the process of determining how to access and use community land. Then there is the question of encroachment on community land by outsiders for uses other than pastoralism, such as large-scale investments in transport corridors, conservation, and energy production (Lind, Okenwa, and Scoones 2020). In instances where pasture access is negotiated, community members seek to benefit from these large-scale projects based on their identity and place (Drew 2020). These dynamics have led to pastoral land tenure systems and institutions being modified to suit emerging needs.

Efforts to alter how individuals and groups access land have implications for people's ability to relate to that land. On one hand, if the law, for instance, individualizes community land and permits privatization and fencing, the rest of the community members are excluded from benefiting from this land. On the other hand, if the law provides for communal ownership and shared equal access then the dynamics of access differ as negotiations for accessing such land are established. For pastoralists, heightened privatization of land and the trend towards sedentism has been found to threaten their access to the land for mobility and pastoral viability (Rutten 1992) with others have highlighted the economic benefits of shifting from communal land (Lesorogol 2008a). Whatever the impact of privatization, such land tenure changes present challenges for pastoralists livelihood and resilience.

This chapter draws on a broad definition of pastoral resilience that recognizes the relational approach and the heterogenous nature of pastoralist communities to highlight the different aspects of resilience within different subgroups in the community (McPeak and Little 2017), in particular women. By focusing on pastoralist women, I highlight how land tenure changes, customary systems, and resource pressures (Pollini and Galaty 2021) influence the ways groups respond to uncertainty and changes in access to land. To understand how land access is gendered, land relations need to be viewed as social relations which have been found to be crucial in determining access (Mackenzie 1990; Berry 1993). This demands a stronger focus on social relations and power dynamics in the society. Questioning these processes would reveal the inclusion and exclusion of segments of the community to land access and their resilience.

Looking at land law reforms and the historical nature of these changes is critical in understanding the characteristics of pastoralist communities that anchor their resilience. Specifically, by studying pastoralists communities as units at the household level while focusing on the pastoralist women in Samburu, this chapter highlights the heterogenous nature of pastoralist communities as well as their multiple forms of resilience.

Data in this chapter is from qualitative fieldwork conducted in the period from 2019 to 2022. Most of the research focused on land law reforms and pastoralist access to land. The analysis of data that informs this chapter is mainly drawn from interviews conducted in 2021 and 2022 that focused on Samburu women's livelihood strategies to understand resilience in the context of changing land laws and looming calls for subdivision of the group ranches situated in Samburu West area.

In the subsequent sections, I start with a discussion of resilience and highlight the theoretical concepts that best suit the issue of pastoralists' resilience in the face of changing land laws based on local Samburu women's ideas, followed by a brief discussion of land in Samburu. Thereafter, I present the findings of the different strategies and resilience of pastoralist women and end with a conclusion on relational resilience in the context of land law reforms among pastoralists.

Pastoralism and resilience thinking

The fate of pastoralism as a viable mode of production given the harsh effects of climate change on pastoralists and their livestock has been questioned for some time (Robinson and Berkes 2010). At the same time, numerous studies have shown evidence of the strength of pastoralist systems to manage variability as well as living with and benefitting from uncertainty (Scoones 2021). There is, in other words, mounting evidence that pastoralists are more resilient than what has commonly been acknowledged (Fratkin 2001; Semplici 2020).

I draw attention to the gendered dimensions of resilience, focusing on how various and dynamic relationships in the pastoralist community shift in response to changing land tenure and how this shapes their resilience. Featuring the gendered nature of resilience reflects the importance of people's experiences and lifestyles, the social actors, ecological actors and how communities are organised (Berkes and Ross 2013; Adger 2000). The recognition of actors, positions, roles and where they belong in the community as intimately connected to daily choices also highlights the need to look at culture in understanding the gendered nature of resilience (Crane 2010; Semplici 2021)

An important dimension of understanding resilience concerns how networks affect resilience, which will be the focus, rather than individual or group abilities to withstand shocks (Konaka and Little 2021). These relations and networks among pastoralists groups are critical in the study of resilience given how central they are for access to land in time of uncertainties and rangeland variability (Scoones 1995; Krätli 2020). Resilience among pastoralists must be understood based on these dynamic environments, where the question of access to land is of paramount importance. While the success of pastoralism has traditionally depended on the mechanisms put in place to address the variability of their environment and access water, grazing land and other resources (Scoones 1995; Krätli and Schareika 2010), threats posed by climate change and land reforms have become increasingly important for understanding pastoralists' resilience (Mwangi 2009). Among pastoralists, climate change dynamics have reduced pasture, water and common pool resources, further complicating access to land and occasioning the need for land use changes.

Under these conditions, land reforms constraining access or altering the customary institutions that have long regulated access, risk having detrimental impacts on pastoral livelihoods that were already struggling due to dwindling resources. These challenges, among others, have stimulated local adaptation in both livestock and pasture management practices, including for example reducing livestock, renting pasture and rotational grazing (Agrawal 2010). There has also been continuous negotiation and reciprocal sharing of grazing areas in the access to land for pastoralists (Ensminger 1996). These adaptation measures have created new challenges of access in the form of a need for pastoralists to embrace the new and changing nature of land tenure and rights.

Changes in how land is accessed and governed have impacted on the coexisting relationships between the private and communal rights of ownership. Land policies in Kenya, including the colonial polices, have long aimed at restricting the mobility of pastoralists and encouraging more sedentary practices such as crop farming. Where grazing was permitted the introduction of grazing schemes limited stock and movement (Lesorogol 2008b). Despite the pushes to limit herding, the ownership and appropriation of land among pastoralist communities has traditionally been based on customary institutions. Customs and arrangements for different pastoralist communities have evolved over time, clarifying and modifying rights for utilizing the land. These include strategies for how land can be used during harsh climatic conditions. Even where customary practices are strong, though, conflicts frequently arise over the usage of land, especially between communally and individually used land.

Land and people in Samburu County

Samburu County is in the northwestern part of Kenya and is approximately 21,000 square kilometers (KNBS 2019). The county neighbours pastoralist dominated counties of Marsabit, Turkana, Isiolo, Baringo and Laikipia. Samburu County can be roughly divided into the highlands, which include the Leroghi (Lorroki) plateau and the lowland plains which consist of different zones with escarpments and mountain ranges. Administratively, the county consists of three sub counties: Samburu East, Samburu West, and Samburu North.

Ecologically, Samburu County is classified as low potential rangeland with minimal annual rainfall. Most of the land is used for pastoralism and the remaining medium potential areas used for agricultural production (Samburu County Government, 2018). In the highland areas the community engages in crop farming with beans, maize and potatoes as the dominant crops. Land in Samburu County is mainly owned as community land (either registered as group ranches or unregistered). Samburu East is mainly comprised of community land registered as group ranches with some land under conservancy management. Samburu West, where this research was conducted, mainly comprises group ranches where some are already subdivided into individual parcels and others were seeking to subdivide due to urbanization and sedentarizing of neighbouring communities and a move towards crop farming (Hassan, Nathan and Kanyinga 2022). Samburu North is mostly communally owned land with group ranches and unregistered land as well.

Samburu County also features rangelands and gazetted forest cover where nomadic pastoralism is the dominant economic activity with many households practicing semi nomadic activities that incorporate crop farming and other forms of wage labour. Common employment includes work in conservancies, police officers, teachers, nurses and staff of the county government of Samburu. Wildlife conservation areas attract tourism activities in the area (Pas 2018)

Demographically the 2019 census in Kenya reported the population of Samburu County as 310,327 with 65,910 households (KNBS 2019). The population is mainly of the Samburu community, which is organised around clans, kinship, age and place, with marked differences in the lifestyles of people in the highlands and lowlands (Pas 2018; Spencer 1965).

Despite differences in the economic activities of the lowlands and highlands a shift to agriculture and trading in local markets has been reported in the highlands (Lesorogol 2008a). Changing land laws and tenure arrangements and the move towards more sedentary lifestyles have affected the roles in pastoral households. Indeed, increased attention to chicken rearing and ready markets in the nearby towns and upcoming shopping centres all contribute to changing gender roles in the community (Wangui 2008; Archambault 2016). As mentioned, pasture and water resources for livestock have been declining in Samburu occasioning the need for pastoralists to move further away with livestock (Pas 2018), yet mobility is not guaranteed due to the changes in land use described above.

Earlier studies in Samburu have focused on the people and their culture as well as their interaction with the environment (Spencer 1973; Fumagalli 1977), including anthropological research about Samburu clans and how labour is organised (Lesorogol 2003; Straight 1997). Furthermore, most of the earlier research on land in Kenya focused on the colonial effects of land expropriation (Okoth-Ogendo 2002; Kanyinga 2000, 2009) as well as the effects of colonial and post-colonial land policies including the establishment of group ranches by the latter (Simpson and Waweru 2021; Fratkin 1994).

According to Samburu County Government (2018) and a review of the Samburu County records, there are 50 group ranches of varying sizes occupying 40 per cent of Samburu's total area. Some of the group ranches had applied for sub-division. Unregistered community land is land that the county government holds in trust, and which local people use for pasture and other purposes. Thus, with changing land laws and tenure arrangements that threaten their livelihoods, pastoralists rely on relational aspects of resilience and reform community networks to support their mode of production (Rodgers 2018). In the following sections I focus on some of the ongoing land questions arising from implementation of the community land act and Samburu women's strategies and resilience.

Dynamics of community land rights and pastoralist women's resilience

There has been much attention paid to communal land rights among pastoralists including the protection of marginalised groups. Mobility and access to land for pasture has been found to be a critical component of supporting livestock in arid and semi-arid lands (Scoones 1995). Communal land specifically enables pastoralists to make use of vast territory by moving the animals and sharing pasture.

Despite the occupation of pastoralists in the vast arid and semi-arid areas, most of which are communally managed, pastoral mobility and communal resource tenure regimes have been identified as the major obstacle to pastoralists' socio-economic development, hampering opportunities for private investment and sustainable resource management (Hagmann 2006). Yet studies in Maasai group ranches in Kenya have shown that the privatization of previously communally owned lands have led to land losses (Kimani and Pickard 1998), including hurried land sales which have left local pastoralists landless (Galaty 2013) and radically altered the power dynamics that affect sharing of subdivided rangelands among pastoralist communities (Jeppesen and Hassan 2022).

In Samburu these changes have been unfolding for decades. Some group ranches in Samburu Central that are considered viable for agricultural production were subdivided soon after they were formed as group ranches in 1968. The remaining group ranches, including some where this study was conducted, are transitioning to comply with Kenya's Community Land Act of 2016. The Community Land Act recognizes community rights to own land and ostensibly safeguards those rights. One of the major issues for Samburu community members is the membership of group ranches.

Whereas previous members of the group ranches were mostly male heads of household, the CLA requires that all adult male and females in the community are recognised and registered as members of the community land. For women in pastoralist communities this has been a major shift towards securing their land rights and equal access to land. However, among the numerous challenges to transitioning group ranches has been resistance to the inclusion of women and young men as community members. Some of this resistance arises due to questions of access to land and membership not being well understood to some of the group ranches (Hassan, Nathan and Kanyinga 2022).

A parallel issue is the subdivision of group ranches which terminate collective community ownership. Samburu women have different experiences of this move towards subdivision depending on their position in the community and their prospects for owning land, if at all. The experiences of Samburu women and the strategies they employ in response to the effects of changing land laws are useful in understanding pastoral women's resilience as a unique category of pastoralists who have been side-lined from land ownership overall.

For this study, the focus is on different views of local Samburu women about debates concerning whether to transit their group ranches to community land or to subdivide into individualized parcels of land for the members and what this reflects about their ideas of resilience and general impacts on pastoral livelihoods. The main contention, based on fieldwork data, are questions around the group ranch register and how it affects women's livelihood strategies. The women's views were quite heterogeneous. As such this chapter presents the concerns women raised in interviews and highlights the ideas around resilience that emanate from the local community perspectives.

The group ranch register is the formal document that enlists members of the group ranch as a registered entity as established by the Land (Group Representative) Act of 1968. Older male members of households were the only registered persons in Samburu when the group ranches were established. The requirements to update the group ranch register would favour inclusion of women in instances where the head of the household had died. This practice varied from group ranch to group ranch, but this is one avenue that women we spoke to in Samburu are pursuing to be included as members. To feature the gendered nature of resilience I now discuss three different strategies women adopted to respond to changes in land tenure: substitution, negotiation, and finance.

Women's resilience through substitution

The questions of who is eligible to be a member of a group ranch elicited different responses in this study. This has become an important question because the current community land law requires that women are recognised as members and as such should be included in the register (GOK 2016). Widowed women navigate these arrangements both formally and informally, as crucial process for their resilience to changing land laws. One of the ways adopted, for example, is to negotiate inclusion by soliciting support from the elders who are custodians of the norms and values in the community. This approach indicated their respect for the existing authority while soliciting their influence to include women in the group ranch registers.

Some women, however, reported having little information about the major changes happening in the community as they are not included in the community meetings where such decisions are made. They indicated that Samburu women are not traditionally allowed to speak in front of elders and as such they have been left out of such public meetings. Some of the women avoided attending these meeting even when they were invited. One critical meeting is the annual general meeting of a group ranch. National government officials attend these meetings to ensure that the group ranch members adopt resolutions to address key land issues in the community. This is the forum where requests are submitted to replace names in the register as well as where decisions to subdivide group ranches are made. Samburu women can still learn about what is happening through their links in the community and families. Most of them draw on the strong social ties they have within the family in which they were born to get information, which is one way they have used to be part of decision making processes. These are tactics Samburu women have created over time to remain relevant in their roles in the community. Pastoralist women deprived of their rights to access and own land utilise the power of these strong ties, given that all decisions are made by elders.

Another group of Samburu women informants highlighted the need to ensure equitable inclusion and sharing of the collectively owned group ranch land. To achieve this, they said they had been subtly nudging their sons to demand their inclusion in the group ranch register. The process of updating the register to include male members of the group ranch has been different but there have been calls to consider the older age sets[1] among the Samburu first as an attempt to harmonise inclusion based on age sets across the different group ranches.

These discussions elicited myriad sentiments, but this chapter will focus on those that demonstrated resilience thinking. Samburu women see the inclusion of their sons in the register as a strategy to increase the amount of clan land that can continue to be used communally after the group ranch is subdivided.

These sentiments express their fear of land sales which have been reported in neighbouring group ranches where land had been subdivided and sold leaving, households with little or no land. One of the key features of pastoralist community resilience is generational inheritance of land. As such Samburu women demonstrate resilience through strategies for safeguarding community-owned land with the aim of ensuring it can be managed and secured for future generations. This group of Samburu women were of the view that including their sons in the group register would help to assure their land access by giving more males in the family a share in decision-making, thus safeguarding the household. In sum, this group of pastoralist women actively drew on their son's power to try to ensure their own access to land.

Women's resilience through negotiation

Recognizing the importance of identity, community, belonging and place (Semplici 2021; Berkes and Ross 2013) as key components of resilience, this section now turns to groups of pastoralist women who have been affected differently by the implementation of community land laws and changing land tenure in the group ranches where they live. The strategies they employed were based in their identity and close connections and enabled them to face the shocks presented to their livelihoods. Specifically,

---

1 In Samburu society, when a male of the community has been circumcised, he joins an age-set comprised of all the young men initiated within a period of about fourteen years. He will maintain a close affinity with these peers until death.

negotiation is a tool employed by women to secure access to land and livestock for their resilience.

At the time of my last phase of fieldwork in Samburu in early 2022, some of the group ranches were grappling with the challenges that would arise from subdividing their collectively owned land and especially the consequences of a process that would result in only some members owning land. There was already alarmed speculation within some households, especially those who were not registered as members. Samburu women shared with me the different strategies they were adopting in the face of these potential changes and challenges.

Close social ties and the long history of families accessing the land leads some households to the strong belief that their family ties and close relations within the Samburu community will enable them to continue to access the land even if they are excluded by the formal processes. One participant said 'Samburu's will remain Samburu's even if we live on trees or in towns we will keep together and share what we have just like we share milk during dry seasons.' The belief in these deep connections reflect community resilience to withstand formal processes and land use changes. This includes the role of the community elders to resolve concerns of landlessness even when they are not directly involved on formal committees to handle group ranch land matters.

The requirement for group ranches to become community land has raised tensions about the fate of families who had migrated to these areas but were never recognised as members in the register. Samburu women in this situation, however noted that whereas it seems as if they would soon be landless, they strongly believed that their clan elders would not allow them to be moved off the land they have occupied. The deep trust in the familial and clan based support they had received in times of crisis reflects the role of solidarity and group-identity in the resilience of pastoralist communities.

Negotiating applies to livestock too. The challenge is that the formal process will rely on the group ranch register, which excludes users of land who have lived in the community for decades. Women from migrant families however still believe that their deep connections and ties within the community will ensure they are still able to share the subdivided land. After discussing the various difficulties that families who have no land in Samburu might face, some women named their brothers as people they would go to to get some livestock to start life elsewhere or even a piece of land from their family's own share after subdivision. Group ranch officials also mentioned this as a consideration for those who will be left landless, those who migrated and were not part of the register.

These strategies are not without challenges. The process of inclusion as members for widows is still stiff and faces delays. Responding to my next question about their plans in case they were excluded, most of them said they know what to do next, they have always had ideas on how to deal with what comes because we know 'our words are our keys' and solutions. This phrase reflects Samburu women's belief that their kinship and their community structures will see them through these changes.

Women's resilience through finance

Another group of Samburu women are those with formal education qualifications who earned wages from employment in government offices or local non-governmental organisations. Like the other women discussed, this group were not included in the group ranch register but through their income they had access to finances to develop the communally owned land that they occupied, either where they married or were born. Three things define their strategies and resilience: creating local women's groups, investing in education, and engaging in income generating activities.

Local women's groups have become an important part of Samburu women's daily lives, as members contribute money to a pool which members can access to invest in small businesses or livestock. These groups are not limited to educated or employed women, but they were the women who formed the groups and assisted in record keeping. They said that the close solidarity and connection felt through women's groups was useful in supporting their households and feeding their children. Apart from wage earnings, other women used the proceeds from the sale of milk and small animals like chickens to sustain their membership in the group.

Providing education to their children was a strategy mentioned by most of the women interviewed. The reasons given to explain why education was important for them as a pastoralist community were linked to the challenges of accessing pasture and changing land tenure. For some, there was no land for children to graze animals and the livestock were also diminishing. Sending them to school was seen as a way to safeguard the community, not only as a direct source of extra income, but also to finance restocking and sustaining pastoralism. The reduced access to pasture due to individualization of land meant that younger Samburu male children who would traditionally have taken care of animals were no longer involved in moving the animals and thus had the chance to attend school.

Those who had access to some income also saw sending their children to school as a form of ensuring that they would one day be able to buy land and thus secure their livestock and households. The participants expressed mixed views on whether they would prefer the group ranches to be subdivided or not. Some were of the view that it would make no difference to them, as they would not be allocated land either way and as such, they had no interest. Others, mostly those who had prospects of accessing finances to purchase land, saw subdivision as opening a window to buy land. In their view, a title deed to a privately owned plot would be preferable to a community land register in which their inclusion was not guaranteed.

Many of the women interviewed also indicated that they were engaged in income generating activities such as keeping small animals. Chicken rearing was the most mentioned, as it is a business in which husbands were not involved, thus giving women full authority to buy and sell. There was a ready market and demand for chickens by town dwellers from neighbouring communities. When asked what they use the money for, they replied that one of the major uses was purchasing pasture in the market and grass for the livestock in the evening. This type of support of the pastoralist system is one of the less well-recognised factors that support the resilience of the pastoralists system. The sale of hay is increasingly a feature of the local markets in Samburu. On market days, men and women approach lorries loaded with hay to negotiate the price for each bunch of hay for the livestock at home. Pastoralist women thus use finance to participate and benefit in the different production systems. These changes have had flow-on effects for household labour which we look at in the next section.

Reduced mobility, changing household labour and resilience

Pastoralists in Samburu have required mobility to access dispersed grazing areas at different times of the year and to deal with recurring droughts. Access to land and movement corridors remain critical to support these movements. Changes in land use, however, have placed constraints on such movement, disrupting previously established systems of access to land. Changing land tenure and subdivided rangelands have blocked livestock movement in parts of Samburu. This includes an increase in large-scale land acquisitions in pastoral areas that have hindered access to grazing land (Lind, Okenwa and Scoones 2020). Some neighbouring communities have adopted agriculture and fenced off land, leaving only road reserves for moving livestock. The growth of towns in pastoralist areas has also impacted access to grazing land, as has the construction of the first tarmac road to Maralal town since independence, which has attracted population growth in some areas.

Access to land and pasture has always influenced how labour is organised among the Samburu. Despite some studies indicating reduced access to pasture soon after Kenya's independence based on fenced off lands in Samburu, the Samburu pastoralists long enjoyed rangeland mobility. The continuing prevalence of herding labour reflected their access to grazing resources (Spencer 1965). Over the years labour was organised through clans who dominated certain areas of pasture and settlements. The division of labour in the clan was further differentiated by age and sex with women and men taking different roles (Spencer 1965; Sperling 1987).

Age as a critical area of organising labour entailed the Samburu age sets which were used to define roles for the males. On the one hand, males in the community were introduced to specific roles as children to look after cows, eventually assuming herding roles outside the homestead. As the younger males assumed these roles the older males significantly reduced their own involvement in herding to engage in tasks of planning and strategizing for access to pasture and water, which included negotiations with landholders (Sperling 1985).

A female's place, on the other hand, was determined by her relationship with the male and position in the age set. Domestic roles were mainly performed by females. Female children were initially introduced to minding goats and then graduated to other roles at home, such as milking and collecting firewood and acacia pods for the animals. The roles of older women also changed significantly as their daughters took over the roles in the homestead.

This brief summary of how labour was generally organised among the Samburu does not provide a complete picture but it reflects the limited role of women in herding outside the settlement area and the critical role of the younger *morans*[2] who would assume responsibility for the livestock over time (Sperling 1987). During drought, the Samburu typically split their herds according to the available forage and the patterns of migration set out for them (Pas 2018).

Significant changes in roles have been occasioned by the reduced mobility in the community as well as the changing social economic demographics in the community which has offered more sedentary lifestyles for the young, leaving older men and women with responsibility for livestock.

Samburu pastoralists have thus had to adjust their mobility strategies, with impacts on the organisation of house labour. Women have found themselves at the centre of both managing livestock at home, as mentioned, and providing alternative income for the household. First, most of the grazing now happens near the homesteads, involving women and younger Samburu males, their children. This is reflected in the efforts the women

---

2 The Samburu community defines a *moran* based on age group and rite of passage. The *morans* 'are the young unmarried men who would at one time have been the warriors of the tribe' (Spencer 1965, ixx).

are putting to secure pasture from outside like buying grass and hay from the market centres. Second the shifting of some household labour to urban centres to earn outside income to support the household has left many women at home taking care of the livestock. Young women are also getting involved in small business – both as owners and employees – resulting from the growing town populations around Maralal, while younger men engage in manual labour and providing motorcycle transport, roles which are different from their usual livestock herding in search of water and pasture.

In response to these changes, Samburu women are now moving with livestock and relocating entire households closer to grazing areas, setting up temporary grazing camps and feeding their families as they take care of animals, a stark contrast to previously, when they would mostly be left behind to await the animals return home to milk and take care of them. The involvement of women in arranging access to pasture is not new but their role is now central. The strategies they employ reflect why pastoralist systems continue to be the most resilient in the rangelands

Who gets employed as a herder has also changed. For decades only male Samburu would be employed by wealthier herders (Sperling 1987). Now, women who have no livestock are also offering herding labour to their neighbours as well as fetching water for livestock and household use. The money earned is mainly used to buy food for the family and hay for livestock at home. The role of income in the resilience of pastoralist women was summed up by a participant who said that 'if you have money, you are okay, you can buy food and you can feed your livestock until the time when the fields are wet, and the grass is back'.

As mobility and labour patterns shift in the community, the changing role of elders is creating tension between the generations. For the traditional custodians of community values who mentored and supervised younger men on migration routes in search of pasture, the change in land uses and the challenges of accessing privatised land has made it difficult for elders to coordinate grazing and access to pasture. Informants complained that the younger *morans* 'move as they wish' and did not take advise from anyone or follow the agreed routes in group herding. Whereas the elders reported feeling helpless about these changes, some of the women reported having decided to accompany the *morans* and guide their search for pasture. One development has been the use of forest resources where animals feed on the leaves of *Rhus natalensis*, a tree locally known as *lmisigiyyo*, and later the women collect the dry branches as firewood for their households. This has reduced the need to move the animals, at least until the community has depleted the available leaves in the forests and move again.

Women's role in providing alternative forms of livelihood by engaging in labour both within and outside the household places them at the centre of pastoralism. Their roles become increasingly important in less stable situations where they fill gaps occasioned by changes in land and labour.

Conclusion

In this chapter I have presented the resilience strategies adopted by Samburu women to adjust to the disruptions and new lifestyles brought about by changing land tenure and heightened subdivision in some of the rangelands in Samburu. This presentation has reflected the options available for pastoralist households and the changing roles that women play.

First, women demonstrated resilience through substitution, where they draw on the influence of their sons and close clan ties to access land when it is subdivided. Second, pastoralist women demonstrated resilience through negotiation based in familial and kinship ties to access land and in some cases livestock through their male relations and clans. Third, pastoralist women create resilience through finance, reflecting the changing nature of pastoral modes of production and the growing markets in Samburu.

These strategies were employed by women from different groups of pastoralists and households. Significantly, women's resilience through substitution was found among the group of pastoralist women who were widowed and whose clan members were in the group ranch register and whose sons could also benefit from the subdivision of land, as well as among married women whose husbands were not eligible for land once the group ranch was subdivided. Additionally, unmarried women and those who migrated to the group ranches but could not claim land in the subdivision would negotiate to get access to land and livestock. Educated women, those who earned wages from employment, took the finance route as they could participate in women's groups, engage in trading and earn money to buy land where possible. Women from poor households sought alternative wage labour to take care of their few livestock and adopted new ways of engaging the customary systems to their advantage.

All three of these examples of Samburu pastoralist women's resilience are responses to reduced mobility and changing household labour wrought by the changing land laws and increased subdivision of land. Women's roles have become more important where mobility has been curtailed.

Critical to pastoralists' resilience is their identity which is deeply tied to a closely knit community and ideas around belonging which are reflected in the tensions around land ownership and access to land. These internal social cultural resources serve to counter the formalization of land and outside demands that threaten the social existence of pastoralist systems. Pastoralist women, I find, draw on their close networks and social capital to diversify their incomes to feed their families and livestock. Identity thus plays an important role in the inclusion of landless women in the community.

Questions of governance and access to land continue to draw attention in the arid and semi-arid areas that provide critical resources for pastoralists. This issue has become even more acute in the rangelands facing big infrastructure projects which entail large scale land acquisitions. Devolved governments in Kenya have also opened previously inaccessible areas, further encroaching on land traditionally occupied by pastoralists. These external threats are quite different from those posed by climate change. While understanding the impacts that these land changes impose on pastoralists' livelihood and resilience is complex, the effects of land tenure changes and the implementation of land laws on pastoralists resilience remains significant. Reflecting on the resilience of the communities faced with such challenges to their livelihoods, there is need to reflect on broader land questions arising from the implementation of the Community Land Act.

This chapter hopes to draw attention and focus on questions of land law reforms as central to enabling or hindering resilience in its different forms and how these effects vary among different groups of pastoralists. Further, the gendered nature of resilience reflects how various relationships among pastoralists fluctuate responding to shifts in the community and land tenure. In this way, resilience is recreated and reconceptualized based on the experiences of pastoralists and the contexts they live in.

Acknowledgments

I wish to thank community members in Samburu and all respondents who contributed to this research. Data in this chapter was jointly collected with my supervisors Iben Nathan and Karuti Kanyinga. Special thanks to Greta Semplici and Shinya Konaka, for their input on the earlier versions of this chapter, their time, advice, and thorough reviews, and to both my supervisors for their reviews, guidance and discussions that have informed my work. This chapter is part of Rahma Hassan's PhD research under the Rights and Resilience Project in Kenya funded by DANIDA (DFC 18-01-KU).

References

Agrawal, A. 2010. 'Local institutions and adaptation to climate change.' In Robin Mearns and Andrew Norton. (eds). *Social Dimensions of Climate Change: Equity and vulnerability in a warming world*. Washington DC: New Frontiers of Social Policy.

Alden Wily, L. 2018. 'The Community Land Act in Kenya: Opportunities and challenges for communities.' *Land, MDPI, Open Access Journal* 7 (1): 1–25.

Archambault, C. 2016. 'Re-creating the commons and re-configuring Maasai women's roles on the rangelands in the face of fragmentation.' *International Journal of the Commons* 10(2): 728–46.

Behnke, R., I. Scoones and C. Kerven. (eds). 1993. *Range Ecology at Disequilibrium: New models of natural variability and pastoral adaption in African savannas*. London: Overseas Development Institute.

Berkes, F. and H. Ross. 2013. 'Community resilience: Toward an integrated approach.' *Society & Natural Resources*. [Online] 26 (1): 5–20.

Berry, S. (1993). *No Condition Is Permanent: The social dynamics of agrarian change in sub-Saharan Africa*. Project MUSE. Madison: University of Wisconsin Press.

Boone, C., A. Dyzenhaus, A. Manji, C. Gateri, S. Ouma, J. Owino, A. Gargule, and J. Klopp. 2019. 'Land law reform in Kenya: Devolution, veto players, and the limits of an institutional fix.' *African Affairs* 118 (471): 215–237.

Bruce, J. and S. Migot-Adholla. (eds). 1994. *Searching for Land Tenure Security in Africa*. Washington DC: The World Bank.

Crane, T. 2010. 'Of models and meanings: Cultural resilience in social–ecological systems'. *Ecology and Society* [Online] 15 (4): 19.

Drew, J. 2020. 'Meanings of place & struggles for inclusion in the Lake Turkana Wind Power Project.' In Jeremy Lind, Doris Okenwa and Ian Scoones. (eds). *Land, Investment & Politics*. Martlesham: Boydell & Brewer. pp. 66–.

Ensminger J. 1996. *Making A Market: The Institutional Transformation of an African Society*. New York: Cambridge University Press.

Fratkin, E. 2001. 'East African pastoralism in transition: Maasai, Boran, and Rendille cases'. *African Studies Review* 44 (3): 1–25. https://doi.org/10.2307/525591

———— 'Pastoral land tenure in Kenya: Maasai, Samburu, Boran, and Rendille experiences, 1950–1990.' *Nomadic Peoples* 34/35: 55–68.

Fumagalli, C. 1977. *A Diachronic Study of Change and Socio-Cultural Process among the Pastoral Nomadic Samburu of Kenya, 1900–1975*. PhD thesis. State University of New York: Buffalo.

Galaty, J. 2013. 'The collapsing platform for pastoralism: Land sales and land loss in Kajiado County, Kenya.' *Nomadic Peoples* 17 (2): 20–39.

Government of Kenya. 1968. *The Land (Group Representatives) Act*. Nairobi: Government Printer.

———— 2010. *The Constitution of Kenya*. Nairobi: Government Printer.

———— 2016. *The Community Land Act*. Nairobi: Government printer.

Hagmann T., 2006. *Pastoral Conflict and Resource Management in Ethiopia's Somali Region*. PhD Thesis, University of Lausanne.

Hassan, R., I. Nathan and K. Kanyinga.2022 "Will community rights secure pastoralists' access to land? The Community Land Act in Kenya and its implications for Samburu pastoralists." *The Journal of Peasant Studies* 1–22.

Homewood, K., P.C. Trench, P. Kristjanson. 2009. *Staying Maasai? Livelihoods, Conservation, and Development in East African Rangelands*. New York: Springer. https://doi.org/10.1007/s10021-001-0045-9.

Jeppesen, M. and R. Hassan. 2022. 'Private property and social capital: Dynamics of exclusion and sharing in the subdivided pastoral rangelands of Kajiado, Kenya.' *Society & Natural Resources*. [Online] 35(1): 92–109.

Kameri-Mbote, P. 2006. 'Women, land rights, and the environment: The Kenyan experience.' *Development* 49(3): 43–48.

Kanyinga, K. 2000. 'Redistribution from above: The politics of land rights and squatting in coastal Kenya.' Research Report No. 115. Uppsala: Nordic Africa Institute.

———— 2009. 'The legacy of the white highlands: Land rights, ethnicity, and the post-2007 election violence in Kenya.' *Journal of Contemporary African Studies: Kenya's Uncertain Democracy: The Electoral Crisis of 2008* 27(3): 325–344.

Kenya National Bureau of Statistics. Population and Housing Census of Kenya, 2019 with *Volume 1A Population Distribution by Administrative Units*. Nairobi: Government Printer.

Kimani, K. and J. Pickard. 1998. 'Recent trends and implications of group ranch sub-division and fragmentation in Kajiado District, Kenya.' *The Geographic Journal* 164(2): 202–213.

Konaka, S. and P. Little. 2021. 'Introduction: Rethinking resilience in the context of East African pastoralists.' *Nomadic Peoples*, Special Issue 25 (2): 166–174.

Krätli, S. (2020). *Valuing variability: New perspectives on climate resilient drylands development.* London: IIED.

Krätli, S., and N. Schareika. 2010. 'Living off uncertainty: The intelligent animal production of dryland pastoralists.' *The European Journal of Development Research* 22 (5): 605–622.

Lesorogol, C. 2008a. *Contesting the Commons: Privatizing Pastoral Lands in Kenya.* Ann Arbor: University of Michigan Press.

———— 2008b. 'Land privatization and pastoralist well-being in Kenya.' *Development and Change* [Online] 39 (2): 309–331.

———— 2003. 'Transforming institutions among pastoralists: Inequality and land privatization.' *Am. Anthropology* 105: 531–541.

Lind, J., D. Okenwa, and I. Scoones. (eds). 2020. *Land, Investment and Politics. Reconfiguring Eastern Africa's Drylands.* Woodbridge: James Currey.

Mackenzie, F., 1990. 'Gender and land rights in Murang'a District, Kenya'. *Journal of Peasant Studies* 17 (4): 609–643.

McPeak, J. and P. Little. 2017. 'Applying the concept of resilience to pastoralist household data'. *Pastoralism: Research, Policy and Practice* 7: 14. https://doi.org/10.1186/s13570-017-0082-4

Mwangi, E., 2009. 'Property rights and governance of Africa's rangelands: A policy overview.' *Natural Resources Forum* 33 (2): 160–170.

Niamir-Fuller, M. 1998. 'The resilience of pastoral herding in Sahelian Africa'. In F. Berkes and C. Folke. (eds). *Linking Social and Ecological Systems: Management practices for building resilience.* Cambridge: Cambridge University Press. pp. 250–284.

Odote, C., R. Hassan, H. Mbarak. 2021. 'Over promising while under delivering: Implementation of Kenya's Community Land Act.' *AJLP & GS* 4 (2): 292-307.

Okoth-Ogendo, H. 2002. 'The tragic African Commons: A century of expropriation, suppression and subversion.' *University of Nairobi Law Journal* 12003(1): 107–117.

Pas, A. 2018. 'Governing grazing and mobility in the Samburu lowlands, Kenya.' *Land (Basel)* [Online] 7 (2), 41–.

Pollini, J. and J. Galaty. 2021. 'Resilience through adaptation: Innovations in Maasai livelihood strategies.' *Nomadic Peoples* [Online] 25 (2): 278–311.

Ribot, J. and N. Peluso. 2003. 'A theory of access.' *Rural Sociology* 2: 153–181

Robinson L. and F. Berkes. 2010. 'Applying resilience thinking to questions of policy for pastoralist systems: Lessons from the Gabra of Northern Kenya'. *Human Ecology* 38: 335–350. https://doi.org/10.1007/s10745-010-9327-1

Rodgers, C. 2018. *Rural, Remote, Raiya: Social differentiation on the pastoralist periphery in Turkana, Kenya*. PhD thesis, Anthropology, University of Oxford.

Rutten, M. 1992. *Selling Wealth to Buy Poverty: The process of the individualization of landownership among the Maasai pastoralists of Kajiado District, Kenya, 1890–1990*. Saarbrücken: Breitenbach.

Samburu County Government. 2018. The County Integrated Development Plan (CIDP).

Scoones, I. (ed). 1995. *Living with Uncertainty: New Directions in Pastoral Development in Africa*. London: International Institute for Environment and Development. https://doi.org/10.3362/9781780445335.

Scoones, I. 1999. Ecological dynamics and grazing-resource tenure: A case study from Zimbabwe. In M. Niamir-Fuller. (ed). *Managing Mobility in African Rangelands. The legitimization of transhumance*. London: Intermediate Technology Publications Ltd. pp. 217–235.

————— 2021. 'Pastoralists and peasants: Perspectives on agrarian change.' *The Journal of Peasant Studies* 48 (1): 1–47. https://doi.org/10.1080/03066150 .2020.180224.

Semplici, G. 2020. *Moving Deserts. Stories of mobilities and resilience from Turkana County, a Kenyan desertscape*. PhD thesis. Oxford Department of International Development, Oxford, UK.

————— 2021. 'Resilience and mobility of identity: Belonging and change among Turkana herders in Northern Kenya.' *Nomadic Peoples* [Online] 25 (2): 226–252.

Simpson, G.L., and P. Waweru. 2021. 'The implausible persistence of pastoralism: Samburu transhumance from their nineteenth-century origins through the period of colonial rule.' *Journal of the Middle East and Africa* 12 (2): 225–249. https://doi.org/10.1080/21520844.2021.1909379

Spencer, P. 1965. *The Samburu: A study of gerontocracy in a nomadic tribe*. London: Routledge and Kegan Paul

————— 1973. *Nomads in Alliance. Symbiosis and growth among the Rendille and Samburu of Kenya*. London: Oxford University Press.

Sperling, L. 1985. 'Labour recruitment among East African herders: The Samburu of Kenya.' *Labour, Capital and Society* 18 (1): 68–86.

————— 1987. *The Labour Organization of Samburu Pastoralism.* PhD thesis, McGill University.

Straight, B. 1997. *Altered Landscapes, Shifting Strategies: The politics of location in the constitution of gender, belief, and identity among the Samburu pastoralists in Northern Kenya.* Ann Arbor: University of Michigan.

Sullivan, S. and K. Homewood. 2003. 'On Non-Equilibrium and Nomadism: Knowledge, Diversity and Global Modernity in Drylands (and Beyond...).' Working Paper No. 122/03. Coventry: CSGR.

Turner, M. 2011. 'The new pastoral development paradigm: Engaging the realities of property institutions and livestock mobility in dryland Africa. *Society and Natural Resources* 24: 469–484.

Wangui, E. 2008. 'Development interventions, changing livelihoods, and the making of female Maasai pastoralists.' *Agriculture and Human Values* 25 (3): 365–378.

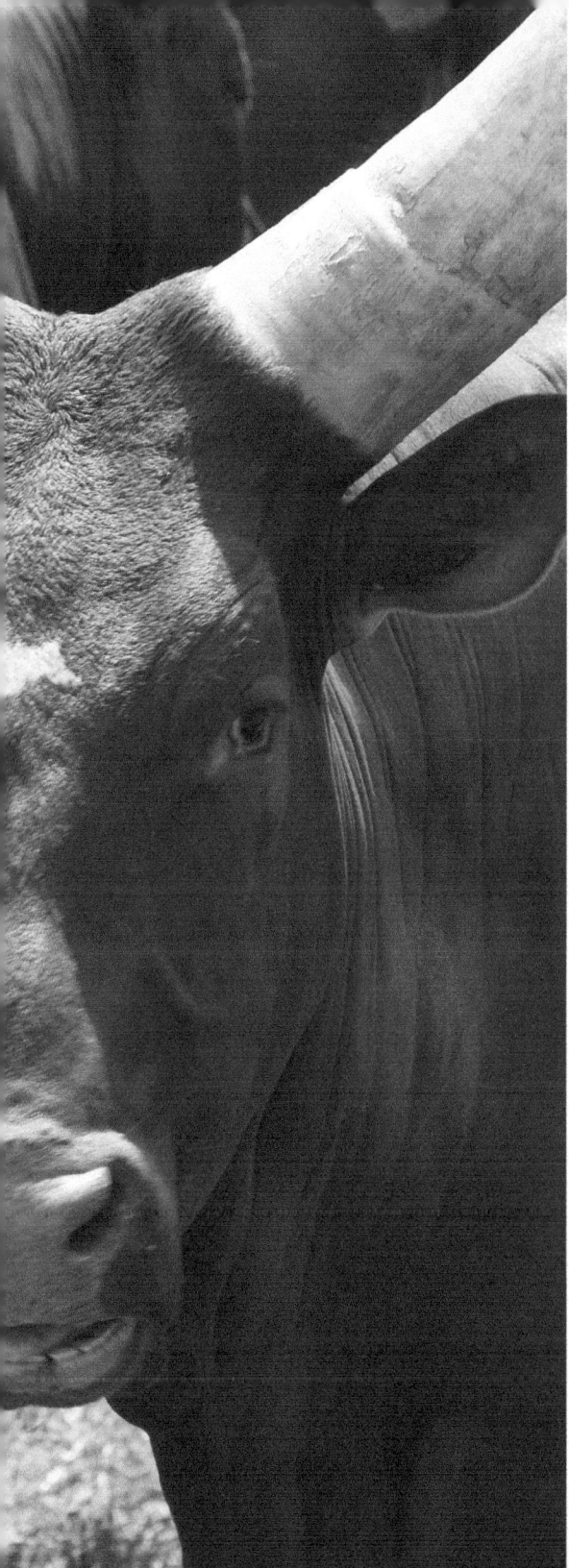

Resilience
of Displaced
Pastoralists
during and
after Conflict

07

Reconsidering the Resilience of Pastoralism from the Relational and Contextual Perspective of Reliability

The case of conflicts between the Samburu and the Pokot of Kenya, 2004–2009

Shinya Konaka

Introduction

This chapter sheds new light on the 'resilience of pastoralists' with an ethnographic case study on a series of conflicts between the Samburu and the Pokot in Kenya, which erupted in 2004 and ended in 2009, exploring the theoretical potentials of reliability pastoralist theory developed by Emery Roe.

Resilience can be generally defined as 'the capacity of a system to absorb disturbance and reorganize while undergoing change so as to still retain essentially the same function, structure, identity, and feedbacks' (Walker et al. 2004). Resilience as an academic line of argument is mainly derived from research in ecology (Holling 1973) and psychology (Garmezy 1971). The scope and study of resilience have been gradually expanded from its roots in the natural sciences to the study of human societies, where the focus has been on 'social and ecological resilience' (Adger 2000; Van der Leeuw and Leygonie 2000; Walker et al. 2004; Folke 2006; Walker et al. 2006).

The application of 'resilience thinking' to the study of pastoralism, especially in terms of rangeland ecology, goes back to the 1980s (Walker et al. 1981; Ellis and Swift 1988), although researchers in these fields preferred, at that time, terms such as 'disequilibrium' or 'new ecological thinking' to 'resilience' (Ellis and Swift 1988; Behnke et al. 1993; Scoones 1995, 2004; McCabe 2004). From approximately the 2000s, researchers began to apply resilience and resilience-related concepts more intentionally to the study of pastoralism with ostensive uses of the term (Niamir-Fuller 1998; Robinson and Birks 2010; Leslie and McCabe 2013; Anderson and Bollig 2018). Although the factors that are considered as outside triggers vary in these discussions, Cervigni and Morris (2016: 46–47) summarised the most important factors as follows: the meteorological factor, caused by short-term weather emergencies or long-term effects of climate change; and the conflict factor, leading to the disruption of livelihood, loss of property, displacement and bodily injury or death. This chapter focuses on the latter.

Unfortunately, the concept of the 'resilience of pastoralists' remains largely unclear and inconsistent. For sure, there are several works that formulate 'pastoral resilience'. For instance, Liao and Fei (2016), in comparing East Africa and Central Asia, proposed indicators of 'pastoral resilience' comprising livestock mobility, land use patterns and livelihood diversification. However, it is important to remember here that resilience was originally developed as a general, universal, and scientifically neutral concept. It has been shown by numerous researchers that resilience is not determined by substantial propensities or traits of certain individuals or systems, but rather is a highly context-dependent or context-sensitive concept (Waller 2001; Ungar 2008; Fletcher and Sarkar 2013; Panter-Brick 2014; Weichselgartner and Kelman 2014; Davies et al. 2015; Chandler and Coaffee 2017; Oliver-Smith 2017). Occasionally, similar lines of argument can be found when discussing 'social resilience',[1] as Keck and Sakdapolak (2013: 14) remark: 'The study of social resilience emphasizes the embeddedness of social actors in their particular time- and place-specific ecological, social and institutional environments. As such, it is a relational rather than an essentialist concept.' Therefore, a more relational and contextual approach to resilience that can embed the resilience concept within the context of pastoralists must be sought by researchers of pastoralism (see 'Introduction').

---

1 Social resilience can be defined as 'the ability of communities to withstand external shocks to their social infrastructure' (Adger 2000: 361).

This chapter intends to expand the scope of resilience with a relational and contextual approach that is relevant to the context of pastoralists[2] by adding the theoretical framework of 'reliability professional' coined by Emery Roe to the study of 'resilience of the pastoralist'. Roe's theory, which is socio-culturally and contextually oriented and politically framed, and which prioritises unpredictability and uncertainty, can provide a keystone aside from the preexisting ecological and psychological theories of resilience. This is not the only or best option to take but simply a possible avenue of research to pursue.

The next section presents the theoretical framework of this chapter, reliability theory. The following four sections mainly describe how displaced pastoralists of Samburu survived under the conflicts between Samburu and Pokot, with special focus on the formation of clustered settlement and inter-ethnic mobile phone networks. The conclusion will be led by the examination of both presented ethnographic cases and the theoretical framework of reliability.

Pastoralists as reliability professionals

The concept of resilience comprises several related concepts, such as unpredictability, uncertainty, bounce-back, threshold, basin attraction, panarchy, transformation and adversity. Above all, addressing the notion of resilience within the context of pastoralism, unpredictability and uncertainty needs to be of primary focus. To this end, the 'high reliability pastoralism' theorised by Emery Roe and colleagues (1998a, 1998b; Roe 2018), is a noteworthy contribution. Although he has rarely explicitly mentioned the concept of resilience in his work, his theory has unexploited potential implications for the 'resilience of pastoralists'. This chapter attempts to harness that potential.

Roe extended the possibility of disequilibrium theory. 'New disequilibrium-based models of ecological dynamics on rangelands enable

---

2 This option originated in the work of Fratkin (1986), who carefully examined the relevance of the application of the biological resilience concept to the Rendille and Ariaal communities of Northern Kenya.

us to see pastoralism as what organization theorists term a high reliability institution' (Roe et al. 1998b: 39). According to Roe et al. (1998a), 'high reliability theory is a relatively recent development of organization theorists interested in how institutions maintain their activities in circumstances where high-cost failure, error, and accidents are very probable'. High reliability organisations studied in the West have included air traffic control systems, nuclear power plants and electricity companies. Roe et al. (1998b) nominated pastoralism as an addition to the list of high reliability organisations by arguing that pastoralism shares most – maybe even all – the attributes of these other organisations.

Although Roe did not give a concise definition of 'reliability pastoralists', the meaning of the concept can be understood with the following three concepts: 'narrative policy analysis', 'risk acceptance strategy', and 'real-time management'. First, Roe (1994: 34) proposed the 'narrative policy analyses'. Policy narratives are 'stories (scenarios and arguments) which underwrite and stabilize the assumptions for policymaking in situations that persist with many unknowns, a high degree of interdependence, and little, if any, agreement'. Therefore, before beginning the analysis of reliability professionals, the process of narrative policy analysis must first reveal the relationship in the broader political context and identify the premise and backdrop of the crisis correctly.

Second, Roe (Roe et al. 1998b, 2018) contrasted the reliability of risk-accepting pastoralists with risk-averting pastoralists. In short, risk-accepting pastoralists primarily accept and take risks, whereas risk-averting pastoralists primarily avoid risks. In this view, the risk-averting pastoralist engages in attempts to avoid or escape the extreme hazards of unpredictability, given that the pastoralist has no control over the probability of these hazards occurring. The pastoralist in search of reliability is, in contrast, actively engaged in ongoing efforts to reduce the probability of unavoidable hazards by managing temporal and spatial diversity in grazing opportunities and diversity in livestock capabilities and response. Both strategies coexist within a pastoral community, but the latter has been scarcely examined in research.

Third, Roe recently elaborated on the theory and extracted the essence of what reliability pastoralists do. He asserted that 'pastoralist systems seek to increase process variance – think, real-time management strategies and options – in the face of high but unpredictable or uncontrollable input variance so as to achieve low and stable output variance' (Roe 2020a: v). 'It is this logic of high input variance matched by high process variance in order to ensure low and stable output variance that characterizes what reliability professionals do' (Roe 2020a: v) (Figure 7.1). This theoretical framework makes it possible to analyse how pastoralists who faced unpredictable and uncontrollable shocks from outside their community have come to ensure their livelihoods through their own improvisational and positive counter-managements. Notably, in this theory, all the management processes of pastoralists are relationally and contextually determined, according to ever-changing outside shocks, and not presupposing any predetermined rules, principles and propensities. In a nutshell, this theory is significant in analysing a positive shock absorbing practice and process of pastoralists.

It must also be noted here that his reliability theory matches the typical definition of resilience: 'the capacity of a system to absorb disturbance and reorganise while undergoing change so as to still retain essentially the same function, structure, identity, and feedbacks' (Walker et al. 2004), although

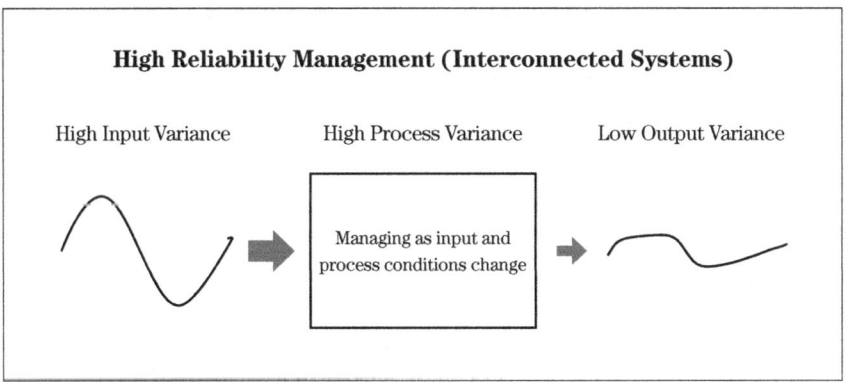

Figure 7.1 High reliability management of infrastructures in terms of input, process and output variance (based on Roe 2020a: 4)

Roe focuses more on the practice and process of absorption rather than the continuity of the system – for him, continuity is a result of this process.

Through interaction with the author, including the examination of this chapter's case study, Roe (personal communication, 2020b see also 2020a: 18) recently defined 'resilience in a pastoralist system' as 'the system's capability in the face of its high reliability mandates to withstand the downsides of uncertainty and complexity as well as [to] exploit the upsides of new possibilities and opportunities that emerge in real time'. Clearly, this definition of resilience in the pastoralist system comprises aspects of the 'reliability pastoralist' and 'real-time management'. Shifting from the African context, insights developed here are in concord with an observation by Reinert and Benjaminsen (2015) of Norwegian Sámi reindeer pastoralism: 'Resilience, like [the] pastoral practice itself, must therefore be understood in open-ended, improvisational and experimental terms' (Reinert and Benjamin 2015: 11).

Several researchers have mentioned Roe's reliability pastoralism theory (Scoones 1999; Krätli and Schareika 2010). However, no such theory has been applied to investigate a specific ethnographical case. If reliability pastoralism theory can be combined with the ethnographic approach of anthropology, it could open possibilities to study the resilience of pastoralists who are in the midst of conflict today.

An outline of conflict and research

This chapter focuses on a case of conflict that erupted in north-central Kenya between the Samburu and the Pokot. The conflict broke out between the two groups in April 2004 and had largely ended by December 2009. Although the conflict was incited by a politician, as I will describe later, it soon escalated to mutual retaliation between the ethnic groups. This chapter does not intend to provide a holistic account of the conflicts between the Samburu and Pokot. The aim of this ethnographic description and analysis is limited to the internally displaced persons of the Samburu, who faced unprecedented crisis when their livelihoods were uprooted after being severely attacked by the Pokot. In the context of a critical conflict situation, resilience specifically indicates their capacity to survive and sustain their livelihood as a pastoral community. Questions then arise: Were the displaced Samburu resilient to the crisis during and after the conflict, and did they succeed in surviving under existential threats? If so, is their survival rooted in the cultural tradition inherited through generations, or outside interventions accompanied by laws, regulations, and guidelines?

Field research on this conflict was conducted intermittently, primarily in five locations in the Samburu, Laikipia, and Baringo Counties in Kenya from 2004 to 2018. This covered 190 incidents and current events and took approximately eleven months of cross-checking the collected information, which was recorded in two languages, Maa and Swahili. Admittedly, the research data have been mainly obtained from the Samburu, Ilchamus and Tugen people, who were the people most severely attacked by the Pokot; thus, it may include biases, although important points were cross-checked directly or indirectly by interlocutors focusing on Pokot. The main research method adopted was a qualitative, in-depth field study comprising semi-structured interviews and participant observation. Under the condition that reliable, official records are unavailable, and media did not report the news material thoroughly, as locals claim, fieldwork was the only available option. The research was conducted by visiting and staying in the same residential areas as the internally displaced persons several times while building rapport. Interviews were mostly conducted collectively, with an average of four collaborators.

The Maa-speaking Samburu are predominantly pastoralists, organised in a dual system of permanent settlements and satellite cattle camps. The Pokot speak the Southern Nilotic Kalenjin language and live as agro-pastoralists who grow crops (rainfed), especially maize, finger millet and sorghum, and who dwell in more permanent settlements. Both ethnic groups raise cattle, goats and sheep, and engage in farming, small businesses and wage work. The conflict led to tremendous damage on all sides, including killings, livestock raiding and the incineration of houses and household items. According to the author's estimate based on fieldwork from 2010 to 2018, the death toll currently stands at 590. Meanwhile, the number of Internationally Displaced People (IDP) generated by the conflict is estimated to exceed 22,000 (IDMC [Internal Displacement Monitoring Center] 2006: 33).[3]

This conflict has been reported and studied by numerous researchers (Straight 2009, 2017; Greiner 2012, 2013; Okumu 2013; Holtzman 2016; Ervin 2020). Aspects of the conflict are described here briefly, since giving full details and reviewing precedent research about this conflict is not the aim of this chapter.

Unravelling the narratives to find the primary cause of conflict

This section attempts to unravel the primary cause of the conflict by referring to 'narratives' on the cause of the conflict. The arguments are not directly connected to the 'resilience of pastoralists' *per se*. However, the premise and backdrops of the crises the pastoralists faced must be clarified before introducing the main arguments. Most Kenyan citizens, including the media and humanitarian/development agencies, attribute the primary cause of the conflict to two pastoralist narratives: the 'traditional livestock rustling narrative' and the 'resource conflict narrative', despite the political backdrops of the conflict, as I will describe. The first – the narrative of livestock rustling – attributes the primary cause of the conflict to traditional cattle rustling activities and hostility among pastoral ethnic groups. Several media outlets have described the conflict using terminology like 'cattle rustlers' and 'ethnic clashes' (cf. Greiner 2013: 217, 232). For instance, a newspaper article represented this conflict as the 'Kosovo of Kenya', where cattle rustling activities have been practised for a very long time.[4]

However, this conflict differs markedly from customary livestock rustling among pastoralists. Traditional livestock rustling is generally thought to be intended to acquire livestock as wealth between ethnic groups, due to deep-rooted hostility. First, in this case, the purpose of the

---

3 The estimation varies from 12,000 to 75,000 (IDMC 2006: 33).

4 A newspaper article, *Daily Nation* dated 11 Nov. 2010.

attack was different from livestock rustling. In 2004, Pokot combatants systematically set fire to the Samburu houses, which was never a common method of livestock rustling attack among pastoralists in this area. Additionally, the Pokot massacred 24 Samburu in 2009, including women and children at the settlement, without stealing the livestock. In fact, this attack occurred after the livestock had been moved to graze in the region where both ethnic groups formerly lived together. This massacre was planned to evict the Samburu from the contested land, an intent far removed from traditional cattle rustling.

Second, although this conflict has been considered an ethnic conflict, this assumption itself is quite dubious. Until the outbreak of the conflict in 2004, the groups were on good terms, without any historical record of large-scale conflicts between them (Greiner 2012: 419, 2013: 227; Holtzman 2016: 134). During the 1998 clash between the Samburu and Turkana ethnic groups, the Pokot provided reinforcements for the Samburu. Although intermarriage and bilingual residents are common, livestock raiding has been quite rare. The IDMC report below clearly represents how local Samburu felt at the onset of the conflict.

> 'Traditionally, the Pokots and Samburus do not fight, so maybe someone is engineering this', stated one member of the Samburu Peace Committee (IDMC 2006: 37).

The second narrative of resource conflict suggests that the environmental repercussions of dry land were the primary cause of this conflict. Certainly, Kenya experienced a serious drought from 2008 to 2009, which adversely affected pastoral communities in drylands. In both the Samburu Central Constituency and Laikipia North Constituency, high mortality rates, particularly for cattle and sheep, were reported (Zwaagstra et al. 2010: 21). It was commonly assumed that the drought made resources, such as pasture area and water, scarce. This caused pastoral people to struggle over these scant resources and even to resort to violence. For instance, a daily newspaper article titled, 'Drought triggers rise in killing: Ranging famine blamed for clash over few resources' describes this conflict within the resource conflict narrative.[5]

However, there is no evidence of such a 'resource-based conflict' – that is, there is nothing to suggest that the struggle over scarce resources, pasture and water led to the conflict between the Samburu and the Pokot. Land ownership of both ethnic groups has been communal. The usufruct rights of land were inclusive of all local residents until the outbreak of the conflict. The Samburu territory is highland savanna at an altitude of approximately 1,500 metres, while the Pokot inhabit low bushlands at an altitude of about 1,000 metres. Seasonally, from July to August, pasture is available in the highlands of Samburu because of the summer seasonal rainfall (*olorikine*). The Pokot, who are looking for sufficient pasture, are allowed to migrate and graze within the Samburu territory. In turn, from October to November, pastures are available in the Pokot lowlands because of the autumn seasonal rainfall (*oltumuren*). The Samburu, who are looking for sufficient pasture, are allowed to migrate and graze within the Pokot territory. This loose system of sharing scarce resources and the mutual aid system of *usufruct*, or right of land, clearly counter the use of resource conflict theory to explain the Pokot and Samburu conflict.[6]

If both narratives are misleading, then what is the primary cause of the conflict? Furnishing an overview, Okumu et al. (2017: 505) state that 'the role of politics and the political elites in these raids are increasingly significant in changing raids in northern Kenya'. Greiner (2013: 236) also suggests that 'much of the current violence in pastoralist areas in Kenya is indeed primarily fueled by politicized dynamics'. On this particular cause of the conflict, interlocutors from all over the area unanimously claim that the conflict was caused by a Pokot politician who inflamed the parochialism of the local people in a bid to garner votes. Since 2000, prior

---

5 *Daily Nation*, 9 Oct. 2009.

6 Holtzman (2017: 43) was also dubious about the resource conflict narratives. Witsenburg and Wario (2009: 536) assert that 'our results clearly show that the intensity of violent conflicts increases during the rainy season as opposed to during dry seasons when resources are scarcer'. See also Van Baalen and Mobjörk 2018: 15.

to the conflict, this politician (we will call him 'X'), gave public speeches promoting the false story that the land occupied by the Samburu had once belonged to the Pokot, and promised that land taken from the Samburu would be distributed to his local supporters as a reward for their support. This information has been confirmed by the tortured captives of the Pokot after the battle, as well as testimony from the Pokot who have Samburu friends and relatives. It is also corroborated by other sources on this conflict as follows.

> Numerous respondents also blamed local government officials for instigating and fueling the conflict between the Samburu and the Pokot. They referred to the alleged financial benefits local government officials receive when conflict takes place (IDMC 2006: 32).
>
> Land ownership, whether communal or individual, provides ready fuel for political incitement and ethnic manipulation by politicians... (Straight 2006: 25).

Another cause of the conflict is related to a conservancy programme proposed by a European resident in 2001. The proposed area for the conservancy is located at the borderland between the territories of the Samburu and the Pokot. First, X was supportive of this programme, which allows local pastoralists to use pasture within the conservancy during the dry season without any fencing. However, after a while, a rumour that the Pokot's rangeland was going to be sacrificed for the conservancy programme circulated among the Pokot. Hearing that, X changed his attitude to use the conservancy programme to incite the Pokot residents by falsely claiming that the Samburu were about to sell the borderland to a European, and planned to occupy it by calling it a new conservancy zone, thus depriving the Pokot of grazing area.[7] Therefore, the factor leading to

---

7 Source: an interview with a Pokot youth on 9 Sept. 2012; a Dorobo elder on 11 Sept. 2012, and a Samburu woman who had lived in Pokot land during the conflict and overheard the conversation of Pokot elders on 10 Sept. 2012. This corresponds to what a Pokot informant told Greiner and Holtzman: 'The Samburu want a conservancy but the area they want to use for it is our land'

the conflict was not the conservancy programme *per se*, but instigation from a politician weaponizing the programme.

Several aspects of 'privatization' (Lesorogol 2008) and 'fragmentation' (Galvin et al. 2008) of land, as well as 'territorialization of ethnicity' (Schlee 2013) can be discerned in the area as well. However, the conflict period was before the implementation of the devolved system of governance in Kenya that commenced in March of 2013. We can safely say that the most prominent factor of this conflict was the 'utilization of land as a tool of instigation' by a politician. These lines of argumentation should not be confused.

In December 2002, X was elected as a Member of Parliament (MP) to represent his constituency. In contrast, the opposing candidate, who insisted on peaceful coexistence, was labelled a coward and thus lost votes.[8] MP X increased his popularity by inciting local administrative chiefs and the local Pokot in continuous acts of violence, including raids as Straight (2009: 25) remarks: 'Political leaders contribute to the cattle rustling as they whip up ethnic sentiments in order to paint themselves as community heroes who deserve to sustain their political leadership.'

In April 2004, MP X was alleged to have supplied the local Pokot with funds, as a patron, along with 500 AK-47 automatic rifles smuggled from Uganda. He ordered the killings of Samburu and raids on their livestock until they gave up the territory. Four locations were attached to one another, and 250 houses were set on fire. The Pokot bribed local police officers with the funds provided by X, and the weapons were used to raid the livestock. Stolen livestock was then sold at the market or butchered, and approximately forty per cent of those livestock sales went to MP X

(Greiner 2012: 420); 'In the views of the Pokot I was interviewing, the goal of this conservancy (which as far as I knew existed only in the minds of Pokot conspiracy theorists) was to alienate the Pokot from their traditional lands around Amaya' (Holtzman 2017: 58).

8 Greiner (2013: 233) notes on the Pokot that 'I was repeatedly told that, particularly during electoral campaigns, peace negotiations are usually avoided by politicians, because any concessions they make are easily used against them by their competitors.'

as repayment for supplying weapons and funds.[9] Livestock sales are used repeatedly to purchase weapons on the black market. This informal accumulation of financial stock can be regarded as a form of 'black capitalism'.[10] This supports Straight's (2009: 25) account: 'Many Samburu I know add that political elites use profits from mafia style livestock raids to finance their campaigns.'

Even though the primary cause of the conflict was evidently the identity politics of MP X (Kaldor 1999), traditional cattle rustling narratives and the resource conflict narrative have played the role of concealing political factors by imputing politicians' war crimes to local pastoralists. The image of warlike pastoralists was convenient for the politician, as Straight (2009: 21) remarks: 'They can also mask the ways in which some elites benefit from the propagation of cultural stereotypes even while deliberately engaging in manipulation of ethnic fault lines.'

This leads us to the next question. Was the pastoral community resilient as it faced crises of survival under uncontrollable and unpredictable types of attack from enemies sponsored by a corrupt politician and officers in a manner never experienced before?

Resilience against weapons proliferation

When the Pokot began to attack the Samburu in 2004 with hundreds of automatic rifles, it was a highly unpredictable and uncontrollable 'input variance' for the Samburu who had few firearms at that time. When the conflict broke out, all the Samburu locals fled to other areas, and large sections of the Samburu-Pokot border area became a no man's land. Giving up the land and migrating to a more peaceful area was one kind of 'risk-averting' strategy that was employed. However, most of the area's pastoralists dared to choose to stay at the battlefront, adopting a 'risk-

---

9 Source: information from a Pokot elder who chatted about the corruption of Pokot with a Dorobo friend [minor ethnic group of the area] in Longewan on 10 Sept. 2012.

10 A similar cycle can be observed in the report of a Pokot informant (Greiner, 2013: 234).

accepting' strategy. After several months, the Samburu IDPs organised their defence so that they could return to their homeland. This was mostly accomplished by gathering in clustered settlements: spontaneous IDP camps that were exceptionally large and fortified. It became a transformed nomadic settlement during the crisis (Photo 7.1).

There were at least ten verified clustered settlements in the county-border area with a total population estimated at 6,700. Clustered settlements may resemble 'ceremonial settlements', which are normally constructed during an age-set ceremony (Spencer 1965). Both, in fact, are called *'lolora'*. However, the clustered settlements formed at the time played several different roles that can be summarised fourfold as follows.

First, all the clustered settlements were constructed at the edge of the Western frontlines of the conflict and were intended to visibly demonstrate the Samburu territory to the Pokot. No resident was confirmed to live beyond the clustered settlement. All the residents of the clustered settlements had dared to choose to live at the frontline even though it was possible to migrate to another area within Samburu territory.

Second, clustered settlements functioned as places for weapons storage. Most of the local police and special forces were bribed by MP X and, therefore, were completely unreliable for Samburu locals. Even the special forces, such as the Anti Stock Theft Unit (ASTU), dispatched to Longewan in April 2005 to secure the Samburu, had been bribed by the

Photo 7.1 A clustered settlement. Photograph by the author, 25 August 2009

Pokot;[11] an officer was in fact witnessed receiving the bribe money from the MP X during patrols.[12] Once, a police officer phoned a leader of a clustered settlement, telling the leader, 'If you could give us more bribe money than the Pokot, we are ready to dispatch troops.' In December 2009, during an operation to disarm the largest clustered settlement, security officers killed one innocent adult man and raped six girls without confiscating any weapons.[13] Villagers had no choice but to ensure their collective protection from enemies through smuggling. Rifles, ammunition, even police uniforms, purchased mostly from police officers, were collected and stored in the clustered settlements.

Third, clustered settlements tend to be large units in the daily subsistence economy, which is based on a collective mode of production. The grazing system used in clustered settlements differs from that used in normally-sized settlements. For instance, in a clustered settlement, all the approximately 3,000 cattle belonging to the settlement are grazed simultaneously. In contrast, in normal settlements small herds and flocks composed from several households are grazed together. Forty grazing guards, mostly young men, scout and patrol the cattle herd, which is followed by fifty herd boys. Four nightwatchmen, each armed with an assault rifle, guard the four entrances to the clustered settlement. The tasks of guarding the daily grazing, grazing-related labour and the night watch are generally allocated according to a daily rotation system. Following Roe's terminology, all the above activities are elements of a process of 'a high input variance' of firearms, and political power has been matched by high process variance in forming an armed clustered settlement collectively.

Fourth, the clustered settlement served as a base of reciprocal exchange and mutual aid for households that lost most of their property during the

---

11 Source: information from a Pokot elder who chatted about the corruption of Pokot with a Dorobo friend in Longewan on 10 Sept. 2012.

12 Source: interview with a Dorobo officer from Longewan who understands both the Samburu and Pokot languages. During a patrol run in an SUV, he accidentally overheard a conversation about delivery of bribe money between Pokot MP X and an inspector of ASTU in the Pokot language, which they mistakenly thought he did not understand.

13 The incident was reported in a newspaper article in *Daily Nation*, 25 Dec. 2009.

evacuation. Livestock was commonly given or loaned to poor households by households inside or outside the clustered settlement. Neighbouring households in the settlement assisted daily in providing food, mostly milk and porridge, to members of poor households who could not afford these essentials. Neighbouring women normally cooperated in building houses and enclosures. Daily commodities were also given or lent to members of poor households who had to leave everything behind during the raid. In 2011, the author conducted a survey that produced an inventory of household commodities, targeting 23 households that had lost all livestock during the conflict (Konaka 2017a and the next chapter). Of all the sources or methods used to obtain household commodities, 'Gift items from local people' occupied the largest percentage (30.9%), while items obtained from aid agencies (mostly The Red Cross) accounted for only 9.2%. In a sense, however, the clustered settlement was itself a case of locally constructed humanitarian assistance.

Notably, the Samburu recognised that the clustered settlement was an unsuitable residential pattern for pastoral production. Pastoralists prefer widely dispersed settlements because pasture is scattered across vast and dry lands. Clustered settlements also tend to exert a strong negative influence on the residents; for example, an infectious disease that causes diarrhea spread in the clustered settlements due to poor sanitary conditions and was exacerbated by aggregation. The environment surrounding the clustered settlements was also damaged by excessive deforestation for firewood, even after collecting firewood had been banned by the village leaders.

Even after the formation of the clustered settlements, the Pokot continued to attack the Samburu. In 2006, even armed clustered settlements became targets of attack. By 2009, most of the clustered settlements were facing an extinction crisis, except for one impregnable clustered settlement. Although eight youths were killed during the fighting there, no livestock were stolen. The armed settlement was the only clustered settlement that was formed without considering clan-phratry affiliations. The tactics employed by the residents were far from traditional cattle rustling. Four retired army officers trained youths (*olmurran*) in modern military training methods, using mobile phones in place of transceivers. Using contributions from residents, they purchased four bazookas from corrupt police officers. Twenty-three trenches were dug around the settlement and defence strategies were prioritised.

In September 2009, 24 Samburu herders, including women and children who were in the grazing camp, were massacred by the Pokot who intended to evacuate the Samburu from the border area. The attack was reported in newspapers and appeared on television news programmes, which invited political pressure from the central government.[14] On the day of the incident, 300 Pokot troops were dispatched to the impregnable clustered settlement. Bazookas were used to counter the attack by Pokot troops, and 120 Pokot youths were killed (102 by a bomb explosion and eighteen by gunshot).[15] As described above, clustered settlements in existential crises illustrate the powerful resilience of pastoralists who have been marginalised from and cannot depend upon the nation-state. Although no militarisation, killings, bribery or torture can be justified, the unreliable and undependable backdrop of all these acts must be considered. Additionally, clustered settlements are completely different from normal or traditional settlements in terms of nomadism, organisation and defence methods.

What, then, was the fate of the clustered settlement after the conflict? Day-by-day, more of its residents returned to the places they had lived before the conflict. All the clustered settlements were dissolved soon after the conflict ended, dispersing the people and the livestock yet again across the vast landscape of the savanna. From a political point of view, it must

---

14 Source: *Daily Nation*, 16 Sept. 2009, supplemented by the sources shown in note 16. See also Greiner (2013: 227).

15 Source: interviews with several Samburu and Pokot locals. The main source is an interview with a Samburu resident of the clustered settlement in Laikipia West Sub-County on 10 Sept. 2010. He was afraid to divulge this information because, up to that point there was no police record of this incident. All the corpses were eaten by hyenas around the area. This information was corroborated by another interview with a Pokot elder in Longewan on 9 Sept. 2012. He said the information was correct and the troop was mobilised mainly from Nasurr and Nkinyang area. A Samburu woman who lived in Pokot land during the conflict also reported that she saw many Pokot mothers crying for their sons that day (an interview in Longewan on 10 Sept. 2010).

have been a strong option for the Samburu, whose bazookas gave them a massive advantage in resuming the conflict and dominating the area. However, no clustered settlers wanted to pursue the conflict; they simply intended to reduce risks and ensure their security. The pastoralists said, 'When living in congestion at the clustered settlement, we see diminishing pasture, increasing ticks, and insanity. That's why we disperse.'[16] They ceased the conflict because they had succeeded in ensuring 'low and stable output variance'. This also demonstrates that in the case of the displaced Samburu, 'warlike' pastoralist narratives are mistaken.

Resilience against information flooding

One of the reasons this conflict escalated rapidly between the two ethnic groups was the use of mobile phones (Konaka 2017b). The role of mobile phone use by Kenyan pastoralists has been discussed by several researchers on herding (Butt 2015), livestock markets (Barrett et al. 2003), nutrition (Parlasca et al. 2019) and veterinary information sharing (Chepkwony 2018). However, little is known about the role of mobile phone use during and after the conflict. Since the mid-2000s, Samburu pastoralists have enthusiastically accepted mobile phones. Nevertheless, their use expanded the conflict rapidly, in fact at a speed never experienced before. Before their introduction, people could only organise combatants by visiting each settlement in person. Pastoralists live scattered over vast areas, and motor transport is not easily available. However, mobile phones enabled the mobilisation of hundreds of combatants at short notice, and battle reinforcements could also be requested if necessary. The mobilisation of combatants is essential because the volume of guns and bullets is a decisive factor in battles. Before the widespread availability of mobile phones, such expansion would have been unimaginable. We must note here that the crisis is not mobile phone use *per se*, but uncontrollable and unpredictable information flooding, which leads to bloodshed. It operated as a strong, uncontrollable and unpredictable 'input variance' for the pastoralists.

---

16 Source: interview with a Samburu elder in Laikipia County on 9 Sept. 2011.

This kind of expansion in the scale of the battle was first exemplified in October 2004.[17] The death toll from both ethnic groups amounted to 53. At the outset, the Samburu were raided, losing 200 cattle and 3,000 goats and sheep. They requested reinforcements from surrounding areas by mobile phone, and 1,000 Samburu combatants wielding assault rifles participated in the battle. It continued from early in the morning until the evening. Finally, the Samburu succeeded in recovering 190 cattle and 3,000 goats and sheep stolen from them.

By 2009, the conflict between the Pokot and Samburu was mostly over. The reason it ended is threefold. First, after the complete defeat during the attack of the impregnable clustered settlement in September 2009, the Pokot recognised that the Samburu were a formidable opponent and feared attacking them again, thus abandoning the invasion of their territory.[18] Second, political pressure was exerted by the internal security minister, who belonged to the same ethnic line as the Samburu, on the Pokot MP who instigated the conflict. Third, the attacks from Pokot had also extended to the Tugen and the Ilchamus, neighbouring ethnic groups on the other side (see Little in this volume). That side of the conflict continued, even after the Pokot abandoned their plans of invading the Samburu.

However, given that mobile phones enabled the rapid expansion of the conflict, subsequent minor skirmishes had the potential to be misunderstood as intentions to reignite a full-scale battle between the two ethnic groups. Peace meetings organised by the national and municipal government have been held many times; however, most of the meetings comprised formal speeches and prayer led by politicians and local officers, remained at the ceremonial level and were thus ineffective. Similarly, an enlightenment campaign to promote peace, called 'Peace Caravan' (cf. Ervin 2020), organised by USAID and touring the borderlands of Samburu and Pokot, seems to be in line with ceremonial actions.

---

17 Source: interview with a Samburu elder, who directly participated in the battle on 21 Feb. 2007, in Loloki; a Samburu elder on 27 Aug. 2009, in Logorate. The information was confirmed by an interview with a Pokot elder on 9 Sept. 2012, in Longewan.

18 Source: interview with a Pokot elder on 9 Sept. 2012, in Longewan.

At a voluntary peace meeting in October 2010, politicians and leaders left early while Samburu and Pokot locals remained to discuss building peace directly for the first time since the end of the conflicts and, through their own efforts, arrived at a solution. They decided to use mobile phones to construct an inter-ethnic information exchange network to ensure peace in the area. They selected six people as inter-ethnic liaisons from each ethnic group and these, in turn, called their counterparts periodically to exchange security information. In the event of any untoward incident, the liaisons would remain in close contact with their counterparts via mobile phone. Nonetheless, it should be noted that, before this meeting, the opportunity for them to communicate rarely existed. To share a communication channel with a dangerous and unreliable enemy was an unimaginable act for both sides, which meant that the pastoralists dared to choose one of their riskiest options. The pastoralists reported that at first, the atmosphere was unfriendly and they felt afraid to communicate until they got used to it.

However, as a result, the mobile phone inter-ethnic information network has proven to be a deterrent to conflict during the post-conflict period. In the event of a minor clash, despite the risk of reinforcements being requested, the cause of the conflict is first investigated, then information is exchanged through the inter-ethnic mobile phone network. Consequently, each minor case is resolved before it can snowball into a larger conflict, thereby significantly reducing repeated retaliatory attacks. These practices are rooted in the improvised operation of local residents; no cultural or traditional roots nor outside intervention into this inter-ethnic mobile phone network can be discerned.

From meetings held between October 2010 and October 2012, the author has determined that the mobile phone inter-ethnic information network was applied in fifteen cases, and it played an important role in conflict resolution in all cases. According to local residents, inter-ethnic conflicts have decreased remarkably since the introduction of the inter-ethnic information network.

For instance, in June 2011, five raiders from the Pokot robbed the Samburu herders of 140 sheep and goats and sixteen calves. After the incident, the Samburu requested via the inter-ethnic mobile phone network that the Pokot search for the stolen livestock. Both ethnic groups cooperated in the search. As a result, all the stolen livestock were discovered and returned to their owners. The mobile phone network has also been combined with the local FM radio network of Samburu. In February 2011, four Samburu robbed a Pokot herder of eleven camels. The Samburu reported the incident using both the inter-ethnic mobile phone network and local FM radio. One Samburu who listened to the notification found the robber. The robber fled, leaving all the camels, which were all returned to the Pokot.

We must note here that countermeasures, such as banning or abstaining from using mobile phones during the conflict could have been a strong option to stop uncontrollable information flooding, which had played a crucial role in the previous conflicts; it was a risky option to communicate directly with the enemies. However, pastoralists did not choose to adopt risk-averting strategies during the peace meeting. Instead, they implemented a risk-acceptance strategy, promoting the utilisation of mobile phones as a tool of direct communication with the enemies that fuelled the conflict. The 'high input variance' of information flooding by mobile phone usage risks has been matched with the 'high process variance' of the inter-ethnic information network. It has thus successfully ensured a low and stable output variance of information flooding.[19]

Conclusion

This research indicates that the premises and backdrops of the 'resilience of pastoralists' must be carefully examined. It reveals that the primary cause of the crises was neither the propensities nor the traits of local pastoralists

---

19 The following peacebuilding effort by World Vision can be supported by the findings of this case study. 'WVK [World Vison Kenya] and another NGO helped people acquire mobile phones so that raids could be reported to authorities and mediation meetings could be facilitated. The provision of mobile phones to peace committee members in the Pokot and Marakwet communities has provided an important institutional mechanism for interrupting the retaliatory cycle of raids and counter-raids between those communities' (Siebert and Epps 2009: 29).

nor environmental and resource factors, but rather political factors outside their community. What led pastoralists to a critical situation was the identity politics of MP X and the interrelated corruption of the state. It has been proposed by numerous researchers that East African pastoralists are far more resilient in general than usually assumed (Niamir-Fuller 1998; Fratkin and McCabe 1999; Fratkin 2001; McCabe 2002; Scoones 2004; Little and McPeak 2014; Anderson and Bollig 2017; Catley 2017; Semplici 2020). This chapter, describing the resilience of displaced Samburu pastoralists facing a critical situation, during and after conflict, is supportive of this view. It suggests that the 'resilience of pastoralists' must be built on the premise that pastoralists are not the cause of crises, but rather find ways of managing variance that do not rely on either state investment or inputs from NGOs and other development agents.

This also shows that considering unpredictability and uncertainty must be prioritised to understand pastoralists who live in marginalised areas in which unavoidable and uncontrollable risk is never the exception, but is rather a day-to-day reality, as illustrated in this case study. Additionally, what pastoralists faced in the Pokot/Samburu clash were new forms of conflict accompanied by new technologies such as automatic rifles and mobile phones, both of which had never been experienced in the western side of Samburu County. Pastoralists who were actively engaged in efforts to reduce risk with a 'risk-acceptance' strategy accepted and reconfigured new technology.

It is incorrect to conclude that the 'narratives' from outside the pastoral community reflect the reality of pastoralist crises, or that pastoralists could manage new types of uncertainties and risks brought in by automatic rifles and mobile phones only with 'traditional culture' or 'indigenous knowledge' inherited over generations (see Roe 2020a: 14). Traditional cultures and indigenous knowledge are not entirely incapable of responding to new types of crises, as is illustrated in the next chapter, but there are certainly some domains where they cannot. We must also note here that the reliability management of the pastoralists does not stem from stable laws, regulations, and guidelines that the nation-state normally presupposes. Given that unpredictability and uncertainty are dominant factors in daily life, coping mechanisms based on stable system-thinking are no longer applicable; real-time improvisation to reduce risk is required. In a nutshell, the case study here illustrates that the factor ensuring pastoralists' livelihood during these crises was an improvised real-time management effort with little reference to outside stable norms or inherited culture.

Therefore, a series of theories developed by Roe is proved to have a definite potential for analysing the resilience of pastoralists under the new, unpredictable situation described here. In the conflict case study, we can observe the 'logic of high input variance matched by high process variance in order to ensure low and stable output variance' through the formation of clustered settlements and inter-ethnic mobile phone networks. As a result, they succeeded in surviving this extremely critical period of conflict under the purview of an unreliable state, which impaired their livelihood. The fact that they sustained themselves based on pastoralism through forming clustered settlements and a mobile phone inter-ethnic network illustrates a different picture of 'the resilience of the pastoralist'. This case illustrates how pastoralists went through the conflict as reliability professionals and provides a new way to think about the 'resilience of pastoralism' that focuses more on risk acceptance and real-time management.

Acknowledgements

I would like to thank all the interlocutors in the East African pastoral societies who collaborated with me. This work was supported by the JSPS KAKENHI Grants JP18H03606, JP25257005, JP16K13305, JP24651275, JP20401010 and Research Funds of the University of Shizuoka. I am very grateful to the anonymous referees of *Nomadic Peoples* and Dr. Greta Semplici for reading my manuscript and for providing significant and insightful suggestions. The early draft of this chapter was presented at the International Workshop 'Thinking Resilience and Development from the "Exceptional" Africa' held in Tokyo on February 1, 2020. I am grateful for Dr. Emery Roe and participants of the panel.

References

Adger, W. N. 2000. 'Social and ecological resilience: Are they related?' *Progress in Human Geography* 24 (3): 347–364. https://doi.org/10.1191/030913200701540465

Anderson. D. M. and M. Bolling. (eds). 2017. *Resilience and Collapse in African Savannahs: Causes and consequences of environmental change in East Africa*. London: Routledge.

Barrett, C.B., J. McPeak, W. Luseno, P.D. Little, S. Osterloh, H. Mahmoud and G. Gebru. 2004. 'Pastoralist livestock marketing behavior in Northern Kenya and Southern Ethiopia: An analysis of constraints limiting off-take rates'. *Economics Faculty Scholarship*. Paper 84, Cornell University, Ithaca, NY. https://doi.org/10.2139/ssrn.611064

Behnke, R.H. Jr., I. Scoones and C. Kerven. (eds). 1993. *Range Ecology at Disequilibrium: New models of natural variability and pastoral adaption in African savannas*. London: Overseas Development Institute.

Butt. B. 2015. 'Herding by mobile phone: Technology, social networks and the "transformation" of pastoral herding in East Africa'. *Human Ecology* 43: 1–14. https://doi.org/10.1007/s10745-014-9710-4

Catley, A. 2017. *Pathways to Resilience in Pastoralist Areas: A synthesis of research in the Horn of Africa*. Boston: Feinstein International Center, Tufts University.

Cervigni, R. and M. Morris. (eds). 2016. *Confronting Drought in Africa's Drylands: Opportunities for enhancing resilience*. Washington: International Bank for Reconstruction and Development, The World Bank. https://doi.org/10.1596/978-1-4648-0817-3

Chandler, D. and J. Coaffee. (eds). 2017 *The Routledge Handbook of International Resilience*. London: Routledge. https://doi.org/10.4324/9781315765006

Chepkwony, R., S.V. Bommel and F.V. Langevelde. 2018. 'Citizen science for development: potential role of mobile phones in information sharing on ticks and tick-borne diseases in Laikipia, Kenya'. *NJAS - Wageningen Journal of Life Sciences* 86–87: 123–135. https://doi.org/10.1016/j.njas.2018.07.007

Davies, J., L.W. Robinson and P.J. Ericksen. 2015. 'Development process resilience and sustainable development: Insights from the drylands of Eastern Africa'. *Society and Natural Resources* 28: 328–343. https://doi.org/10.1080/08941920.2014.970734

Ellis, J.E. and D.M. Swift. 1988. 'Stability of African pastoral ecosystems: Alternate paradigms and implications for development'. *Journal of Range Management* 41: 450–459. https://doi.org/10.2307/3899515

Ervin, G.M. 2020. 'Now we sleep without our shoes… The story of the Laikipia Peace Caravan'. In S.L. Connaughton and J. Berns. (eds). *Locally Led Peacebuilding: Global case studies*. Lanham: Rowman & Littlefield. pp.47–59.

Fletcher, D. and M. Sarkar, 2013. 'Psychological resilience: A review and critique of definitions, concepts, and theory. *European Psychologist* 18 (1): 12–23. https://doi.org/10.1027/1016-9040/a000124

Folke, C. 2006. 'Resilience: The emergence of a perspective for social–ecological systems analyses'. *Global Environmental Change* 16: 253–267. https://doi.org/10.1016/j.gloenvcha.2006.04.002

Fratkin, E. 1986. 'Stability and resilience in East African pastoralism: The Rendille and the Ariaal of northern Kenya'. *Human Ecology* 14 (3): 269–86. https://doi.org/10.1007/BF00889031

Fratkin, E. 2001. 'East African pastoralism in transition: Maasai, Boran, and Rendille cases'. *African Studies Review* 44 (3): 1–25. https://doi.org/10.2307/525591

Fratkin, E. and J.T. McCabe. 1999. 'Introduction' to Special Issue of Nomadic Peoples: East African pastoralism at the crossroads. *Nomadic Peoples* 3 (2): 5–15. https://doi.org/10.3167/082279499782409442

Galvin, K.A., R.S. Reid, Jr., R.H. Behnke and N.T. Hobbs. (eds). 2008. *Fragmentation in Semi-Arid and Arid Landscapes: Consequences for human and natural systems*. Dordrecht: Springer. https://doi.org/10.1007/978-1-4020-4906-4

Garmezy, N. 1971. 'Vulnerability research and the issue of primary prevention'. *American Journal of Orthopsychiatry* 41 (1): 101–116. https://doi.org/10.1111/j.1939-0025.1971.tb01111.x

Greiner, C. 2012. 'Unexpected consequences: Wildlife conservation and territorial conflict in Northern Kenya'. *Human Ecology* 40 (3): 415–425. https://doi.org/10.1007/s10745-012-9491-6

Greiner, C. 2013. 'Guns, land, and votes: Cattle rustling and the politics of boundary making in Northern Kenya'. *African Affairs* 112 (477): 216–237. https://doi.org/10.1093/afraf/adt003

Holling, C.S. 1973. 'Resilience and stability of ecological systems'. *Annual Review of Ecology, Evolution, and Systematics* 4: 1–23. https://doi.org/10.1146/annurev.es.04.110173.000245

Holtzman, J.D. 2016. *Killing your Neighbors*. Berkley: University of California Press. https://doi.org/10.1525/california/9780520291911.001.0001

IDMC (Internal Displacement Monitoring Center). 2006. *I am a Refugee in My Own Country*. Geneva: Internal Displacement Monitoring Centre.

Kaldor, M. 1999. *New and Old Wars: Organized violence in a global era*. Cambridge: Polity Press.

Keck, M. and P. Sakdapolrak. 2013. 'What is social resilience?: Lessons learned and ways forward'. *Erdkunde* 67 (1): 5–19. https://doi.org/10.3112/erdkunde.2013.01.02

Konaka, S. 2017a. 'Articulation between the material culture of East African pastoralists and non-food items of humanitarian assistance'. *African Study Monographs, Supplementary Issue* 53: 53–67. https://doi.org/10.14989/218915

Konaka, S. 2017b. 'Reconsidering the spatiality of nomadic pastoralists in East African pastoral society'. *Senri Ethnological Studies* 95: 279–292. http://doi.org/10.15021/00008587

Krätli, S. and N. Schareika. 2010. 'Living off uncertainty: The intelligent animal production of dryland pastoralists'. *European Journal of Development Research* 22: 605–622. https://doi.org/10.1057/ejdr.2010.41

Leslie, P. and J.T. McCabe. 2013. 'Response diversity and resilience in social-ecological systems'. *Current Anthropology* 54 (2): 114–143. https://doi.org/10.1086/669563

Lesorogol, C.K. 2008. *Contesting the Commons: Privatizing pastoral lands in Kenya*. Ann Arbor: University of Michigan Press. https://doi.org/10.3998/mpub.300488

Liao, C. and D. Fei. 2016. 'The operationalization of resilience in pastoral regions in East Africa and Central Asia'. In D. Chandler and J. Coaffee. (eds). *The Routledge Handbook of International Resilience*. London: Routledge. pp. 198–210.

Little, P.D. and J.G. McPeak. 2014. 'Resilience and pastoralism in Africa South of the Sahara'. In S. Fan, R. Pandya-Lorch and S. Yosef. (eds). *Resilience for Food and Nutrition Security*. Washington, DC: International Food Policy Research Institute. pp. 75–82.

McCabe, J.T. 2002. 'The role of drought among the Turkana of Kenya'. In A. Oliver-Smith and S. Hoffman. (eds). *Culture and Catastrophe*. Santa Fe: School of American Research Press. pp. 213–236.

McCabe J.T. 2004. *Cattle Bring Us to Our Enemies: Turkana ecology, politics, and raiding in a disequilibrium system*. Michigan: University of Michigan Press. https://doi.org/10.3998/mpub.23477

Niamir-Fuller, M., 1998. 'The resilience of pastoral herding in Sahelian Africa'. In F. Berkes and C. Folke. (eds). *Linking Social and Ecological Systems: Management practices for building resilience*. Cambridge: Cambridge University Press. pp. 250–284.

Oliver-Smith, A. 2017. 'The concepts of adaptation, vulnerability, and resilience in the anthropology of climate change'. In H. Kopina and E. Shoreman-Ouimet. (eds). *Routledge Handbook of Environmental Anthropology*. London: Routledge. pp. 116–136. https://doi.org/10.4324/9781315768946-17

Okumu. W. 2013. *Conflict over Ltungai Conservancy: A case of fatal competition over grazing land and water among the Samburu and Pokot in north-western Kenya*. Paper presented at The International Conference on 'Large Scale Agricultural Investments in Pastoral Lowlands of the Horn of Africa: Implications for Minority Rights and Pastoral Conflicts.' Institute of Peace and Security Studies, Addis Ababa University, Addis Ababa.

Okumu, W., K.N. Bukari, P. Sow, and E. Onyiego. 2017. 'The role of elite rivalry and ethnic politics in livestock raids in Northern Kenya'. *Journal of Modern African Studies* 55 (3): 479–509. https://doi.org/10.1017/S0022278X17000118

Panter-Brick, C. 2014. 'Health, risk, and resilience: Interdisciplinary concepts and applications'. *Annual Review of Anthropology* 43: 431–48. https://doi.org/10.1146/annurev-anthro-102313-025944

Parlasca, M. C. and O. Mußhoff, and M. Qaim. 2019. 'Can mobile phones improve nutrition among pastoral communities? Panel data evidence from Northern Kenya'. *Agricultural Economics*. 51: 475–488. https://doi.org/10.1111/agec.12566

Reinert, H., and T.A. Benjaminsen. 2015. 'Conceptualising resilience in Norwegian Sámi reindeer pastoralism'. *Resilience* 3 (2): 95–112. https://doi.org/10.1080/21693293.2014.988916

Robinson L.W. and F. Berkes. 2010. 'Applying resilience thinking to questions of policy for pastoralist systems: Lessons from the Gabra of Northern Kenya'. *Human Ecology* 38: 335–350. https://doi.org/10.1007/s10745-010-9327-1

Roe, E. 1994. *Narrative Policy Analysis: Theory and practice*. Durham: Duke University Press. https://doi.org/10.1515/9780822381891

Roe, E., L. Huntsinger and K. Labnow. 1998a. 'High reliability pastoralism'. *Journal of Arid Environments* 29: 39–55. https://doi.org/10.1006/jare.1998.0375

Roe, E., L. Huntsinger and K. Labnow. 1998b. 'High-reliability pastoralism versus risk-averse pastoralism.' *Journal of Environment & Development* 7 (4): 387–421. https://doi.org/10.1177/107049659800700404

Roe, E. 2018. *Except-Africa: Remaking development, rethinking power*. London: Routledge. https://doi.org/10.4324/9781351289887

Roe, E. 2020a. *A New Policy Narrative for Pastoralism? Pastoralists as reliability professionals and pastoralist systems as infrastructure, STEPS Working Paper 113*. Sussex: STEPS Centre.

Roe, E. 2020b. *Resilience Isn't What You Think*: https://mess-and-reliability.blog/2020/02/24/resilience-isnt-what-you-think/

Scoones, I. (ed.) 1995. *Living With Uncertainty: New directions in pastoral development in Africa*. London: International Institute for Environment and Development. https://doi.org/10.3362/9781780445335

Scoones, I. 1999. 'New ecology and the social sciences: What prospects for a fruitful engagement?, *Annual Review of Anthropology* 28: 479–507. https://doi.org/10.1146/annurev.anthro.28.1.479

Scoones, I. 2004. 'Climate change and the challenge of non-equilibrium thinking'. *IDS Bulletin* 35 (3): 114–119. https://doi.org/10.1111/j.1759-5436.2004.tb00144.x

Schlee, G. 2013. 'Territorializing ethnicity: The imposition of a model of statehood on pastoralists in Northern Kenya and Southern Ethiopia'. *Ethnic and Racial Studies* 36 (5): 857–874. https://doi.org/10.1080/01419870.2011.626058

Semplici, G. 2020. *Resilience in Action: Local practices and development/humanitarian policies. A review of resilience in the drylands of Turkana*. London and Nairobi: EU Trust Fund for Africa (Horn of Africa Window) Research and Evidence Facility.

Siebert, J. and K. Epps. 2009. *Addressing Armed Violence in East Africa: A report on World Vision peacebuilding, development and humanitarian assistance programmes*. Mississauga: World Vision Canada.

Spencer, P. 1965. *The Samburu. A study of gerontocracy.* London: Routledge. https://doi.org/10.1525/9780520337091

Straight, B. 2009. 'Making sense of violence in the "badlands" of Kenya'. *Anthropology and Humanism* 34 (1): 21–30. https://doi.org/10.1111/j.1548-1409.2009.01020.x

Straight, B. 2017. 'Uniquely human: Cultural norms and private acts of mercy in the war zone'. *American Anthropologist* 119 (3): 1–15. https://doi.org/10.1111/aman.12905

Ungar, M. 2008. 'Resilience across cultures'. *British Journal of Social Work* 38: 218–235. https://doi.org/10.1093/bjsw/bcl343

Van Baalen, S. and M. Mobjörk. 2018. 'Climate change and violent conflict in East Africa: Integrating qualitative and quantitative research to probe the mechanisms'. *International Studies Review* 20 (4): 547–75. https://doi.org/10.1093/isr/vix043

Van der Leeuw, S.E. and C.A. Leygonie. 2000. 'A longterm perspective on resilience in socionatural systems'. Paper Presented at the Workshop on System Shocks–System Resilience, Abisko. pp. 22–26. https://doi.org/10.1142/9789812701404_0013

Waller, A. 2001. 'Resilience in ecosystemic context: Evolution of the concept'. *American Journal of Orthopsychiatry* 71 (3): 290–297. https://doi.org/10.1037/0002-9432.71.3.290

Walker, B.H., D. Ludwig, C.S. Holling and R.M. Peterman. 1981. 'Stability of semi-arid savanna grazing systems'. *Journal of Ecology* 69: 473-498. https://doi.org/10.2307/2259679

Walker, B.H., C.S. Holling, S.R. Carpenter and A. Kinzig. 2004. 'Resilience, adaptability and transformability in social-ecological systems'. *Ecology and Society* 9 (2): 5. https://doi.org/10.5751/ES-00650-090205

Walker, B.H., and D. Salt. (eds). 2006. *Resilience Practice: Building capacity to absorb disturbance and maintain function.* Washington, DC: Island Press.

Walker, B.H., J.M. Anderies, A.P. Kinzing and P. Ryan. 2006. 'Exploring resilience in social-ecological systems through comparative studies and theory development: Introduction to the special issue'. *Ecology and Society* 11 (1): 12. https://doi.org/10.5751/ES-01573-110112

Weichselgartner, J. and I. Kelman. 2014. 'Geographies of resilience: Challenges and opportunities of a descriptive concept'. *Progress in Human Geography*: 1–19. https://doi.org/10.1177/0309132513518834

Witsenburg, K.M. and R.A. Wario. 2009. 'Of rain and raids: Violent livestock raiding in northern Kenya'. *Civil Wars* 11 (4): 514–38. https://doi.org/10.1080/13698240903403915

Zwaagstra, L., Z. Sharif, A Wambile, J. de Leeuw, M.Y. Said, N. Johnson, J. Njuki, P. Ericksen and M. Herrero. 2010. *An assessment of the response to the 2008–2009 drought in Kenya: A report to the European Union delegation to the Republic of Kenya.* ILRI, (International Livestock Research Institute) Nairobi: Kenya. https://cgspace.cgiar.org/bitstream/handle/10568/2057/assessment_drought_2010.pdf?sequence=3&isAllowed=y

CHAPTER

08

Contextualizing Resilience to Material Culture of Pastoralists and Humanitarian Assistance in Northern Kenya[1]

Shinya Konaka

Introduction

This chapter explores the contextualization of the resilience concept in relation to the material culture of East African pastoralists and humanitarian assistance through an ethnographic case study of Samburu, Tugen, and Ilchamus in Northern Kenya, displaced by conflict with the Pokot from 2004. For this study I employed the 'relational approach', which we proposed in the introduction to this volume.

Although the material culture of pastoralists plays an important role in sustaining livelihoods during emergencies, and is inseparable from humanitarian assistance, its significance for resilience has not been fully explored, and to some extent remains unquestioned. From a classical anthropological and archaeological perspectives, material culture refers to the physical aspects of human culture mainly embodied as artefacts. It is typically reflected in subsistence goods, household commodities, clothing, and dwellings. In ecology, resilience is most commonly understood as 'the capacity of a system to absorb disturbance and reorganize while undergoing change so as to still retain essentially the same function,

---

1 This chapter partially includes revised content previously published as Konaka (2017b, 2019, in press).

structure, identity, and feedbacks' (Walker et al. 2004: online). From this perspective, physical aspects of human-environment interactions and reactions are crucially important, and in turn so is the material culture that provides the intermediate realm, where interactions and reactions between human and environment occur.

Interstate and intrastate conflicts have frequently occurred in East African countries, particularly since the end of the Cold War, and large numbers of pastoralists have become refugees or internally displaced persons (IDPs). Even if we limit the argument to the material aspects, this type of displacement differs substantially from the normal nomadic movements of people in search of pasture and water for themselves and their animals. Fundamentally, nomadic pastoralists take all of their belongings with them when they move in search of water and pasture, but displaced pastoralists are typically forced to abandon their property when they unexpectedly evacuate. Displaced pastoralists therefore need basic commodities such as shelters, cooking and eating utensils, clothing, and food to survive.

Humanitarian assistance in East Africa has a rather long history. Since the Ethiopian famine of 1983, international aid organizations, such as the United Nations World Food Programme, the Food and Agricultural Organization, and USAID in cooperation with the national and municipal government have provided humanitarian assistance, mostly as food aid, to East African pastoral societies in times of famine. Typical humanitarian assistance to the area residents has been assumed to be food aid to combat famine. Basic commodities other than food are categorized as 'non-food items' in the humanitarian assistance practice (Sphere Project 2011) and are essential for the survival of pastoralists following a crisis.

Shifting our perspective from the humanitarian donors to the recipients, non-food items have been termed as 'material culture' in studies of the culture of East African pastoralists. There have been many studies on the traditional material culture of East African pastoralists, most of which have been in archaeology and anthropology (Robbins 1973, ; Hodder 1982, ; Kurita 1983, ; Larik 1985, 1986, 1987; Moore, 1995; Prussin 1995, 1996; Kassam and Megerssa, 1996; Bianco, 2000; Kratz and Pido, 2000; Grillo, 2014).

Notably, pastoralists' material culture has undergone considerable transformation since the colonial and postcolonial periods in East Africa. One important turn was sedentarization, or settling, predominantly from the early to mid-twentieth century, which transformed a mobile material culture into something more permanent (Prussin 1995, 1996). Another important turn was the introduction of the market economy, evidenced by the establishment of livestock markets, especially in the late twentieth century after market privatization (Konaka 1997; McPeak and Little 2006). Pastoralists sold their livestock and bought manufactured products with the profits. For instance, the traditional goat leather skirt was replaced by synthetic fibre clothes imported from Southeast Asia and traditional wooden water containers were replaced by factory-made polyethylene water containers.

Hence, the material culture of East African pastoralists is no longer 'pure', traditional, mobile, natural, or subsistence-oriented. Rather, it can be characterized as an articulation of indigenous material culture with market commodity culture (Konaka 2017a). However, little attention has been paid to the articulated interrelations of local and universal material cultures. This chapter interrogates material culture not as a purely traditional culture, but as the ongoing dynamics of such interrelationships. Taking a 'relational approach' as we proposed in the Introduction, material culture is reconfigured to refer to the incessant dynamic interrelations and interactions between 'human and things' as well as 'community and external actors', not objects *per se*, nor limited to notions of distinct traditions.

However, non-food items provided through humanitarian assistance have never been addressed in studies of the material culture of contemporary pastoralists and vice versa. From the perspective of aid recipients, both introduced non-food items and traditional material culture are equally important because both are part of their livelihood without qualification. The 'relational approach' relativizes both the humanitarian perspective and the material culture of pastoralists and places them horizontally on the same line.

Furthermore, while preceding research on the material culture of pastoralists has focused on routine circumstances in peacetime, the material culture of critical situations and uncertainty such as conflicts or disasters has received little attention, especially with regard to pastoral communities. This is even though pastoral communities are often represented as living in permanent crisis due to recurrent droughts and low-intensity conflicts, and even though uncertainty and unpredictability are central concepts in the research on pastoralists (Scoones, 1992; Krätli, 2017; Roe, 2020). Moreover, research has tended to conceptualize material culture as discrete objects without questioning their interrelationships with broader contexts such as human bodies or livestock. Thus, greater focus on the context of risk, uncertainty, unpredictability, mobility, and broader interrelationship are needed in the study of pastoralists' material culture.

Based on these concerns and orientations, this chapter interrogates the resilience of new material culture, which has been recreated as a complex of pre-existing material culture and newly introduced non-food items of humanitarian assistance in East African pastoral societies with case studies of three ethnic groups of northcentral Kenya: Samburu, Tugen, and Ilchamus. Typically, the resilience of material culture can be observed when the normal material culture is affected by disruption or destruction through natural or socio-cultural factors – mostly displacement. Therefore, the resilience of material culture will be questioned in terms of the relationships and contexts of sustainable livelihood of displaced pastoralists: What material goods do displaced pastoralists possess? What items did they carry with them in flight?

To this end, a survey to inventory the possessions of pastoral IDPs was conducted by the author. The interpretation of the survey data and contextural analysis illustrate the process of recovering household commodities among displaced pastoralists. In addition, a comparative analysis of pastoral ethnic groups reveals the culturally defined 'minimum set of possessions' that are distinctive to each ethnic group. The sets are the most indispensable items they narrowly brought when fleeing, and are presumed to reflect the most valued commodities among pastoralists. The backdrops of the minimum set will be analysed with relational ontology of pastoralists. Finally, implications from this case study for the humanitarian and development policies will be proposed.

Outline of the research

Conflict and displacement

This section outlines the backdrops of internal displacement induced by a conflict that erupted in north central Kenya, mainly between the Pokot and its neighbouring ethnic groups: the Samburu, Tugen, and Ilchamus.

All four ethnic groups included in this research are livestock keepers. The Maa[2] speaking Samburu, typically living in highland savanna at an altitude of approximately 1,500 meters, are predominantly pastoralists and the most mobile of the four groups, organized in a dual system of permanent settlements and satellite cattle camps. The Pokot and the Tugen speak the Southern Nilotic Kalenjin language[3] and like the Ilchamus (Maa speaking), grow crops (rainfed), especially maize, finger millet and sorghum, at an altitude of about 1,000 meters. Ilchamus and Tugen have more permanent settlements (Hodder 1982: 16–17) while the Pokot are more like the Samburu in this respect.

Details of the conflict were discussed in the previous chapter. Although it rapidly escalated in mutual retaliation, the conflict was triggered by a Pokot politician inciting Pokot youth to attack neighbouring ethnic groups. This conflict inflicted tremendous damage on all sides, including killings, raided livestock, and the torching of houses and household items.

The conflict between the Pokot and the Samburu broke out in 2004 and, according to my survey, the death toll currently stands at 590. The number of Samburu it displaced is estimated to exceed 22,000 (IDMC 2006: 33). This conflict has been reported and studied by numerous researchers (Ervin 2020; Greiner 2012, 2013; Holtzman 2016; Okumu 2013; Straight 2009, 2017).

---

2 A Maa dialect of the Eastern Nilotic Maa language (Sommer & Vossen 1993).
3 The label 'Kalenjin' however is a recent construct; see Lynch (2011).

During the 2000 electoral campaign, prior to the conflict, this politician (we will call him 'X'), gave a public speech claiming that the land occupied by the Samburu had once belonged to the Pokot, and promising that land taken from the Samburu would be distributed to his local supporters as a reward. In the following months he increased his popularity by inciting local administrative chiefs and local Pokot to continuous acts of violence, including raids. In December 2002 he was elected as a Member of Parliament (MP). Prior to the instigation to violence by X, Samburu and Pokot had maintained good relationships and shared grazing land for years, particularly during periods of drought.

According to my interlocutors, during the conflict X supplied his local supporters with hundreds of automatic rifles smuggled from Uganda and encouraged and sponsored their raids. Allegedly, money was also provided to bribe local police officers. More weapons were purchased from the sale of raided livestock.

When the conflict broke out, Samburu and Pokot fled to other areas and large sections of the Samburu-Pokot border area became no-man's land. After several months, the Samburu IDPs started to organize their defence so that they could return to their homeland. This was mostly done by gathering in clustered settlements: spontaneous IDP camps, exceptionally large and fortified. There might have been help from national and international aid organizations, but it was rare. In conditions of conflict, clustered settlements are both the first line of defence and strongholds of survival. There were at least 10 verified clustered settlements in the district-border area in 2010, with a total population estimated to number about 6,700.

By 2009, the conflict between Pokot and Samburu was mostly over, following political pressure from the central government as well as the strong reaction by the Samburu themselves, especially one particular clustered settlement where people had purchased four bazookas (allegedly from corrupted officers in the Kenyan police).

In the meantime, the attacks from Pokot had extended also to the Tugen and the Ilchamus in Baringo County. That side of the conflict continued even after the Pokot abandoned their plans of invading the

Samburu. Between 2005 and 2013, 22 Tugen and 10 Ilchamus had been killed, while some 10,000 had been displaced. Unlike the nomadic Samburu, Tugen and Ilchamus resided in dispersed settlements and did not form clustered settlements.

Research methodology

Two research methods were used in this research: the participant observation and qualitative analysis of anthropological research and a more quantitative survey of household possessions. Similar surveys have been conducted with East African pastoralists (Robbins 1983; Kurita 1983; Moore 1996), but not with pastoralists IDPs. Survey data were analysed with reference to additional interviews and my experience of participant observation among East African pastoralists since 1990. The survey targeted Samburu, Tugen, and Ilchamus IDPs who had been most seriously damaged by the conflicts described in the previous section, people who had lost most of their belongings. My sample included 47 households: 23 Samburu, 10 Tugen, and 14 Ilchamus. For the Tugen, the sampled households were selected from among the poorest in the area. In addition to the survey, 7 qualitative interviews were conducted, focused on material culture, thus bringing the total number of main interlocutors to 54.

In the case of Samburu and Ilchamus, this included all the households in the area of my fieldwork, who had lost all of their livestock. Livestock are pastoralists' most important possessions in terms of property, subsistence, market income, bride price, social network, prestige, and symbolic meaning. Completely losing one's livestock means losing all these things. However, most studies on pastoralists have assumed livestock holdings or ownership and little is known about ex-pastoralists who have lost all their livestock.

The survey inventoried all items within the home and surrounding area including clothing, except those items belonging to members of the household who were not present at the time of the survey. For a comprehensive understanding of material culture, both commodities obtained at market and traditional subsistence goods are inventoried. No livestock was found. First, each possession owned by each household at the time of the survey was examined, photographed and recorded. Second, semi-structured interviews were conducted in relation to each commodity and related issues. Third, the data were cross-checked and supplementary interviews were conducted. The interviews were mostly conducted with women, generally in charge of household possessions. Sometimes they were accompanied by their husbands and children. The research was carried out directly in Northern Maa dialect (Samburu and Ilchamus) and Swahili (Tugen). The correct understanding was checked through collaboration with local research assistants.

Overview of the survey

The household-possessions survey with Samburu IDPs was conducted in August and September 2011, three years after the conflict ended. These people had been displaced at different times between 2005 and 2009. Three years later, between August and September 2014, the survey was repeated, targeting the same Samburu households. This survey method aimed to demonstrate the process of recovering household commodities among the pastoral IDPs during the three years between surveys, which expectedly reflects the resilience of material culture of pastoralists. With Tugen IDPs, the survey was conducted in September 2013, approximately one year after their displacement. The survey with Ilchamus IDPs was conducted in September 2014. Most of this group was displaced in 2005 and then again in July 2014. The number of inventoried items was 567, 218, and 156 for the Samburu, Tugen, and Ilchamus, respectively. The average number of possessions per household was 24.7 for the Samburu (567 items over 23 HH), 21.8 for the Tugen (218 items over 10 HH), and 11 for the Ilchamus (156 items over 14 HH). The items taken when fleeing accounted for 24% of all items for the Samburu, 51% for the Tugen, and 45% for the Ilchamus. The higher number of items amongst the Samburu, despite their more mobile lifestyle, is likely to be due to the fact that they had more time to flee. The time taken to flee has not been strictly measured, but the attacks on Tugen and Ilchamus were the first unexpected ones on the ethnic group, which prevented them from fleeing. In contrast, the Samburu had already been attacked by the enemy several times before the incident, providing them the time to prepare for fleeing.

Recovering commodities among Samburu pastoral IDPS

Outline of the process of recovery

The process of recovering household commodities among the Samburu pastoral IDPs was analysed using the survey data in 2011 and 2014. In 2011, the average number of commodities per household was 24.7; in 2014, it was 30.0. The average increase was 10.3 and the average decrease, which was mainly due to exhaustion, was 6.0. The net increase in commodities between 2011 and 2014 was 4.7 (119%).

Before the conflict, an exhaustive household survey was conducted in 2003 on two households of Samburu which provides us with a pre-conflict baseline for comparison. The average number of household commodities was 212.5. Assuming the sample is representative of the average household, the Samburu pastoral IDPs had a mere 11.6% in 2011 and 14.1% in 2014 of the average number of commodities per household. This finding suggests that the Samburu pastoral IDPs survived with far less than the average household. In sum, the results suggest that the pastoral IDP households would be unlikely to successfully acquire commodities up to the level of the average household for many years after evacuation. However, the results also found that the Samburu pastoral IDPs were able to make slow progress in recovering household commodities.

Composition of commodities

To demonstrate what items displaced pastoralists possess, the composition of Samburu pastoral IDPs' household commodities was analysed; Table 8.1 shows the composition in 2011. The largest percentage was clothing, at 39.3%, followed by cooking and eating utensils (22.0%) and ornaments

(14.1%). Table 8.2 shows the composition of the increase of the same household commodities from 2011 to 2014. The increase rate for clothing was 60.1%, amounting to an increase of 134 items compared with 2011. The increase rate for cooking and eating utensils was 42.4% and the increase rate for water containers was 81.0% (from 21 items in 2011 to 38 items in 2014). Among the commodities, clothing had the highest number of items in both data sets and the percentage of clothing was higher in 2014 than 2011. Therefore, clothing was the most necessary commodity for the pastoral IDPs during the process of recovering household commodities.

Clothing seems to be the dominant commodity because of the socio-cultural context of Samburu and the following conditions. First, Samburu households frequently borrow and lend cooking and eating utensils, so individual households do not need to acquire all their own cooking and eating utensils. Relatively poor households borrow from their neighbours. However, lending and borrowing clothing outside a household is not

Table 8.1 Composition of Samburu pastoral IDPs' household commodities in 2011

Composition	Real number	Percentage
Clothing	223	39.3%
Cooking and eating utensils	125	22.0%
Ornaments	80	14.1%
Mat	27	4.8%
Water container	21	3.7%
Stool	20	3.5%
Hatchet	20	3.5%
Milk container	18	3.2%
Building material	10	1.8%
Stick	4	0.7%
Flywhisk	4	0.7%
Staff	2	0.4%
Bedclothes	2	0.4%
Lamp	2	0.4%
Others	9	1.6%
Total	567	100.0%

Table 8.2 Composition of the increase of Samburu pastoral IDPs' household commodities from 2011 to 2014

Composition	Real number	Percentage
Clothing	134	56.5%
Cooking and eating utensils	53	22.4%
Water container	17	7.2%
Ornaments	15	6.3%
Milk container	4	1.7%
Stool	3	1.3%
Building material	3	1.3%
Mat	2	0.8%
Rope	2	0.8%
Others	4	1.7%
Total	237	100.0%
Staff	2	0.4%
Bedclothes	2	0.4%
Lamp	2	0.4%
Others	9	1.6%
Total	567	100.0%

acceptable. Even touching the clothes of a member of the other sex is prohibited by pastoral community norms. Clothing is specific to the individuality and dignity of the persons affected.

Second, food and clothing are in the category of consumable goods. Cooking and eating utensils are necessary items, but once they were acquired at the time of evacuation, they did not need frequent replacement. Among this impoverished group of people, clothes are of poor quality and tend to wear out in a short time. Therefore, clothes can be regarded as consumable items, similar to food. In addition, in the Samburu community, clothes are gift items similar to food. Clothing also serves as a means of social communication; for example, to propose marriage, a suitor must give a cloth to the intended wife's father along with some food. In the humanitarian assistance framework, clothing is categorized as a non-food item. However, in the Samburu community, clothes are consumable items more similar to food than to non-food items.

Another important commodity is the water container. The household survey data found that the percentage of water containers increased by 81.0% between 2011 and 2014, the highest increase among all commodities. Pastoralists live in a dry climate and water containers are essential to survival. However, the durability of water containers is low under the harsh conditions of their use. Because of the need for frequent replacement, water containers are also categorized as a consumable commodity for the pastoralists.

Methods of obtaining commodities

The survey data were analysed to elicit how displaced pastoralists have recovered the possessions. Table 8.3 shows the methods used for obtaining household commodities in 2011. 'Gift items from local people' including items purchased with gift money, were the most common (30.9%), followed by 'carried items' at 24.3% and 'purchased items' at 16.9%. These three ways to obtain household commodities comprised 72.1% of the ways to do so. Items obtained from aid agencies accounted for only 9.2%, suggesting that the pastoral IDPs acquired relatively few non-food items from humanitarian assistance after evacuation.

Table 8.4 shows the Samburu pastoral IDPs' methods of obtaining household commodities from 2011 to 2014. Newly obtained items in this period amounted to 237. In 2014, about 74.7% of newly obtained commodities were 'purchased items,' which accounted for the largest percentage, followed by 'gift items from local people' (18.6%). No items were obtained from aid agencies.

The results of the two surveys were quite different. After evacuation, the pastoral IDPs had obtained household commodities mostly as gifts or purchases. Three years later, the pastoral IDPs obtained their commodities mostly by purchasing them. The 'purchased items' markedly increased from 96 in 2011 to 273 in 2014 – a 184.3% increase. In contrast, commodities obtained as 'gift items from local people' increased from 175 in 2011 to 219 in 2014 – only a 25.1% increase.

Generally, pastoral IDPs have little in the way of cash money. All the 'purchase items' reported in the survey were purchased with wages from casual labour. Several years after evacuation, some of the pastoral IDPs started working as wage labourers as, for example, night watchmen, agricultural workers, or charcoal burners. After several years, relatively affluent neighbours began to offer manual work to the pastoral IDPs such

Table 8.3 Methods used for obtaining household commodities by Samburu in 2011

Methods	Real number	Percentage
Gift items from local people	175	30.9%
Carried items	138	24.3%
Purchased items*	96	16.9%
Self-made items	55	9.7%
Items obtained from aid agencies	52	9.2%
Found items	51	9.0%
Total	567	100.0%

*Purchased with wages from casual labor.

Table 8.4 Samburu pastoral IDPs' methods for obtaining new household commodities from 2011 to 2014

Methods	Real number	Percentage
Purchased items*	177	74.7%
Gift items from local people	44	18.6%
Self-made items	14	5.9%
Found items	2	0.8%
Total	237	100.0%
Found items	51	9.0%
Total	567	100.0%

*Purchased with wages from casual labor.

as carting water, collecting firewood, and building traditional houses, all of which had previously been categorized as activities conducted by unpaid mutual aid. This change meant that the pastoral IDPs no longer relied on neighbouring and free commodities through sharing, and they had to earn money through self-help efforts. Before the conflict, pastoralists who had lost all their livestock after the drought sought casual labour as employed herdsmen. The recent emergence of casual household labour seems to be derived from the employment system of herdsmen. However, it differs from that of the herdsmen in that the wages are too small for reacquiring household livestock. The survey results also found counterevidence to the image of pastoralists as stereotypically dependent on mutual and humanitarian aid. The pastoral IDPs needed to be self-reliant and independent, even from their neighbours.

Relationship between local donors and pastoral IDPs

The survey data were analysed to reveal the relationships between local donors of commodities and the pastoral IDPs. Table 8.5 shows relationships in 2011. The 'wife's brother' was the most common type of donor, at 36.6%, and the 'wife's sister' was 23.4%, the second-highest percentage. Combined, they accounted for about 60% of local donors. Table 8.6 shows relationships with local donors of newly obtained household commodities from 2011 to 2014. The 'wife's sister' was reported as a donor by 36.4% of the pastoralist households; this was the largest percentage.

This finding suggests that female relationships are most important between local donors and pastoral IDPs. The reason for this is likely related to the residential pattern of Samburu, which is regulated by the rule of paternal exogamy. After marriage, wives usually leave their parental homes and reside in their husbands' homes. Consequently, a married woman's brothers and sisters live farther from her than do her husband's brothers and sisters. When a man is displaced, his brother's household is also very likely to be displaced, whereas his wife's brothers' and sisters' households are relatively less likely to be affected and more likely to be safe. During peaceful times, a wife's brother or sister will often visit her household for socializing and will bring gifts. Additionally, they usually give the gift to the wife, because wives are expected to keep and manage the household commodities. Therefore, female relationships are predominant among relationships between local donors and pastoral IDPs, suggesting that it is important to assess the importance of women's relationships as key element of resilience during humanitarian crises.

The uses of distributed goods of Tugen

During the period when the survey data were being collected, shortly after their evacuation, observations were made of uses of distributed goods among Tugen. The Red Cross distributed tents, water containers, saucepans, knives, dishes, cups, spoons, blankets, and packets of maize

Table 8.5 Relationships between local donors of commodities and Samburu pastoral IDPs in 2011

Relationship	Real number	Percentage
Wife's brother	64	36.6%
Wife's sister	41	23.4%
Clanmate	15	8.6%
Wife's mother	9	5.1%
Husband's brother	7	4.0%
Neighbors	7	4.0%
Husband's mother	6	3.4%
Daughter	6	3.4%
Husband's sister	4	2.3%
Husband's maternal uncle	4	2.3%
Friend	4	2.3%
Wife's father	2	1.1%
Wife's paternal uncle	2	1.1%
Wife's maternal grandmother	1	0.6%
Daughter's husband	1	0.6%
Son's wife	2	1.1%
Total	175	100.0%

Table 8.6 Relationships of local Samburu donors of newly obtained household commodities from 2011 to 2014

Relationship	Real number	Percentage
Wife's sister	16	36.4%
Husband's mother	4	9.1%
Son	4	9.1%
Wife's mother	4	9.1%
Clanmate	4	9.1%
Husband's brother	2	4.5%
Husband's friend	2	4.5%
Suitor to husband's sister	1	2.3%
Husband's sister	1	2.3%
Father (within household)	1	2.3%
Brother of wife's mother	1	2.3%
Wife's friend	1	2.3%
Election campaigner	1	2.3%
Wife's brother	1	2.3%
Daughter	1	2.3%
Total	44	100.0%

flour to the agro-pastoralist Tugen IDPs in November 2012. The IDPs used the distributed items from humanitarian agencies in their own ways. Tents were one of the distinctive aid items. Four household cases regarding tents distributed by the Red Cross in November 2012 are of interest. In the first case, after the tent was worn out (December 2012), the household constructed a traditional-style hut using the tent material (Photo 8.1). After the tent was worn out in the second case (June 2013), the household constructed an enclosure of young goats and sheep with the tent material (Photo 8.2). The third household constructed a new house when their tent wore out in February 2013 and used the tent material as a bed sheet (Photo 8.3). Last, the fourth household constructed a grazing-camp-style

Photo 8.1 Traditional house of Tugen, Baringo County, Kenya, September 8, 2013 (photographed by the author)

Photo 8.3 Bed sheet made from a tent, Baringo County, Kenya, September 10, 2013 (photographed by the author)

Photo 8.2 An enclosure of young goats and sheep, Baringo County, Kenya, September 9, 2013 (photographed by the author)

Photo 8.4 Grazing-camp-style temporary hut, Baringo County, Kenya, September 10, 2013 (photographed by the author)

temporary hut with the tent material when it was no longer usable as a dwelling (January 2013) (Photo 8.4).

The tents were distributed to assist the IDPs' construction of temporary shelters. However, these four cases indicate that the tent material was used in many unexpected ways. In fact, most of the pastoralists and agro-pastoralists have the skills to construct shelters from the vegetation around them, and tent material does not fit their construction methods. Nevertheless, the pastoralists and agro-pastoralists used the tent material for other purposes as described above. Thus, tent material is useful to the pastoralists, but not just for its original purpose; they adapt it to meet their needs.

Material culture of displacement

Material culture of displacement

This section focuses on the material culture of displacement in East African nomadic communities based on the inventory survey findings. 'Material culture of displacement' refers to the interrelationship between humans and things that emerges when people are displaced or prepare for displacement for various reasons: conflict, environmental aggravation, government policies of forced migration, and so on. Mostly, material culture of displacement can be observed through the things people brought when fleeing. Working among the Awá hunter-gatherers in Brazil, González-Ruibal et al. (2011) noted that making and using arrows is indissolubly woven with the self of Awá men who were displaced from their land, radically altering their livelihoods. However, studies on this subject are rare and none, to my knowledge, has looked at pastoralists.

To consider material culture of displacement of nomadic pastoralists, their cultural interest in livestock requires specific attention. Notably amongst pastoralists, livestock is a most valued asset but in case of conflict the fact that it is easy to move and to sell makes it an attraction for attackers. In most cases, people left it behind. They left as quickly as possible, taking with them just a few objects. In other words, the items they bring with them when fleeing are considered to embody the ultimate value of pastoral society that is more valuable than livestock. Re-analysing their household possessions of the inventory survey with the same data set revealed what items they carry with them when fleeing and leaving livestock behind.

Items carried when fleeing

This section started from a survey on the material culture of pastoralist IDP households. Most households that participated in the survey had been unable to carry money and had almost no money at the time of the survey. Normally, Samburu keep their money in the large tin box called *sanduku* (Swahili) locked with a padrock. It made it difficult to carry during the sudden attack.

Clothing and accessories comprised a significant proportion of their possessions: 95% for the Samburu, 80% for the Tugen, and 87% for the Ilchamus. Nearly all of these items were simply what people had been wearing at the time of their displacement. This corroborated the general impression that people had taken hardly anything with them when they fled. Items other than clothing and accessories amounted to 21 for the Samburu, 22 for the Tugen, and 6 for the Ilchamus.

In the case of the Samburu, these items of carried objects comprised milk containers (29%), hatchets (24%), stools (19%), and rugs (10%) (Table 8.7). The Tugen, mainly carried livestock milk containers (50%), ropes (14%), spatulas (14%), and milk container fumigators (9%) (Table 8.8). Store-bought utensils for eating and cooking comprised 44% of all items by the Ilchamus, with stools, spatulas, stirring rods, and water vessels each constituting a small percentage (Table 8.9).

Households that carried hatchets had done so to protect themselves from enemy attacks. The other items are subsistence goods of each ethnic group (Photo 8.5). Livestock milk containers are used to express or preserve milk. They are hollowed from *Commifora Africana* for the Samburu, *Psiadia punctulate* for the Tugen. Stools are used to sit on when having a guest over or doing work hollowed from *Commifora Africana*; rugs are used for sleep cut from cowhide; ropes to attach livestock and to carry firewood or food formally cut from cowhide or hide of eland currently stitched with nylon fibers extracted from sack; spatulas for cooking; stirrers for livestock milk containers to fumigate livestock milk

Table 8.7 The composition of items other than clothing and accessories carried by the Samburu during evacuation [23 households]

Composition	Real number	Percentage
Livestock milk containers*	6	29%
Hatchet	5	24%
Stool*	4	19%
Rugs*	2	10%
Cooking and eating utensils	2	10%
Water vessel	2	10%
Total	21	100%

*Ominous items to leave for the Samburu.

Table 8.8 The composition of items other than clothing and accessories carried by the Tugen during evacuation [10 households]

Composition	Real number	Percentage
Livestock milk containers*	11	50%
Rope*	3	14%
Spatula*	3	14%
Milk container fumigator	2	9%
Stirring rod	1	5%
Hatchet	1	5%
Whisk	1	5%
Total	22	100%

*Ominous items to leave for the Tugen.

Table 8.9 The composition of items other than clothing and accessories carried by the Ilchamus during evacuation [14 households]

Composition	Real number	Percentage
Cooking and eating utensils	4	44%
Stool*	1	11%
Spatula	1	11%
Stirring rod	1	11%
Water container	1	11%
Other items	1	11%
Total	9	100%

*Ominous items to leave for the Ilchamus.

and stirring rods for porridge-like foods. Spatulas, stirrers, fumigators are made of various plants. All items were produced within the ethnic group with local plant or livestock materials except stools made by the Dorobo (a minor ethnic group) that are obtained mostly through barter exchange with livestock. These results show some commonality across the different groups in the sample in the type of items carried when fleeing.

Minimum set of possessions

At the next stage of the survey, the people in the same data set were interviewed about their possessions, to understand why those particular items had been chosen at the time of displacement and which ones they perceived as particularly necessary (Table 8.10). The items listed here (perceived as necessary) are mostly consistent with the items carried by IDPs when fleeing, as revealed in the survey of material culture of each ethnic group. The concordance rate of the items perceived as necessary and items actually carried is 76.9%. Items that correspond with the survey results of material culture comprised those that each ethnic group had

| Milk Container   Stool   Rug |

| Rope   Stirring Rod   Fumigator of Milk Container |

Photo 8.5 Items carried by the pastoral IDPs, Samburu and Baringo County, Kenya, from 2011 to 2014 (photographed by the author)

ceased to use with the arrival of the market economy. Although these items have little monetary value, they play a role as subsistence goods for daily life within each ethnic group as I described in the previous section.

The Tugen case give us a clue why these items were carried. They place livestock milk containers, spatulas, and stirring rods in a particular place in their houses (Photos 8.6 and 8.7). They reported that those items are intentionally bound together and specifically placed in the house so that they can easily be carried out in case of displacement. When an emergency arises, this enables them to flee with these necessary items. It is not for simply reducing the risk of forgetting items when moving camp frequently, since the items continue to be bound together in settled houses. In each ethnic group, items that were perceived as necessary to carry when fleeing are also considered indispensable in the nomadic life of their ethnic group.

Photo 8.6 Livestock milk containers placed at a particular place in the house, Baringo County, Kenya, September 8, 2013 (photographed by the author)

Photo 8.7 Spatulas, and stirring rods placed at a particular place in the house, Baringo County, Kenya, September 10, 2013 (photographed by the author)

Table 8.10 Items generally perceived as necessary to carry by each ethnic group during evacuation

Group	Items
Samburu	Livestock milk container, rug, stool, fire rod, arrow, roof sheet.
Tugen	Livestock milk container, rope, insect repellent rod, spatulas, stirring rod.
Ilchamus	Livestock milk container, stool.

Based on interviews with informants of each ethnic group

In other words, to survive critical situations, such as conflict and drought, each ethnic group seems to have defined what could be termed as the 'minimum set of possessions' – a minimum set of items that should be carried when fleeing.

Network of relationship between humans and things

However, the role of these items as subsistence goods does not fully explain why the IDPs in each of the ethnic groups chose to bring them when fleeing. First, although each ethnic group also uses modern household items made in factories such as cups, cookpots, and knives, and no less practical and important than subsistence goods in their everyday lives, IDPs tended not to carry these items with them. Furthermore, the Samburu and Tugen IDPs had fled while carrying livestock milk containers even though almost no households were able to carry milk-able livestock with them. In fact, at the time of the survey, the livestock milk containers of these households had long been empty and were not used for their original purpose. Additionally, many of the other items they had brought with them could easily have been replaced in the IDP shelter; as long as plant or livestock materials were freely available at the destination, that was mostly possible. Therefore, the reason for carrying these items cannot be explained solely in reference to their functional roles. However, as an old Samburu woman told me, these items are more important than livestock.

> If the enemy attacks, you must flee with those items you have prepared at the time of your circumcision. When we hear the first gunshot, we grab those items first, never mind if the enemy takes your livestock away. Your items are the first things you must flee with. They are more important than livestock. (author interview with a Samburu elderly woman in March, 2016)

Most of the Samburu men and women commonly share the sentiments of the above statement, which was commonly heard in the material culture survey. Each ethnic group has specific items that they believed would bring misfortune to their owners if they were left behind when fleeing. For the Samburu, these items comprised livestock milk containers, rugs made of livestock hide, and stools. Livestock milk containers are made by the mother of boys who are to be circumcised. When the boy marries, his wife is told to cherish the livestock milk container as if it were her husband. If the husband then dies, his livestock milk container is abandoned in the bush. In the case of women, a livestock milk container is made for the wedding of a daughter who will be married. On her wedding day, the livestock milk container is smeared with red clay and then tied to her back to represent the baby she will give birth to in the future. It is believed that if a woman leaves her livestock milk container behind while fleeing, she will not conceive. Like the husband's livestock milk container, hers is also abandoned in the bush if she dies.

Rugs made of livestock hide are laid under the body of a person during circumcision, whether male or female. Since the rug is stained by the blood of the person undergoing the surgery, it is said to represent their body. The hides of animals slaughtered for a specific ceremony are treated in the same manner. If its owner dies, the rug is abandoned in the bush, like the livestock milk container. Stools are owned by the family head and used to welcome guests, for shaving the head of a person to be circumcised, and for celebrations. The stool is owned by the family head and is passed from the father to the first-born son.

For the Tugen, the items they believed would bring misfortune to their owners if they were not brought with them when fleeing comprised livestock milk containers and ropes. As among the Samburu, if the owner dies, his livestock milk container is abandoned. Only special rope made from the cow hide slaughtered for circumcision ceremonies are carried. Normal ropes were not carried when fleeing. It is believed that if the livestock milk container and the rope are left behind in the village or stolen by the enemy, their owner will never be able to obtain property in the future.

For the Ilchamus, the items they believe will bring misfortune to their owners if not carried when fleeing comprise rugs made of livestock hide. For them, leaving such rugs behind in the village when fleeing and allowing them to be burned by the enemy means a physical threat to its owner. If the owner is a woman, it is believed that she will not be able to conceive. In the Ilchamus community, if a person dies, his/her corpse is laid over the rug and buried.

Among all three ethnic groups, those items will bring misfortune if they are not brought in any cases, whether they are fleeing or just moving their camp, and whether they are forgotten, lost or broken. A Samburu woman told me that they are always careful not to have lost them within their houses, checking daily. In particular, if an enemy were to find those items after they flee, it would cause the worst misfortune. If a family was unable to bring these items with them during an enemy attack, they might return later to seek them. There are reported cases of people risking their lives to return to their former houses to search for these items and bring them back to their current shelters. In sum, all three ethnic groups have strong prohibitions against leaving such items behind.

Minimum set as a body

Our discussion has hitherto revealed that the minimum set of possessions carried by IDPs when fleeing is strongly linked to their owners' bodies. This is most notably the case for the Samburu, for whom the livestock milk container and the rug made of livestock hide are not mere items, but an integral part of the body of the owner. As an elderly Samburu woman told me:

> We, women, make a milk container as a body and proper name of our son who is supposed to be circumcised. You see my three sons all have their milk containers each stored. Wives should cherish these as their husbands' bodies. That is the reason he never leaves his milk container behind in his village when fleeing. If he does so, it seriously threatens the health of his body, due to illness, accident, or homicide. If a woman leaves her milk container behind when fleeing, it means her body is also spoiled. In that case, she will never bear a child again with her spoiled body. (author interview with an elderly Samburu woman in March, 2016)

In observing that these possessions are a part of someone's body, we may appreciate how leaving them behind when fleeing must be avoided.

As mentioned, the items carried by households that have been forced to abandon their livestock when fleeing embodies greater social value than livestock, even in pastoral communities where livestock is obviously highly valued. This value associated with the minimum set of possessions is neither monetary nor utilitarian: these possessions are seen as (quite literally) part of their owners' bodies.

Thus, the minimum set of possessions is an inseparable part of the owner's body. Does minimum set of possessions represent or symbolize certain hidden meanings? The informants told me that they carry the items, because they are part of the body, not because they 'symbolize' or 'represent' something. The important point here is that they never said 'because they symbolize the body,' although they have the equivalent vocabulary to 'symbolize.' This leads us to discard the distinction between the subject and object: between the somebody who represents something from the something which is represented by somebody. In this regards, ontological view of things gives us a clue. Ontologically speaking, things are treated as *sui generis* meanings (cf. Henare et al. 2007: 3–4). If so, we must say that the minimum set of possessions itself enunciates meanings. The importance of these items is not because they 'represent' or 'symbolize' certain hidden meanings that need to be 'interpreted' or 'decoded.' At the least, the pastoralists did not take such a view.

Thus, the minimum set of possessions for each ethnic group must be understood not only in terms of their role as subsistence goods, or their symbolic meaning, but through the ontological relationships people form with such objects within each ethnic group. As archeologists González-Ruibal et al. (2011: 2) put it, 'ontology implies a relation between humans and things that is prior to symbolization and therefore deeper and less obvious for the human actor.'

Konaka (2018) discusses in depth how East African pastoralists view the relationship between livestock and wild animals. Here, I will only mention how pets are treated when fleeing. In all three ethnic groups, cats and dogs raised as pets also play a role as a work-animal, such as watchdogs and sheepdogs. Amongst the Ilchamus, leaving behind a pregnant dog or a dog that has just delivered puppies while evacuating is believed to bring misfortune to its owner. According to several Ilchamus informants, this is because 'the dog is the eyes of human beings' and must not be left behind. In the words of a young Ilchamus woman:

> Dogs are your eyes. With dogs you can see even at night. Dogs are the eyes watching ahead of your own eyes. If you leave dogs, particularly delivered puppies when fleeing, it is a very bad sin. Your livestock might be eaten by wild animals and your children might die after birth. (author's interview with a young Ilchamus woman, March, 2016)

Like livestock milk containers and rugs, the dog is considered a part of the human body, and thus seen as inseparable from their owners.

In other words, the reason the minimum set of possessions and pets embody an ultimate value to the nomadic community is because they are not simple possessions or animals, but are understood as constituting part of their owners' bodies.

The minimum set and mobility of pastoralists

This section comprises an attempt at an ontological reflection on the East African pastoralists with special reference to mobility. East African pastoralists have been increasingly settling down. Many have modified their dwelling practices from moving camp every few months to a dual system with a more durable home moved every few years, combined with a mobile camp where most of the livestock is kept. However, the mobile lifestyle still constitutes the core of their material culture, which is immediately apparent in the material culture of displacement due to conflict. Anderson (2002) illustrated how mobility was common among Samburu, Tugen, and Ilchamus in the nineteenth century. It can be generally presumed that in an era when East African pastoralists moved frequently with their livestock and a small number of subsistence goods, the human body was inseparable from their livestock and subsistence goods, which together constituted the basic, migratory unit of movement of nomadic pastoralism.

Hazama (2015) argues that the identity of the Karimojong-Dodoth, in Uganda, is defined not in terms of a distinction between humans and non-humans, but in terms of a unit comprising humans and livestock and distinct from all the rest. He asserted that 'from the infancy, through the life cycle, the residential area [of humans and livestock] has been a place that establishes their [multispecies] identity as a member of the community and promotes socialization, because it functions as a core of the interaction between human and livestock' (Hazama 2015: 247). Thus, humans and livestock are intertwined in a common identity.

Similarly, the identity of IDPs in this research cannot be defined in terms of humans and non-humans, but as a unit consisting of humans and a minimum set of possessions on one hand, and all the rest on the other. For the pastoralists in the research, humans are unimaginable without this minimum set of possessions (fleeing with possessions, even in the face of danger), while possessions without humans are equally unimaginable (only items significant for humans are viewed as possessions). The continuous relationship between humans and things is akin to the relationship between humans and livestock.

The important point here is that the demarcating line of self-identification goes beyond the boundary of human skin and the external world. The nomadic life, in which humans as well as livestock and objects are always subject to movement, generates this body that is unrestrained by the skin. In their societies, because the self and livestock are intertwined in this manner, livestock is not an object; rather, it is a part of the self.

It is well known that East African pastoralist communities decorate themselves with extravagant – at least from the perspective of the outside observer – physical ornaments (Hodder 1982), which signifies that physical extension objects, such as beads used for embellishment, are remarkable among them. Furthermore, body modifications, such as piercing earlobes and removing teeth, which are widely observed in East African pastoralist communities, signify that human bodies are remodeled as objects. Therefore, in East African pastoralists communities, which presuppose instability and nomadism, the boundary between a human body and objects/possessions dissipates as the objects become included in the human body and vice versa. It can therefore be assumed that mobility of pastoralists is the main factor that has given rise to minimum set of possessions.

Conclusion

This chapter illustrated what possessions displaced pastoralists have and how displaced pastoralists recovered their possessions with inventory survey data. It also revealed what items they carried with them when fleeing. In conclusion, these research findings have implications for assessing humanitarian assistance frameworks for East African pastoralists in terms of resilience. The contextual and relational approach suggests that both stereotypical conceptions of material culture and the naive assumption that humanitarian assistance can be provided free from friction with local contexts are misleading. While global organizations provide 'non-food items' as humanitarian assistance to those affected by crisis based on established standards, which are in turn based on humanitarian principles supposedly common to all human beings, more attention must be paid to the complexity of material cultures among the recipients of assistance. This research observed neither purely universal humanitarian assistance nor purely indigenous material culture among the three ethnic groups surveyed. The contextual and relational approach urges us to see the interactions between universal and local contexts of materiality and beyond the conventional local-universal dichotomy. This approach could open a new horizon beyond stereotypes and prejudices towards recognizing the resilience of pastoralists and thereby improving humanitarian assistance frameworks to support pastoral IDPs.

For instance, it would be incorrect to assume that resilient nomads do not suffer from displacement. Most of the pastoralists in this research had more permanent social relationships in residential areas than had been common in the area before the mid-twentieth century. Certainly, the resilience of pastoralists is heavily dependent on their social relationships with neighbours. Therefore, we should not overlook the plight of displacement of nomadic pastoralists. It would also be mistaken to assume that pastoralists are overly dependent on mutual aid and humanitarian assistance, as even their neighbours expected pastoral IDPs to be self-reliant and independent earners. Therefore, humanitarian assistance should consider creating economic opportunities that support continuous self-help efforts among pastoral IDPs based on social relationships.

These findings suggest that a redefinition of the distribution frameworks of humanitarian assistance is required, at least regarding pastoral IDPs, as clothing and water containers are rapidly consumed and require frequent replacement. Clothing is important to maintain the pastoralists' dignity. Thus, the more frequent provision of clothing and water containers would enhance the material resilience of pastoral IDPs.

The belief that male relationships to patrilineal kinsmen are the most important element of resilience was contradicted by the data. It is reasonable that female relationships among pastoralists should be assessed as potential sources of resilience. If humanitarian agencies distributed emergency relief goods primarily to women, then it is more likely that the relief goods would be redistributed extensively and effectively.

The idea that relief goods are always used for their original purposes was overturned. Pastoralists and agro-pastoralists have used tent materials for other uses – for example, as construction materials for traditional-style houses, as enclosures for young goats and sheep, as bed sheets, and as grazing-camp temporary huts. This finding suggests that distributing humanitarian assistance items is useful for pastoralists, but not only for their original purposes: they also serve other needs. Humanitarian assistance items provide pastoralists with an opportunity to improve their lives in unexpected ways and, at the very least, it is recommended that humanitarian agencies recognize this potential and embrace another possibility for enhancing resilience.

Finally, this chapter has demonstrated that in East African pastoralists' relational ontology, humans are not fundamentally distinct from their objects and livestock. The analysis of the material culture of displacement with both qualitative and quantitative data revealed a 'minimum set of possessions' which is a part of the human body and closely related to the mobility of pastoralists.

From the humanitarian assistance perspective, the material culture of displacement described in this chapter can be seen as an important means of survival nurtured by the victims themselves independent of the assistance of external societies. For people who have lost everything, to move towards recovery by restoring their own physical image, the minimum set of possessions that has been integrated with their bodies provides a starting point of resilience that cannot be overemphasized. The informants of all ethnic group reported that they feel relieved, self-confident, good fortune, and even the completeness of self, if they can successfully carry those minimum set items when fleeing. Even if they have lost all their livestock, the minimum set of possessions can catalyze the restoration of the extended physical image and the recovery of their self-esteem over the course of time. This role cannot be substituted by tents or utensils distributed through humanitarian support. This aspect of material culture seems to have been historically nurtured as a provision of local resilience with cultural dignity against displacement caused by recurrent drought and livestock raiding – and has proved to still be functioning today. It should also be noted that resilience is deeply rooted in their mobility. I believe that these findings provide a key to rethinking humanitarian assistance in East African pastoralist communities whose livelihoods are dominated by uncertainty and unpredictability towards new frameworks rooted in cultural dignity.

Acknowledgements

I would like to thank all the interlocutors in the East African pastoral societies who collaborated with me. This work was supported by the JSPS KAKENHI Grants JP18H03606, JP25257005, JP16K13305, JP24651275, JP20401010 and Research Funds of the University of Shizuoka. The early draft of latter sections of this chapter was presented at the Panel: LL-NAS07 < order OR stability > Working with Pastoral Systems in a 'Messy' World at CASCA/IUAES2017 Conference in Ottawa. I am grateful for the organizer, Dr. Saverio Krätli and participants of the panel.

References

Anderson, D. 2002. *Eroding the Commons: The Politics of Ecology in Baringo, Kenya 1890–1963*. Oxford: James Currey.

Bianco, B. 2000. 'Gender and material culture in West Pokot, Kenya'. In D.L. Hodgson. (ed). *Rethinking Pastoralism in Africa: Gender, culture & the myth of the patriarchal pastoralist*. Oxford: James Currey. pp. 29–42.

González-Ruibal, A., A. Hernando, and G. Politis. 2011. Ontology of the self and material culture: Arrow-making among the Awá Hunter–Gatherers (Brazil). *Journal of Anthropological Archaeology* 30: 1–16. https://doi.org/10.1016/j.jaa.2010.10.001

Greiner, C. 2012. 'Unexpected consequences: Wildlife conservation and territorial conflict in Northern Kenya'. *Human Ecology* 40 (3): 415–425. https://doi.org/10.1007/s10745-012-9491-6

———— 2013. 'Guns, land, and votes: Cattle rustling and the politics of boundary making in Northern Kenya'. *African Affairs* 112 (477): 216–237. https://doi.org/10.1093/afraf/adt003

Grillo, K. 2014. 'Pastoralism and pottery use: An ethnoarchaeological study in Samburu, Kenya', *African Archaeological Review* 31(2): 105–130. https://doi.org/10.1007/s10437-014-9147-6

Hazama, I. 2015. *The Coexistence Logic of Pastoral World: An ethnographic study of Karimojong-Dodoth*. Kyoto: Kyoto University Press [in Japanese].

Henare, A., M. Holbraad, and S. Wastell. 2007. 'Introduction: Thinking through things'. In A. Henare, M. Holbraad and S. Wastell. (eds). *Thinking through Things: Theorising artefacts ethnographically*. London: Routledge. pp.1–31.

Hodder, I. 1982. *Symbols in Action: Ethnoarchaeological studies of material culture*. Cambridge: Cambridge University Press.

IDMC (Internal Displacement Monitoring Center). 2006. *I am a Refugee in My Own Country*. Geneva: Internal Displacement Monitoring Centre.

Kaldor, M. 1999. *New and Old Wars: Organized violence in a global era*. Cambridge: Polity Press.

Kassam, A. & G. Megerssa. 1996. 'Sticks, self, and society in Booran Oromo: A symbolic interpretation'. In M.J. Arnoldi, C.M. Geary, and K.L. Hardin. (eds). *African Material Culture*. Bloomington: Indiana University Press. pp. 145–166.

Konaka, S. 2017a. 'Introduction: The articulation-sphere approach to humanitarian assistance to East African pastoralists', *African Study Monographs, Supplementary issue* 53: 1–17. https://doi.org/10.14989/218918

———— 2017b. 'Articulation between the Material Culture of East African Pastoralists and Non-food Items of Humanitarian Assistance', *African Study Monographs, Supplementary Issue* 53: 53–67. https://doi.org/10.14989/218915

————— in press. 'Material culture of displacement: Ontological reflections on East African pastoral internally displaced persons', *Japanese Review of Cultural Anthropology* 23(2).

————— 2018. 'Livestock as Interface: The Case of the Samburu in Kenya', In I. Tokoro and K. Kawai. (eds). *An Anthropology of Things*. Melbourne: Trans Pacific Press. pp.241–257.

————— 2019. 'Savanna Ontology: Material culture of displacement of East African pastoral societies', In I. Tokoro and K. Kawai. (eds). *An Anthropology of Things 2*. Kyoto: Kyoto University Press. pp.103–119 [in Japanese].

Krätli, S. 2017. 'Pastoral localization of humanitarian aid: The need to re-qualify the pastoral context'. *African Study Monographs, Supplementary Issue* 53: 141–146. https://doi.org/10.14989/218909

Kratz, C. & D. Pido 2000. 'Gender, ethnicity and social aesthetics in Maasai and Okiek Bead-work'. In D.L. Hodgson. (ed). *Rethinking Pastoralism in Africa: Gender, culture & the myth of the patriarchal pastoralist*. Oxford: James Currey. pp. 43–71.

Kurita, K. 1983. 'Material culture of the Pokot in Kenya: With special reference to circulation of articles'. *African Study Monographs* 3: 87–104. https://doi.org/10.14989/67989

Larik, R. 1985. 'Spears, style, and time among Maa-speaking pastoralists'. *Journal of Anthropological Archeology* 4: 206–220. https://doi.org/10.1016/0278-4165(85)90003-0

————— 1986. 'Age grading and ethnicity in the style of Loikop (Samburu) spears'. *World Archaeology* 18(2): 269–283. https://doi.org/10.1080/00438243.1986.9980003

————— 1987. 'The circulation of spears among Loikop cattle pastoralists of Samburu District, Kenya'. *Research in Economic Anthropology* 9: 143–166.

Lynch. G. 2011 *I Say to You: Ethnic politics and Karenjin in Kenya*. Chicago: Chicago University Press.

Moore. H. L. 1995. *Space, Text, and Gender: An anthropological study of the Marakwet of Kenya*. Cambridge: Cambridge University Press.

Okumu. W. 2013. *Conflict over Ltungai Conservancy: A case of fatal competition over grazing land and water among the Samburu and Pokot in North-western Kenya*. Paper presented at the International Conference on 'Large Scale Agricultural Investments in Pastoral Lowlands of the Horn of Africa: Implications for Minority Rights and Pastoral Conflicts.' Institute of Peace and Security Studies, Addis Ababa University, Addis Ababa.

Prussin, L. (ed.) 1995. *African Nomadic Architecture: Space, Place, and Gender*. Washington: Smithsonian Institute Press.

————— 1996. 'When Nomads Settle: Changing technologies of building and transport and the production of architectural form among the Gabra, the Rendille, and the Somalis'. In M.J. Arnoldi, C.M. Geary and K.L. Hardin. (eds). *African Material Culture*. Bloomington: Indiana University Press. pp. 73–102.

Robbins, L.H. 1973. 'Turkana material culture viewed from an archaeological perspective', *World Archaeology* 5(2): 209–214. https://doi.org/10.1080/00438243.1973.9979567

Roe, E. 2020. *A New Policy Narrative for Pastoralism? Pastoralists as reliability professionals and pastoralist systems as infrastructure*, *STEPS Working Paper* 113. Sussex: STEPS Centre.

Scoones, I. (ed). 1995. *Living With Uncertainty: New directions in pastoral development in Africa*. London: International Institute for Environment and Development. https://doi.org/10.3362/9781780445335

Sommer, G. & Vossen, R. 1993. 'Dialects, sectiolects, or simply lects? The Maa language in time perspective.' In T. Spear and R. Waller. (eds). *Being Maasai: Ethnicity and identity in East Africa*. London: James Currey. pp. 25–37.

Sphere Project 2011. *Humanitarian Charter and Minimum Standards in Humanitarian Response*. Hampshire: Hobbs the Printers.

Walker, B., C.S. Holling, S.R. Carpenter and A. Kinzig, 2004. 'Resilience, adaptability and transformability in social-ecological systems'. Ecology and Society 9 (2): 5. https://doi.org/10.5751/ES-00650-090205

Man-animal
Social Relationship
as the Source
of Resilience

Itsuhiro Hazama

Introduction

Marshall (1983) defined *citizenship* as a status given to a full member of a community. Those with citizen status are equal regarding the rights and obligations associated with that status. Ideally, universal equality is achieved through citizenship. However, the realities of citizenship have developed differently across space and time. For example, the Greek concept of policing and rights depends on the exclusion and suppression of slaves, women and subjects of the Empire (Lazar 2013). This heritage persists in the uneven distribution of the benefits of citizenship in the modern state. Rosaldo (1994) noted that citizenship produces a progression from second-class to full citizenship; second-class refers to those structurally marginalised within the state system. Sassen (2002) distinguished between a citizen who is not authorised but is approved and a citizen who is authorised but not approved, and defined the latter as a full citizen who is not yet recognised as a political entity. Many groups fall into this category, including indigenous peoples and immigrants (Lazar 2013).

In 2000, the pastoralists in the Karamoja region in north-eastern Uganda experienced disarmament and the prohibition of nomadism through forced sedentarisation. Mwangu (2014) reported that this policy had become synchronous with land-grabbing, further exposing already vulnerable Karamoja pastoralists to marginalisation as second-class citizens. From the colonial period onwards, pastoralism in Karamoja and other parts of East Africa has been stereotyped as inefficient and outdated (Hazama 2012; Ohta 2009). Against a backdrop of development policies that contain human rights abuses, pastoralists' agency has been denied (HRW 2007). Their status as unapproved citizens has justified a mission of pastoralist enlightenment and edification from the colonial period to the postcolonial period (Mamdani, Kasoma and Katende 1992; Walker 2002). This study presents an ethnographic case study considering resilience as the ability to regain well-being which may have developed uniquely in the human-animal relationships of pastoralist societies, referring to how Karamoja pastoralists resist oppressive policies and organise citizenship-related practices.

Kirksey and Helmreich (2010) argued that animals exist at the periphery of anthropology as human food, as part of the landscape and symbols. However, ethnography, once limited to human subjects, has been transformed into interspecific ethnography, focusing on symbiotic connections, interconnections and non-hierarchical alliances between humans and non-human animals (Kirksey and Helmreich 2010). From the perspective of interspecific parity, the pastoral world is compatible with it. In the pastoral world, animals and humans are persons with alterity. This perspective contrasts with Western relationships with animals as human property, products to sell, sources of power and sometimes resources to preserve. The state sits outside natural society because the etic relationship between colonial and post-colonial state principals and animals excludes an emic understanding. The view of the modern state is that animals do not have independent agency, do not deserve collective rights and are of little value outside of their service to humans.

First, this study explores how Dodoth refugees from Karamoja, under poor state governance, construct citizenship autochthonously to deal with such key issues as the protection provided by the local community on behalf of the poor state – the safety net. In places where nation-state citizenship is dysfunctional, the concept of citizenship should not be limited to nation-state citizenship; the concept *per se* should be flexed and re-examined as a more comprehensive concept.

Second, there has been a rapidly growing movement in recent years to establish inclusive citizenship and achieve equality in relations (Nyamnjoh 2006; Clifford 2012; Kymlicka 2018). This is due to a mounting awareness and criticism of the history of systematic exclusion of certain subjects (e.g. people with intellectual disabilities) from traditional political processes, using the ability to manipulate language and logic as indicators. This study adopts an 'ontological framing' to question scientific epistemology which reproduces the human/nature dichotomy (Konaka and Little 2021), shows that the circle of inclusion of citizenship includes non-human entities, and sheds light on another perspective of how modern Western citizenship-based political culture has been biased towards linguistic agency (Steiner 2013).

This study is based on twenty years of research with pastoralists in the Karamoja region of Uganda from 1998 to 2017, particularly among two pastoralist communities: the Karimojong, who mainly live in the Moroto and Napak districts, and the Dodoth, who mainly live in the Kaabong district. Here, it focuses especially on the Dodoth in north-eastern Uganda. The primary data presented were obtained from participant observations in *manyatta* and kraal of the Dodoth family with whom I have lived, as well as unstructured and semi-structured interviews with their residents, refugees and key informants. These interviews focused on longer-term research issues related to an anthropological understanding of intrinsic logic in problem-solving in a low-intensity conflict and reassessment of the value of local practices for alternative development. In December 2012, 77 people outside of my host family were interviewed in Kopus about their interactions with the military after the disarmament, its impact on their lives and how they dealt with it in their families. Most of the interviews were conducted in the Dodoth language. However, the interviews with the key informants were conducted in either English or the Dodoth language. Participant observation of the Dodoth refugees was conducted in Kabekenyang on the border between Uganda and South Sudan, a hillside north-west of Mt. Zulia in 2007. Due to time and access constraints, this study does not sufficiently reflect the perspectives of the military and the central government.

Structural violence against nomadic pastoralism

The Dodoth are part of the Nilotic peoples. Their language belongs to the Eastern Nilotic language family. Their society is based on a symbiosis with domesticates in one of Africa's harshest environments. They live adjacent to the Karimojong, Jie and Teso in Uganda; the Turkana in north-western Kenya; and the Jiye, Nyangatom and Toposa in South Sudan. These groups formed a single *Ateker* political community 200 years ago that subsequently separated into different groups because of conflicts over animal raiding. Since then, these groups have been in dizzying flux, ranging from the peaceful sharing of rangelands to raiding feuds (Lamphear 1976).

Uganda's central government has had many forms, including colonial, post-independence, socialist, neoliberal, populist, and authoritarian. In the modern nation-state, central governments with a teleology that ones requires satisfying an infinite appetite for revenue have unified administrative, economic, and cultural standards by pacifying and subordinating non-national spaces and citizens while imposing social control (Scott 2009). Although ethnicity in East African pastoral society is characterised by flexibility of attribution, the concept of the territorial state is now embedded in East Africa; however, the concept of a two-dimensional surface area enclosed by a boundary remains foreign. Administrative divisions and grazing areas were established during the colonial period when tribal boundaries were imposed by the government (Gray 2000). The compartmentalisation and territorialisation of ethnicity resulted in the creation of tribal territories, supported by a framework of ecology-driven pasture conservation and the appointment of chiefs to divert control away from collective local decision-making (Schlee 2013).

Since the earliest colonial period, the modern state has been opposed to the pastoralists' ecological logic and politics. The key to Uganda's colonial policy was control over pastoralist mobility and raiding practices. With the decentralisation of state power in the twenty-first century and turning the gaze of surveillance on remote border areas of Uganda, Kenya and South Sudan, international concerns have focused on how power operates in the so-called failed and fragile states of South Sudan and Somalia. Asad (2004: 279) wrote that the margins of these nations are so unstable that state law and national order must be continuously re-established. The twenty-first century pastoral societies in East African territorial states are the core units that clarify the mechanisms – and limits – of modern politics and power.

Since 2001, severe and complicated armed conflicts and intervention by the international community have made this explicit. External intervention in armed conflict in Karamoja reflects global governance measures against transboundary terrorism networks. Immediately after the 9/11 terrorist attacks, the failed states were designated as hotbeds of terrorism. The disorder of north-east African countries, including Sudan, was emphasised; Western governments and international organisations drafted and implemented policies that included anti-terrorism activities and social development. A joint statement from bilateral Uganda–Sudan talks, hosted by the United Kingdom, and involving the United Nations and European non-governmental organisations (NGOs), positioned pastoral societies in Karamoja as rebellious groups equivalent to the Lord's Resistance Army (LRA), an anti-government organisation located in southern Sudan and northern Uganda. A cooperation target was set to eliminate the arms trade beyond the border to pastoral areas.[1] As part of its International Development Agency mission, the United States has also addressed transboundary violence in the border areas of Kenya, Uganda, Ethiopia and Sudan (i.e. the entire *Ateker* group residential area). They have directly requested President Museveni to continue supporting the disarmament of Ugandan pastoralists (Knighton 2003). Financial aid for anti-terrorism

---

1 Cattle raiding, a local arms race, the proliferation of small firearms promoted by 'cross-border crime networks', inappropriate security maintenance and insufficient national security were identified as the main factors stimulating violent conflict. The disarmament of pastoralists was part of a broader plan for regional disarmament.

social development has flowed into Uganda, which is praised for its anti-terrorism policies, and the World Bank has supported defence expenditure on disarmament. International politics accept the Ugandan government as a single entity with established social order.

Social development in Karamoja was combined with security operations, which had the ironic result of adding state violence to the mix. Domestic and international development activities accompanied by disarmament have rapidly increased in this area. For example, the number of health centres near settlements had almost doubled from twelve in 2002 to 22 in 2010 (Karamoja Data Centre 2002, UNOCHA 2010). The Karamoja Integrated Disarmament and Development Programme was remarkable for the number of military personnel involved, the size of the budget, length of the implementation period and the impact on the everyday lives of the pastoralists. The government pursued the recovery of law and order centred on sedentarisation and disarmament through compulsory detention for animals and humans at military kraals installed adjacent to the army barracks. This system is called protected kraal.[2] However, Western donors, United Nations agencies and NGOs who support the militarisation and authoritarianism of the central government are complicit in the violations of pastoralists' human rights. For example, my Dodoth host family continued to practice nomadism between kraals from 2006 to 2008 and engaged in battle on five occasions with the Ugandan People's Defence Forces (UPDF), which sought to enforce their sedentarisation in a protected kraal. After being bombed by UPDF battle helicopters, the family members heard their head saying, 'let animals be watched by the government'[3] and decided to move to a protected kraal in 2008.

---

2 There seems to be regional differences in military-related terminology. Stites and Akabwai, who conducted a survey in the district where most Karimojong and Jie reside, wrote that a kraal alongside an army barracks is called a 'security kraal' (2009). In Dodoth, the English-speaking heads of local administration and military commanders used 'protected kraal' instead of 'security kraal'.

3 Interview, Dodoth elder, Lonyia (Kalapata), 5 Aug. 2009.

Table 9.1 shows the results of a medical history survey conducted in 2011 among the Dodoth. In this survey, family patriarchs and married women were interviewed about their own and their family members' episodes of locally recognized diseases. The 10 major diseases are listed here in descending order by the number of people affected by the diseases. The data is based on the recall method, and although there are some effects of misremembering, the overall trend can be seen. The number of cases of *anakakinet* increased more than threefold in the 1990s and again in the 2000s. *Anakakinet* is caused by eating only hard grain sorghum and corn as aid food, with little or no milk or butter. It suggests the spread of aid food and the shortage of livestock products, which means the destabilization of pastoral livelihoods. The number of cases of *etukuri*, in which the sound of gunshots or exploding bombs affects the heart, and *ngikerep*, a disease of insanity, also increased sharply in the 2000s. Both diseases have a specific pathology that results in a bizarre state of consciousness. This suggests that in the 2000s, the Dodoth society suffered from serious conflicts triggered by structural violence, and that psychological tensions among the people were on the rise.

Table 9.1 Decadal change in major diseases

Local Name of a Disease	Symptom	1970-79	1980-89	1990-99	2000-09	Total
Etid	Swelling of the pancreas and liver	17	26	29	37	109
Lokou	Severe headaches	18	20	31	34	103
Erogo	Weakness, hypother-mia, diarrhea	18	24	17	22	81
Anakakinet	Constipation, leg and back pain	2	3	▲11	▲31	47
Lomagal	Rib pain	7	9	14	17	47
Ekwam	Jaundice, lethargy	9	11	12	15	47
Sil	Vascular pain	5	8	15	18	46
Ngikerep	Madness	2	3	8	▲18	31
Puulu	Rash, measles	4	7	11	5	27
Etukuri	Palpitations, fainting	0	0	0	▲8	8

▲ indicates that there has been a more than twofold increase compared to the previous decade. Figures are total number of persons.

Citizenship of a living being

In December 2006, Dodoth kraals were attacked by Turkana and the Ugandan army. The Dodoth managed to escape fifty kilometres northward to pastureland on the South Sudan border. Together with the Toposa and a different Turkana group, they constructed eight kraals that shared a single enclosure, enabling them to see each other. This group is called an *akigunya* (neighbourhood group) and functions as a joint camp of around 4,600 people, including refugees simultaneously fleeing from *manyatta* to escape everyday harassment and cordon-and-search by the army. More than half of the Dodoth became refugees during this period and fled to border areas with South Sudan, such as Nyuukuj, Kabekenyang and Lotukei, escaping repression by the military.[4] Meetings are held in the camp every morning and evening.

There are eight elected councillors to manage everyday camp issues and two representatives responsible for encouraging discussion and guiding the group towards a consensus. A shift system is operated wherein individuals from each kraal guard grazing herds during the day and in the camp overnight. There are shepherding groups, armed escort groups and night watchman groups; all pastoralists voluntarily participate as representatives, councillors or members of herding and security management organisations. Their share of the labour and security risks are equal. Representatives do not eloquently express sublime ideals or enlightenment about the virtues of citizens but play a coordinating role, helping others share a common narrative by resolving discord through humour or the remembrance of past events. There is no counterpart to the controller in this system, which emphasizes consensus building based on horizontality rather than top-down organisational hierarchy. Social life proceeds in a structure that is solely horizontal in the exchange of ideas and not coercive. Everything that happens in the operation and decisions of the community can happen naturally. Put another way, things in the community proceed in an interchangeable form, in the recognition that there are no non-exchangeable elements. For example, to overcome a potential enemy

---

4 Interview, Dodoth elder, Lopech (Kalapata), 1 Aug. 2008.

whom a traditional healer had prophetically dreamed of encountering on grazing routes, all meeting participants agreed to sacrifice a black sheep. A girl from the herd's milking group burst into tears, 'Do not kill *kirion* [the sheep's name, based on its black colour],' while others praised the owner for generously offering the sheep. Dodoth representatives uttered metaphors during this discord, such as 'There is no need to surround the small warrior [the girl who cries for the sheep] and confiscate (*aremun*) animals.' The place filled with laughter, implying that there was no choice but to postpone the sacrifice as it had similarities with military cordons. The pastoralists sought an alternative pasture ground and eventually established a different grazing route.

The collaborative camp system seeks to guarantee healthy minds and bodies through the formation, maintenance, and expansion of webs of mutual aid, and traditional welfare citizenship (as defined by Isin and Turner 2007) is practised.[5] Individuals without livestock and property, those with disabilities and the elderly can maintain a basic quality of life by borrowing lactating cows and receiving gifts through begging and food aid (cereal flour). Physical and social conditions stand as absolute disadvantages in the lives of the same fellow human beings, as the misery of the weak. This is especially true in the difficult living conditions of evacuation. Even more than in normal life, pain, hurt, and weakness convey a clear sense of the dire need for someone's tender care. Therefore, being with a defenceless human being can lead to the conviction that a being so dependent on others must itself be a sign of a good and precious existence. Disability and weakness can extend the value of human existence. Camp members with friends and relatives in places where food aid is distributed share information about food aid at the meetings. Dodoth camp members crossed the national border on foot to obtain the food distributed at a Turkana place near the Kakuma refugee camp. Inhabitants in the target area around Kakuma received international food aid every one to two

---

5 There are increasing problems reported in South Karamoja of chronically high, acute malnutrition and growing inequity of livestock ownership (Stites and Howe 2019).

months in 2007. Women and boys walk to the Kenya–Uganda border to ask for food aid from friends. Every individual returns with twenty kilograms, sometimes shared with those with disabilities, their families and the elderly in a collaborative camp.

These citizenships in cluster settlements autogenously constructed by Dodoth refugees can be evaluated as nodes connected to humanitarian assistance. Konaka (2016) concluded that, in a pastoral society embroiled in direct and structural violence, external societies should provide humanitarian assistance to internally displaced persons (IDPs) by adjusting to the citizenship developed by pastoral communities to ensure collective survival. IDPs are subject to the jurisdiction of the state in which they reside. However, a state-initiated citizenship approach cannot ensure their protection. Recognising the provisional role of citizenship practised by the Dodoth refugees may enable cooperation between international and local communities for effective humanitarian assistance during emergencies.

The collaborative camp created by Dodoth refugees is a substitute for the dysfunctional citizenship of the nation-state, satisfying the requirements for both republican citizenship, which emphasises the equal obligations of its members, and liberal citizenship, which emphasises the equal rights of its members. A collaborative camp enables the survival of pastoralists. Their moral interdependence is not based on a fixed, consistent traditional idea of social norms and obligations as a fixed member of a belief community (Thompson 1971). Instead, citizenship is practised across permeable ethnic boundaries within non-homogenised spaces of ethnicity. Pastoralists do not view the categorical unit of ethnicity as absolute and have shaken off the collective and binary thinking of 'them and us,' 'enemy and friend.' A natural empathy for the act of being alive separates the sphere of life from the political sphere, which is established by representations of the other, and creates a break in a single ethnic group. The evacuation of the Dodoth from state control into a collaborative camp led to and was supported by the construction of a new community as an attempt to determine what alternative forms of mutual behaviour were possible.

A natural empathy and resilience

Something – a living being – is struggling to survive. This commonplace reality brings a fresh sensation, a surprise to pastoralists, as in the practice of citizenship in a collaborative camp. It is also a moment of awe at the world in which one lives. There is no reason why 'something' should be limited to human beings, especially in the perceptions fostered in natural societies. In the pastoral world, animals evoke a natural empathy, or empathy as a living being, for the commonplace act of being alive.

Living with herds

Among East African pastoralists, animals are both raw materials and labour force. Animals are an extended body of pastoralists. In Dodoth, less than 20% of the plants growing on the savanna are edible to humans, but animals eat more than 60%. Each herd produces three litres of milk per person every day. The plants and trees become the bodies of the animals, and their bodies become the bodies of the pastoralists. In Karamoja, cattle, donkeys, goats and sheep have wealth and exchange value; cattle and donkeys are also working animals. They are not treated as vacant things to be manipulated at will. Animals have customary rights, can participate in ceremonies, and have the power to curse. When animals and people move from settlements to kraals during the dry season, they are blessed by elders. For the Dodoth, animals are an irreplaceable point of identification with their ancestors because who we used to be includes coexistence with animals. For pastoralists, animals embody the habitus of people as active members of society. Interspecific interaction is necessary for everyday herding. Each animal has an individual name, gender, emotions, wisdom, personality, and personhood (Hazama 2015).

Dodoth shepherds identify animals individually based on recognition of body colour, ear shape, and horn shape. There is a vocabulary of about 40 words each to describe the body colour, ear shape, and horn shape, which can be combined to categorically identify more than 1,000 animals. The memorized individuals are arranged structurally in the herder's mind, which includes dead and migrated animals. Pastoralists use this mental phylogeny to confirm the presence or absence of individuals during herding. Animals also identify humans as individuals, and the Dodoth see this as essential to the establishment of herding. Prospective shepherd boys are instructed to lead the herd to the watering hole so the animals learn to identify him. Animals prefer the cloth of the shepherd which is soaked with the scent of the family and the herd's milk, and especially the cows, who, as soon as they smell an unrecognizable person, emit a deep, low voice and try to flee.

Animal behaviourist Walther uses the term 'anima-morphism' to describe the module of understanding others that operates when nonhuman animals creatively interpret and construct the world (Walther 1991). In the workings of the mind called anthropo-morphism, humans understand other animals by ascribing human-like characteristics to them. Even on the side of non-human animals, Walther points out, they understand and act by incorporating others of different species into the same species relationship. From the human side, anthropomorphism is cast onto the animal other. The non-human animal side also throws its own anima-morphism as a module for understanding the other. In Dodoth lands, pastoral life is built on the foundation of this kind of intertwining. Biopsychologists Kaminski and his colleagues concluded in experiments at a research facility in the United States that goats do not walk in the direction of a human gaze (Kaminski et

Table 9.2 Pastoralists' calling sounds for goats and their responses

Sound	Meaning	Number of correct responses within 5 seconds (responses/call)
Ngaai	Come here.	78/111
Ai	Come back to the herd.	30/52
Tsutsui	Keep moving	71/122
Iwa	We're on the move. Gather around.	44/52
Pochu	Come to the water.	23/25
Kwaa	Stop suckling/Don't kick/ Don't move.	43/47
Iikiu	Calm down and let your kid suckle your milk.	74/77
Mee	Come and get your milk.	52/54

al. 2005). In Dodoth, cattle and goats not only stare at the point on which the shepherds have their eyes set, but they also widen their nostrils to smell and move further in that direction. Between animals and pastoralists among the Dodoth, there is joint attention clearly established, in which one's interest and attention are directed to what others are interested in and paying attention to.

During herding and milking, the pastoralists make calling sounds using tongue and gums, whistling, and vocalising. They can explain the meaning of the sound, i.e., the response they expect from the focal animal. A total of 17 different sounds were identified for goats and sheep. The most frequently heard sounds, together with the goats' responses, were observed, and on average, more than 70% of the goats responded as expected.

When a shepherd calls a personal name to cattle, the surrounding unaddressed individuals are almost completely unresponsive, whereas the focal individuals respond 60–100% accurately. The cattle are clearly aware of their names. Animals are not a silent set that is unilaterally controlled by humans. In grazing and milking, herders watch the behaviour and gestures of individual animals and vocalize instructions. Individuals that are called adjust their behaviour and mesh well with the pastoralist's instructions. Pastoralism is supported by the delicate power of interactions among animals.

Collective imagination for resilience

Individual animals have interspecific personhood because they are citizens of pastoral societies alongside humans. An expression of animal-inclusive citizenship is often observed at peace meetings (*ekisil apukan*) between the government and pastoralists. The Dodoth often consulted with the Karimojong, Turkana, Jie, district governors, council members and NGOs.[6] For the pastoralists, the focus was the insistence that the freedom and peace of animals be restored. Subjectivities were raised on behalf of the animals,

---

6 NGOs include Kaabong Peace and Development Agency, Kotido Peace Initiative and Lotus Kenya Action for Development Organization.

such as 'Animals must not be confined in one place by the government' and 'The heart of a cow detached from the family of humans and other animals is filled with suffering. Stop raiding the cow from the owner.' Statements from the pastoralists revealed that abnormal animal behaviours are a form of resistance, and animals have the agency to change disposition through the (empathic) attitude of pastoralists. This is expressed in the following case study.

In a protected kraal in Kurao, shepherds milk the animals early each morning. On the morning of 23 September 2011, a bull named Inyang made a short and low bellow at the pen's edge. Simultaneously, 200 cattle began to make sounds, jumped, rushed about, pushed, and thrust each other with their horns. Under pressure, the enclosure fence collapsed. The shepherds and young people calmed Inyang and the other animals. One young man called Lour explained the reason for the disturbance: 'Inyang remembered the smell of Namchele, and his grief was transmitted to the herd. They tried to commit suicide by grieving for their companion's death.' Namchele was a mature female shot dead by a soldier four nights earlier when she moved outside the fence. Animals frequently attempt to escape during the night because the fixed kraal is uncomfortable. Lour, other family members, and his friends call this *awi angikeyain ka echoto* or a military shithole camp. Soldiers often kill and eat such animals, saying, 'Animals must be housed within the fence at night, or the owner will be punished for neglecting his duty.' The herd's behaviour was caused by bereavement. When explaining the animals' actions, the pastoralists use 'we' instead of 'they.' People have realized that while they thought they were tracing back the history of the incident to find the cause of the animals' abnormal behaviour, they were eventually shedding light on the present circumstances of the humans.

Pastoralists claimed to abolish the military kraal regime by acting as one with their co-citizen animals, who share their lives and feelings. For the Dodoth, Karimojong and Jie, rejection of state control is a translation of animal resistance (*akitepeg*) to the military kraal regime. Thus, the pastoralists recognised the following behaviours:

Goats and sheep jump on the fence. Animals lie down, refusing to eat grass... The cows ignore the shepherd ... the animals drag themselves.

Cows and goats hide ... and try to escape. Bulls pulling an ox-plough ... will not move. Donkeys who tow building materials ... collapse, and rampage, and destroy your luggage and will bite you.

Animals refuse to return to the protected kraal... Even if you hit a cow ..., the cow ... breaks through the fence. Following the sudden command of the military to move inside, the animals ... fight with people... The cows become ferocious, especially when they hear explosions or gunshots.

When a cow owner is charged with the possession of a gun, the cows ... are left in the enclosure where they continue to vocalise and protest.

These behaviours were referred to by Scott (1985) as weapons of the weak. In small villages in Malaysia, farmers opposed to compulsory rural development practiced soft resistance such as pretending to be stupid, not taking orders, slogging, walking with a limp, not working, taking breaks, voicing complaints, stealing, destroying, running away, and using violence, to keep their rhythm of rural communal life close to nature. Although rarely organised, this kind of 'art of life' can be seen in the everyday behaviours of the oppressed and is sometimes effective (Matsuda 1999). In pastoral societies, an attitude of casual non-obedience is not limited to humans. To resist the protected kraal system, animals adopt the art of life of the oppressed. By interpreting animal behaviour, pastoralists recognise the political subjectivity of animals and represent them as co-citizens. Animal resistance and death cross species through their interdependence with humans and generate objections to modern civil society. Yet, at least in the perception of the pastoralists, the animal's voices were not heard by the state side. Instead, in meetings with the army, pastoralists' speaking for animals activated human-animal relations as the oppressed and began to strengthen cross-ethnic solidarity.

This system of protected kraal, which the pastoralists call 'military camp' or 'military shithole camp' or just 'shit camp' was terminated in 2014. One of the major turning points is the meeting between the state and the pastoralists in which the pastoralists confirmed their common understanding of the act of being alive and experienced the inability to transmit their voice to the state. In the following, I describe this gathering. There is a strong voice in Western academia that it is not right to Westernize what is non-Western, and that the translation should emphasize the otherness of the original. Here, however, I translate with emphasis on readability:

> The August 2013 meeting, in the shade of an acacia tree in Loyoro, the populated area with local administrative organs on the boundary of each pastoralist group, was attended by three governors, the commander of a division of tens of thousands of infantrymen, their men, and about 100 people each from Dodoth, Karimojong, Turkana, and Jie. The pastoralists sat on handmade palm-sized stools on the ground. The guest officials sat on wooden couches or plastic stools, facing the pastoralist groups.
>
> After the elders seated in the front row of the pastoralists stood up and delivered a diplomatic address to the state, Kayo, a Dodoth elder, began talking. 'Since the military camp was built, our animals have been unable to move their home at all. We are suffering in a swamp of their shit. By *we*, of course, I mean the animals and the people,' he said in a voice like a thunderclap.
>
> He paused until his words sank into the audience, and then lowered his voice. 'When the pastoralists want to move, they are ordered to write a letter to the governor or the army and make a request. But that is foolishness. Who else but us people of animals would know all about a safe and scenic place for our animals to settle down, and have a deep understanding of their needs?'

An elderly servant of Dodoth who speaks English, bent down and whispered into the division commander's ear. The division commander raised his left hand in mid-sentence, shooing the interpreter away, crossing his arms, closing his eyes, and began to be silent as hard granite during Keyo's speech. He was looking like sleeping.

Keyo's nephew, Lopech, spoke up. 'What the elders are saying is right,' he softly told the congregation. 'The animals that have lost their freedom are refusing the present living condition. Not a single animal has accepted the military camp. It is obvious. We can see it. When we go herding, we feel like being cattle. Because we have cattle in our blood. Those cattle are complaining of suffering.'

His younger brother, Lour, turned his back on the guests and spoke to the pastoralists. 'The camp is in the most terrible state,' he said in a voice seething with frustration. 'Cattle refuse to lie down in the shit camp. No one wants to sleep on their own shit, not so? The goats and sheep feel so bad and unbearable of it that they jump over the fence. Do these people understand?' he said, and lifted his chin towards the big guys' seats.

Another young interpreter told the guests a boldly abbreviated English translation in a mechanical voice. 'Their property is suffering from deteriorating sanitary conditions.' The seats on the state side appeared to be lined with cold stones surrounded by a hint of scorn.

A young man from Jie sat up. He turned back to official guests too and said, 'We used to live comfortably with the cows until the government built army barracks. When we are with cattle in the pasture, we feel like we are holding a baby. I wonder if the men would understand if I tell them this,' he said, and then paused. 'In the very calm wind and light, the cows are grazing, defecating in harmony. They are just being themselves. I was with the cows then.'

Another young man from Jie stood up. 'Listen,' he said, facing the pastoralists' side, 'our animals are dying one by one. Do you know what that means? Their friends, their children, their mothers are gone, their grief biting them become unbearable, and the happiness of the animals is being undermined. What do you know is their happiness?'

In response to his question, a voice from the audience responded, 'To be with friends and eat together.' Another pastoralist said, 'To be next to a friend,' and Lopech said with a deep voice, 'To sleep huddled together.'

One of the Dodoth elders sitting next to Keyo answered, 'To bathe in the sand,' while a Turkana elder whispered hoarsely, 'To feel the breeze and light,' and a Toposa elder said almost simultaneously, 'To smell and lick each other's anus and back.'

'Listen, gentlemen,' Loul said, turning to the officials, 'The military camps you are forcing on us are a death trap for our animals. We can't afford to be in such places anymore. For example, the donkey, which would not be daunted by even a slight burden, falls down quickly when an empty backpack is placed on its back, as if it is saying "I can't be of any more use to you." They ruin any other tools or possessions that humans have, including timbers, jerrycans. They kick and bite us when we try to make them stand up. We have scars inflicted by them in our whole body.'

A young Karimojong half-rose to his feet and responded, 'My cows are so angry at you, they won't listen to a shepherd calling your name.' The place is attacked by a great burst of laughter. The division commander remained as still as a stone with his eyes closed. He may have been buried in full-blown sleep.

From somewhere on the pastoralist side, a voice said, 'The cows and goats are finding their current life so hard that they hide where no one can see them and sneak away from the herd.'

'Oh my father, it's the same at my place.' Some utterances over-lapped. The crowd groans and there is in an uproar. The older and younger interpreters raised their hands to signal to the pastoralists to quiet down.

Another man from Karimojong stood up and said, 'At midnight, the cows start to taunt us. They break through the fence, run around half-crazed, and break their own legs.'

The man next to him made a terrified expression and said, 'Something similar happens when soldiers fire their guns. Our cows get paralyzed with fear, and then they turn into like insane buffaloes, stabbing anything that catches their eyes.'

The pastoralists laughed wildly and maliciously, mimicking burned-out cows. The guests ignored them completely.

The buzz of excitement and anger gradually subsided, and when it was completely quiet, Loul said, 'We don't want to live a tortured life with no peace, no comfort, or even a moment of safety.' Then, 'by *we*,' he added, 'I mean the animals and the people.'

What is most striking here is that animals have the sense to perceive the simple happiness of living and existing in this world, that, goats, cattle, donkeys, and sheep do not have any species-specific form of happiness, but only the simple happiness of what we already know, that is, the fulfilment of the basic conditions of life, such as eating and sleeping, the companionship of familiar friends, communion with nature and the body, and that they complain about the lack of such happiness and helplessness as something unfair and difficult to accept. Resilience is the ability to regain well-being, and animals teach pastoralists what well-being is. The pastoralists confirm to each other that they share a natural empathy for the act of being alive, and that such empathy cannot be shared with the state side.

Immediately after this meeting, pastoralists started visiting each other across ethnic groups almost daily and holding frequent peace meetings from below by utilizing local networks of elders among several ethnic groups, which provided an opportunity for pastoralists to share information and negotiate the launch of a self-help policing system that crossed ethnic and national borders, leading to a revival of nomadic pastoralism in 2014. Resilience is the ability to regain wellbeing, and animals have given pastoralists collective imagination to regain well-being. The pastoralists who participated in these peace meetings called each other 'people of animals' and criticized the government for being outside the pastoralist world and oblivious to the fertile intertwining world of living creatures. At the beginning of every peace meeting, a song was sung: *We will live with herds and animals, Oh, army of presidents, you go, you may live on millions of shillings*. This is a personal song (*emong*) composed in prison by a young Dodoth man named Lodul, who was arrested for fighting the government against the forced confinement of his families' cattle-herd in the protected kraal. Ultimately, the pastoralists succeeded in reviving nomadism, using this autonomous peace meeting as a starting point to create a security system that transcended national borders and ethnic groups.

Self-help security systems

In this section, I will demonstrate how animal-initiated pastoralist solidarity led to the creation of social change.

Citing Uganda as an example, Mamdani called the modern state in Africa a bifurcated state, identifying urban areas as societies ruled by civil laws created in a colonialist space, while in rural areas customary law thrives under the indirect rule of the state. He showed that the duality of the urban citizen and rural subjects under the rule of a king or chief has been a curse of the modern African state (Mamdani 1996). The pastoralists of Karamoja have never had kingdoms or chiefdoms as a system of governance. Instead, they share the history of East African pastoral societies where social groups are not fixed units defined by a single cultural and ethnic tradition (Schlee 1998; Lamphear 1993; Sobania 2011). Even in their mediation with contemporary governments, they bridged the rupture with urban citizenry through local negotiations and the horizontal production of face-to-face identities. With a nodal point of solidarity and a lifestyle that understands animal resistance and deeply values the lives of animals (Tshimba 2013), as nomadic pastoralists, they formed a coalition transcending national and ethnic boundaries from below.

The pastoralists, who resisted military-controlled kraals, continued to promote the voice of animals at the inhabitants' meetings and, in 2011, convinced the state to agree to review kraal movement restrictions. Pastoralists had only to send a letter of application to the district governor and the army to obtain permission to move kraal. However, the military refused these requests and the route, as well as the start and end times of day-trip herding, remained under military control. In August 2014, pastoralists regained their right to herding and movement self-determination. The decisive moment in the negotiations that changed government attitudes came when a security organisation combined community hegemony and external hegemony in a pluralist manner.

When the last remaining group of Dodoth in South Sudan returned to the Ugandan side of the border in August 2014, the Dodoth, Karimojong, Turkana and Jie established a joint kraal community on pasture land on the ethnic boundary. The pastoralists, including Toposa, organised 'traditional' search teams at each sub-county level where members of the same ethnic group pursued and reclaimed raided animals. The armed conflict stemmed from the escalation of raiding between pastoralists dispersed across Kenya, Uganda and South Sudan. Thus, limiting the scope of raiding deterrence to within Uganda was unlikely to effectively resolve the conflict. The northeast Ugandan escarpment is a wet savannah that provides suitable pasture land. Neighbouring pastoralists, including Toposa and Turkana, have a strong desire to share this land. The pastoralist groups, including the Karimojong, Jie, Dodoth, Turkana and Toposa, displayed solidarity across national boundaries to improve security and ensure pasture sharing. The self-help security system crossed national and ethnic boundaries and became the basis of a new mixed pastoral community system. Through a forum involving district-level local governments in Uganda, Kenya and South Sudan, as well as NGOs, agreements were made to trial a variety of mechanisms. These agreements included pastoralists reporting the proceedings of cattle-herders' meetings to the military every two weeks, participating in district-sponsored joint meetings to establish search teams for raided animals every three months and developing a system to coordinate search team activities with the government and NGOs.

When organising the search teams, pastoralists recalled the former indigenous self-governing collective practices. For example, the Dodoth referred to the practice and collective memories of their attacks on armed groups in 1988 when stolen animals were returned to their owners from different groups. The 1988 event occurred when a subset of Dodoth raided large herds of cattle belonging to the Karimojong, who had formed a collaborative kraal with the Dodoth. The Dodoth believed that recapture of the herds that their subset raided was essential for restoring security and continued pasture sharing. So, they gathered friends and young people from neighbouring villages to track the herds' hoofprints. Over 400 armed pastoralists arrived at the raiders' base. A meeting was held in the surrounding pasture ground to plan the attack and clarify the role of each participant. The attack was successful with the recovery of all the Karimojong cattle.[7]

The search team that the pastoralists organised to end the military kraal regime was called *atukot* or crowd because it was organised autonomously against raiders from within the same ethnic group. The word 'crowd' is an old term that refers to the counterraid of 1988 by former shepherds. The word was reused here as a label to cognitively establish the search teams' new institutional and social organisation of the search team. Social and political change generally occurs at a pace that exceeds the capacity of

---

7   Interview, Dodoth elders, Lokong, Loita, and Nangor (Kalapata), 1 Jan. 2015.

existing institutions and organisations to respond (Lund 2006). The new system of articulating a search team across national and ethnic borders with external hegemony was a counterstrategy that pastoralists improvised to revive nomadism in suitable pasture lands. To execute this strategy under new circumstances, the pastoral community selected from stocks of existing symbols and practices that could be used both internally and externally, and could be combined in new ways to embody a new system of thought and practice.

During the meeting with the government, all pastoralist representatives, including those from the Dodoth, structured their behaviour when speaking. They delivered individual praise for the government's disarmament and social development policies. Even shepherds who had lost family and friends in the complex violence immediately after returning their guns to the government said, 'Guns take away our lives, and that is something that prevents us from living together' and 'The government trying to eliminate them is good.' People put aside the history and logic of disobedience and resistance and behaved as citizens devoted to the state community, creating a powerful tool to ingratiate themselves with the state. Pastoralists respected the dignity of state representatives during the meeting and completely ignored them in their absence, such as in the collaborative kraals formed in border areas where national rule does not reach. Through the practice of multiple parallel structures and alternative mechanisms (e.g. plural collective decision-making, life security and communities), the concept and rationality of the unity of a singular state system can be controlled.

Conclusion

In the formation of politics, economics, culture and society, the state's anti-pastoralist policies and denial of the pastoralists' equality effectively meant that the state was no more than an animal raider of a different form (Knighton 2003). The state does not merely physically remove the animals. The state forces pastoralists to sever the qualitative relationship between animals and humans. When forced disarmament and sedentarisation were violently imposed with international cooperation, the Karamoja pastoralists did not appeal to normative ideas of human-centred citizenship but developed their own animal-inclusive citizenship practices from their position as second-class citizens. Animals resisted the lethal military camps, using the weapons of the weak, and pastoralists translated the subjectivity of animals. Animal resistance generates objections to sedentary society and the state through the pastoralists who form a symbiotic circle with animals, while the animals' voices are not heard by the granite-like state side. The agents of the state are outside of the pastoralist world, unaware of the intertwined living things that make the world fertile.

Tim Ingold (1990) points out that anthropologists have succeeded in carving humans out of nature and creating a disciplinary identity distinct from biology, in exchange for perpetuating the human-nature divide that has been a catastrophic consequence of Western civilization. The urgent task of contemporary anthropology is to reconnect human beings to the organic continuum of life to overcome the crisis of civilization. Ecological anthropology, which has achieved much in the study of rural societies in Africa, has focused on the resilience of rural societies surrounded by abundant nature, and has shown that the resilience of such societies is richly embedded in the human-nature relationship and in human society. In the case of the Dodoth, the crisis was not the result of a natural disaster but was a human-/state-made crisis based in the idea and practice of sedentarisation. The social sense of animals gave pastoralists insight to the origin of well-being and a creative solution to overcome the crisis. The resilience of pastoralist society is embedded in human-animal social relationship.

Acknowledgements

This work was supported by the JSPS [18H03606, 17H04538], and Joint Research: South Africa–Japan 'Human Resilience in the Face of Man-made and Natural Disaster in Japan and South Africa: Ethnographic Perspective'.

References

Asad, T. 2004. 'Where are the margins of the state?' In V. Das and D. Poole. (eds). *Anthropology in the Margins of the State*. Santa Fe: School of American Research Press. pp. 279–288.

Clifford, S. 2012. 'Making disability public in deliberative democracy'. *Contemporary Political Theory* 11 (2): 211–228. https://doi.org/10.1057/cpt.2011.11

Gray, S. 2000. 'The experience of violence and pastoralist identity in Southern Karamoja'. In G. Schlee and E.E. Watson. (eds). *Changing Identifications and Alliances in North-East Africa*. New York: Berghahn. pp. 73–100.

Hazama, I. 2012. 'The sequence of disarmament operations in Karamoja, Northeastern Uganda'. *Asian and African Area Studies* 12 (1): 26–60 (in Japanese).

———— 2015. *Logic of Coexistence in a Pastoral World: Ethnography of the Karimojong and Dodoth*. Kyoto: Kyoto University Press (in Japanese).

Human Rights Watch (HRW). 2007. *'Get the Gun!' Human rights violations by Uganda's National Army in law enforcement operations in Karamoja Region*. New York: Human Rights Watch.

Ingold, T. 1990. 'An anthropologist looks at biology.' *Man* 25 (2): 208–229.

Isin, E. and B. Turner. 2007. 'Investigating citizenship: An agenda for citizenship studies.' *Citizenship Studies* 11 (1): 5–17. https://doi.org/10.1080/13621020601099773

Kaminski, J., J. Riedel, J. Call, and M. Tomasello. 2005. 'Domestic goats, Capra hircus, follow gaze direction and use social cues in an object choice task.' *Animal Behaviour* 69 (1): 11–18.

Karamoja Data Centre. 2002. 'Field Survey.' http://www.karamojadata.org/index.htm (accessed 21 March 2012).

Kirksey, E. and S. Helmreich. 2010. 'The emergence of multispecies ethnography'. *Cultural Anthropology* 25 (4): 545–576. https://doi.org/10.1111/j.1548-1360.2010.01069.x

Knighton, B. 2003. 'The state as raider among the Karamojong: "Where there are no guns, they use the threat of guns"'. *Africa* 73 (3): 427–455. https://doi.org/10.3366/afr.2003.73.3.427

Konaka, S. 2016. 'Citizenship of internally displaced persons in Africa – The case of pastoral societies in East Africa'. In A. Nishikida (ed.), *Citizenship of Migrants/Refugees*. Tokyo: Yushindo (in Japanese). pp. 60–80.

Konaka, S., and P. Little. 2021. 'Introduction: Rethinking resilience in the context of East African pastoralism.' *Nomadic Peoples* 25 (2): 165-180.

Kymlicka, W. 2018. 'Connecting domination contracts'. *Ethnic and Racial Studies* 41 (3): 532–540. https://doi.org/10.1080/01419870.2018.1389968

Lamphear, J. 1976. *The Traditional History of the Jie of Uganda*. Oxford: Clarendon Press.

———— 1993. *The Scattering Time: Turkana responses to colonial rule*. Oxford: Clarendon Press.

Lazar, S. (ed.). 2013. *The Anthropology of Citizenship: A reader.* Oxford: Wiley-Blackwell.

Lund, C. 2006. 'Twilight institutions: Public authority and local politics in Africa'. *Development and Change* 37 (4): 685–705. https://doi.org/10.1111/j.1467-7660.2006.00497.x

Mamdani, M.M. 1996. *Citizen and Subject: Contemporary Africa and the legacy of late colonialism.* Princeton: Princeton University Press

Mamdani, Mahmood M., P.M.B. Kasoma, and A.B. Katende. 1992. *Karamoja: Ecology and history.* Kampala: Centre for Basic Research Working Paper No. 22.

Marshall, T.H. 1983. 'Citizenship and social class'. In D. Held et al. (eds). *States and Societies.* Martin Robertson, in association with The Open University. pp. 248–260.

Matsuda, M. 1999. *The Resisting City: From the world of Nairobi immigrants.* Tokyo: Iwanami Shoten (in Japanese).

Mwangu, A.R. 2014. 'Land-grabbing in Uganda: Are pastoralists second class citizens?' In G.B. Mulugeta. (ed.). *A Delicate Balance: Land use, minority rights and social stability in the Horn of Africa.* Addis Ababa: Institute for Peace and Security Studies. pp. 192–223.

Nyamnjoh, F.B. 2006. *Insiders and Outsiders: Citizenship and xenophobia in contemporary southern Africa.* Dakar/London: CODESRIA/Zed. https://doi.org/10.5040/9781350220775

Ohta, I. 2009. 'Pastoralists are proficient in cultivating positive social relationships: Case of the Turkana in Northwestern Kenya'. *Mila (NS)* 10: 24–38.

Rosaldo, R. 1994. 'Cultural citizenship and educational democracy'. *Cultural Anthropology* 9 (3): 402–411. https://doi.org/10.1525/can.1994.9.3.02a00110

Sassen, S. 2002. 'The repositioning of citizenship: Emergent subjects and spaces for politics'. *Berkeley Journal of Sociology* 46: 4–25.

Schlee, G. 1998. 'Some effects on a district boundary in Kenya'. In M.I. Aguilar (ed.), *The Politics of Age and Gerontocracy in Africa: Ethnographies of the past & memories of the present.* Trenton: Africa World Press. pp. 225–256.

Schlee, G. 2013. 'Territorializing ethnicity: The imposition of a model of statehood on pastoralists in northern Kenya and southern Ethiopia'. *Ethnic and Racial Studies* 36 (5): 857–874. https://doi.org/10.1080/01419870.2011.626058

Scott, J. 1985. *Weapons of the Weak: Everyday forms of peasant resistance.* New Haven, CT.: Yale University Press.

———— 2009. *The Art of Not Being Governed: An anarchist history of upland South-east Asia.* New Haven, CT.: Yale University Press.

Sobania, N. 2011. 'The formation of ethnic identity in South Omo: The Dassenech'. *Journal of Eastern African Studies* 5 (1): 195–210. https://doi.org/10.1080/17531055.2011.544542

Steiner, G. 2013. *Animals and the Limits of Postmodernism.* New York: Columbia University Press.

Stites, E. and D. Akabwai. 2009. *Changing Roles, Shifting Risks: Livelihood impacts of disarmament in Karamoja, Uganda*. Medford, MA.: Feinstein International Center, Tufts University. https://doi.org/10.1017/S0022278X18000642

Stites, E. and K. Howe. 2019. 'From the border to the bedroom: Changing conflict dynamics in Karamoja, Uganda'. *Journal of Modern African Studies* 57 (1): 137–159.

Thompson, E.P. 1971. 'The moral economy of the English crowd in the eighteenth century'. *Past and Present* 50: 76–136. https://doi.org/10.1093/past/50.1.76

Tshimba, D. 2013. '"Our" cows do matter: Arguing for human and livestock security among pastoralist communities in the Karamoja Cluster of the Greater Horn of Africa'. *IHL Paper Series* 1 (1): 1–17.

UNOCHA (United Nations Office for the Coordination of Humanitarian Affairs). 2010. *Basic Services Accessibility Atlas*. Geneva: UNOCHA.

Walker, R. 2002. 'Anti-pastoralism and the growth of poverty and insecurity in Karamoja: Disarmament and development dilemmas.' *Report to DFID East Africa*.

Walther, F. R. 1991. 'On herding behavior.' *Applied Animal Behaviour Science* 29 (1–4): 5–13.

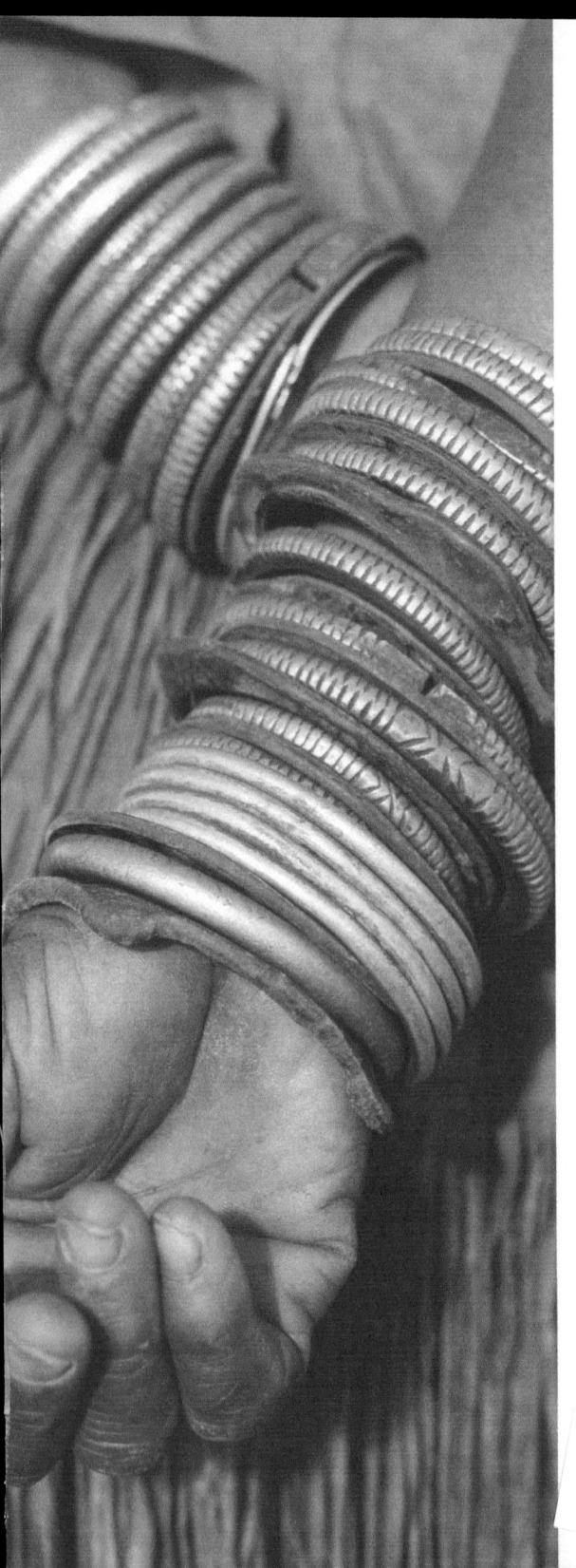

Co
Pe
o
a

10

Resilience through (Im)mobility and Patience: Kel Tamasheq in Bamako

Giulia Gonzales

In Sirakoro, a peripheral new neighbourhood of Bamako, known as 'the Tuareg neighbourhood', a family from Kidal, in northern Mali, holds a naming ceremony for a new-born baby.[1] Zeinabou, my host during fieldwork, a Kel Tamasheq woman around her mid-fifties and a long-term resident of the Malian capital, originally from the region of Timbuktu, is getting dressed for the event. The mother of the new-born baby invited her to participate in the ceremony, as both are from the Kel Ansar *tewsit*.[2] Geographically located in the region of Timbuktu, where they established their political influence since the eighteenth century (Norris 1975 : 139), the Kel Ansar claim to descend from those who welcomed the Prophet in Medina. Today, they are known for their marabouts, Islamic clerics with healing powers. In Bamako, Kel Ansar started circulating as early as the 1950s, establishing

---

1 Seven days after a child is born, the child's name is chosen, frequently in the presence of the imam, and publicly announced during the naming ceremony. It signals the baby's entrance into society (Scheele 2012: 217).

2 Pl. *Tewsiten*. A socio-political relational group, organised around notions of lineage descent and social hierarchies, galvanising identity affiliation (Diallo 2019: 87). With the concept of tribe, also fundamental to the organisation of social groups (Bonte and Ben Hounet 2009: 13), the notion of *tewsit* does not represent reified realities but unfolds in the course of history and covers political issues in its own right when it is mobilised around solidarities and affiliations. Kel Ansar also refers to a larger federation (*ettebel*) enclosing other *tewsiten* than the Kel Ansar.

connections with the administrative and civil apparatus of the state (e.g., customary officers, teachers). By circulations, I mean movements back and forward of people and their goods across the Saharan-Sahelian space, which encloses Bamako.

Zeinabou knows that her participation in the naming ceremony is expected, as is the ongoing reconstitution of the *tewsit*'s links in urban areas. Social gatherings are crucial to ensure a sense of group continuity. At the same time, for Zeinabou, the ceremony is an opportunity to mingle with other families, such as the *tewsit* of the baby's father, the Ifoghas, another important political group in the region of Kidal, which consolidated its influence between the nineteenth and twentieth centuries (Boilley 1999a). The neighbourhood of Sirakoro, since its establishment, is increasingly associated with Kidal's Kel Tamasheq, as many of them are concentrated there. So, participating in a ceremony in this neighbourhood, with members of other Kel Tamasheq families, whether from Kidal or elsewhere, provides an important occasion to keep up to date with recent circulations and arrivals to the city. And this is a valuable opportunity, especially for someone like Zeinabou who always looks for new social and political links to incorporate into her humanitarian and cultural projects, dedicated to raising awareness of the Kel Tamasheq's situation in the north of Mali.

The naming ceremony is also an opportunity for Aishata, a cousin who is temporarily staying at Zeinabou's place, and who is now in the courtyard, filling a big black plastic bag with veils and other accessorizes. She will use the ceremony to conduct her *petit commerce*, an informal type of commerce done in private spaces where goods (veils, jewellery, beauty products, etc.) are sold and advertised. At the naming ceremony, Aishata aims to sell her products, but also to attract novel affiliated costumers. Getting in the car, Zeinabou takes the front seat, next to the driver, a younger Kel Tamasheq, and checks WhatsApp messages while Aishata loads the big black sack in the trunk. It is almost 8.30 in the morning and off we go. We need to stop elsewhere on our way to the naming ceremony.

Introduction

In this chapter,[3] I would like to ponder how practices of mobility and immobility, what I refer to as (im)mobility, become resourceful strategies for a nomadic pastoralist population to embrace emerging possibilities in the urban space. These strategies of switching from mobility to immobility and vice versa can be connected to 'collective values'. In the case of Kel Tamasheq, I use the phrase 'keep on going' to encompass how their shared, collective values shape their strategies of embracing and actively engaging with an ever-changing political, environmental and socioeconomic context. In particular, I emphasize how these collective values are embodied in the concept of tazidert, or patience, which encompasses numerous sociocultural and religious dimensions of shared values. As it emerged from the fieldwork that informs this chapter, patience gives the necessary strength to 'keep on going'. I intend to nuance definitions of resilience that generally focus on mobility as the crucial element of resilience by widening the lenses of analysis and considering not mobility *per se*, but the ability to switch between mobility and immobility depending on the context. Accordingly, I argue that resilience resides in, first, the ability to switch between mobility and immobility, and, second, associated collective values. In this case, I emphasize patience, a religious and sociocultural value that is framed as a crucial trait of collective identity.

Resilience refers to "the capacity to create a fundamentally new system when ecological, economic, or social structures make the existing system untenable" (Konaka and Little 2021: 167). It counters strict binaries (local/global, human/natural, and culture/environment) that produce reductionist understandings of social and environmental phenomena, with the aim of governing them. These binaries separate objects of research and then neglect to analyse the relationships involved (Chandler in Konaka and Little 2020: 69). In this chapter, I argue that the binary mobility/immobility should also be reconsidered, conceiving them not as

---

3 This paper was written in the context of PASTRES (Pastoralism, Uncertainty, Resilience: Global Lessons from the Margins), ERC Advanced Grant (n 740342). I thank all the people who welcomed me in Bamako.

discreet states of being, but as processes that intersect in the composition of everyday practices, trajectories, and imaginings of collective life. Moving and staying put are crucial strategies for responding to volatility and embracing novel possibilities in everyday life, and as such are crucial for resilience. At the same time, resilience generates, and equally derives from, a sense of continuity to collective identity. This continuity needs to be understood in the context of historical transformation, by preserving a sense of sameness (Schlee 2017) that is entangled with moral values. This continuation is embodied by the sociocultural and religious concept of *tazidert*: to be patient is to acknowledge what life, and God, has given you and to be able to embrace it – by switching between mobility and immobility, in this case – and so to be actively engaged in life's changes. Resilience emerges in this reproduction, by transformations, of practical and representational dimensions.

Empirically, I will draw on the example of some Kel Tamasheq (aka Tuareg) families, a nomadic population, which has relied on pastoralism, caravanning, and raiding, in the Saharan drylands of northern Mali. Despite a contested history with the state, which has structurally marginalised nomadic pastoralist populations, Kel Tamasheq are now moving in large numbers to Bamako, the capital in the country's south. Due to long-term socioeconomic, political, and environmental transformations, Kel Tamasheq started perceiving Bamako as a place for potentially novel opportunities and diversification. This chapter analyses Kel Tamasheq (im)mobilities and narratives of being in Bamako as ongoing (re) articulations of Kel Tamasheq's flexible adaptation to the urban space. By flexible adaptation I mean not only how Kel Tamasheq are adjusting to the urban environment, but also how they are becoming active agents, contributing to practices of place-making as well as processes linked to broader regional trends (McDougall and Scheele 2012).

Settlement is a strategy of socio-spatial mobility, characteristic of both nomadic and sedentary environments (Grémont 2011: 181). Settlement in Bamako creates opportunities to expand and steer local and translocal comings and goings (Belalimat 2019: paragraph 48) in pursuit of individual and collective socioeconomic and political improvement. Moving to and staying put in Bamako helps to foster the "right types of connections" (Scheele, 2012: 201), crafting new ones, while nurturing or discarding older ones.

To discuss (im)mobile practices in Bamako, I examined visiting patterns among families in the urban space. Through these visiting patterns, (im)mobilities disclose the extension and organisation of people's relationalities, and the ways in which Kel Tamasheq dwell in the urban space. At the same time, Kel Tamasheq's collective enclosure of Bamako needs to be integrated into collective identity narratives, which is achieved through the concept of patience, as a collective sociocultural and religious moral value. If "resilience rests in the capacity for change, remaining open and as flexible as possible in decisions and practices [and] in the mobility of being, employed as mechanism to navigate change and different cultural environments" (Semplici, 2021: 248), I suggest that the combination of practices of (im)mobilities of Kel Tamasheq in Bamako and their representations of their nomadic pastoralist and Muslim identity as patient contributes to a broader definition of resilience, where the point is not to be mobile or immobile, but to embrace a practical and conceptual openness to life's changes. To support my argument, I will first present the debate around current patterns of (im)mobilities in the Saharan-Sahelian context.

Shifting terrains: Patterns of circulations and fixations in the Sahara-Sahel

Literature on Kel Tamasheq mobilities has focused mainly on pastoral, seasonal, and forced migration (Bourgeout 1990; Claudot-Hawad 2002; Diallo 2018; Grémont 2014; Oxby 2014). These studies on Kel Tamasheq movements emphasize emic understandings of (im)mobility that do not prioritise place over movement in the making of Saharan-Sahelian social realities (Boesen et al. 2014). Rather, mobile patterns in response to contextual volatility and the ongoing care of social connections coalesce in Kel Tamasheq mobile identities and practices. Based on a rationale of mobility and stasis, what Rossi defines as kinetocracy, a "power [which] comes to be expressed primarily as control over one's own and other people's movements" (Rossi 2015: 149), Kel Tamasheq mould their socio-moral stances, craft political alliances, and ensure the continuation of their livelihoods (Oxby 2014: 118–121).

Mobility is key in navigating environmental fluctuations, and access to resources has been explained in terms of (social and spatial) flexibility and fuzzy boundaries in pastoral practices (Homewood 2008: 3). However, shifts in land tenure systems throughout the continent, associated with colonial and postcolonial centralisation of powers, recent privatisation, and market integration of pastoral products (Homewood 2008; Kohler 2020; Nori 2022), have influenced pastoral and related flexibilities and (im)mobilities. As a result, nomadic and pastoral flexibility to engage with variabilities has been limited, severely undermining their well-being. The compelling drive to act, sometimes violently, against such odds, caused by environmental as much as by political issues, "lies not only in the place of land as a basic need, but in the question of when, where and how social actors and groups are able to act collectively and violently on this claim, and to connect to other agendas that go beyond the question of access to and control over land" (Chauveau et al. 2020: 13, my translation). Throughout the Sahel, both pastoralist and sedentary people are attempting to secure their emplacement with claims of autochthony (Geschiere, 2005: Kohler, 2020: 98; Lentz, 2013; Poudiougou and Zanoletti, 2020), stressing "the importance of being the first to arrive" (Kohler, 2020: 47).

A parallel can be drawn to Kel Tamasheq's recent history of migration to Bamako. Historically, Kel Tamasheq's political and socio-economic networks did not extend into today's southern Mali, as they were not present in the southern regions. Decolonisation (in 1960) did not provide a bridging moment that might have unified southern Mali, around the capital, Bamako, and the north of the country, where populations, including Kel Tamasheq, became marginalised from the state's structures (Sèbe 2015). This division was long lasting: during and after the Sahelian droughts of the 1970s and 1980s, which disrupted pastoralist livelihoods, only a minority of Kel Tamasheq migrated to Bamako. The majority preferred to move north towards the Maghreb in response to political, socio-economic, and environmental variabilities. As already mentioned, many effective strategies for tackling variabilities were undermined by colonial and postcolonial powers which greatly affected the political and socioeconomic arrangements of the region, especially by defining and governing populations and processes through binaries (sedentary/nomadic, agriculturalist/pastoralist). In these circumstances, Bamako and southern Mali remained unappealing migratory destinations for Kel Tamasheq.

The situation changed slightly after the second Kel Tamasheq rebellion against the Malian state in the early 1990s. Following processes of democratization and decentralization, Mali integrated ex rebels into its civic and military apparatus while a small urban-based elite of Kel Tamasheq occupied prestigious positions within the state. This created a shift in Kel Tamasheq's migratory processes towards Bamako: those few families that were already in Bamako invested in land and constructed houses in the capital, which facilitated kin and friends' arrivals and circulations through the city. Prospects for jobs, to build networks with non-profit organisations to support development projects in the north, to access sanitary services, or to generally improve life's prospects, drew growing numbers of Kel Tamasheq to Bamako. Nevertheless, most Kel Tamasheq remained marginal to the state and politically disenfranchised, while iterative and extensive pastoralism continued to be penalised by international and national livestock markets and commodification systems. Simultaneously, integration into a globalised world translated into significant capital flows into the region (through tourism, humanitarian aid and smuggling), and the benefits were captured very unevenly by different actors in society, fostering inequalities (ibid.).

In 2012, Kel Tamasheq led another rebellion against the state (Lecocq et al., 2013). They chased the Malian state out of northern Mali, proclaiming the region as the independent state of Azawad. Violence broke out and people fled to neighbouring countries (in Bamako, Kel Tamasheq and Arabs, racially perceived as white, left for fear of retaliation); a *coup d'état*, sparked by the government's inappropriate handling of the situation, led to institutional chaos; regional armed Muslim movements consolidated their power in the north of Mali; regional and international military interventions, following the paradigm of 'the war on terror', began in 2013. Armed Muslim movements were dispersed in the region and in 2015, the government, pro-independence and pro-state movements signed the Algiers Accord (Dida Badi 2013; Lecocq et al. 2013; MNLA 2012). Refugees returned (UNHCR 2015), but the situation on the ground remains extremely volatile, with entrepreneurial proliferations of violence (Poudiougou and Zanoletti 2020). Meanwhile, Bamako once again attracts growing numbers of Kel Tamasheq (Gonzales, forthcoming) in hope of a better life and gaining support for their claims of autochthony in northern Mali, it also triggers feelings of estrangement and alienation due to its conflictual history.

Dwelling in Bamako is part of a broader nomadic and pastoralist tendency towards employing a certain degree of 'being fixed' in space. According to Grémont, Kel Tamasheq's spatial fixation follows a rationale of mobility and stasis, which comprises a connubium of greater patterns of mobilities, while at the same time, a relative anchoring in a place where socio-economic and political connections are consolidated (2011: 181). These dynamics have been observed mainly in rural settlements and villages, the so-called process of *villagisation* (Lunacek 2020), while Kel Tamasheq rural-urban migration has been overlooked. Moving through and within Bamako enables consolidation of social connections and access to resources that are not necessarily located in the capital city, but that can be unlocked by mobilising contacts there. Flexible practices of (im)mobilities can thus be seen as responses to volatilities and increased political and environmental variabilities.

Travelling with Zeinabou: Reflecting on (im)mobilities through visiting

To frame (im)mobile habits, I lean on daily visiting patterns and the concept of *tbushak*. The latter refers to the period when a married woman goes to visit her family, passing from one nomadic camp to another, weaving together genealogical, political and geographical dimensions of mobile identities (Bonte 1986). In Bamako, *tbushak* acquires novel meanings: it is not exclusively a gendered practice, and it concerns shorter visits, even of a couple of days. This resignification mirrors changes in patterns of (im)mobilities (Grémont 2014), where spatial mobilities, settled presence and an overall increase in exchanges embody urban and extra-urban visits, while movement remains a daily strategy to tackle variabilities, including nurturing or discarding connections. Analytically, visiting practices interlace with social ceremonies and celebrations of life events, such as the naming ceremony described at the beginning of the chapter.

What follows is a continuation of the introductory vignette, describing a typical day travelling with Zeinabou and providing the frame for understanding how urban patterns of movement are structuring Kel Tamasheq's adaptation to, and inclusion of, Bamako within their socio-economic and political spheres. It then provides a ground to reflect on the patterns of presence and flexible adaptation of Kel Tamasheq in and with Bamako, their incorporation of the city into their regional networks, by accessing the many opportunities that the city offers and by participating in co-produced patterns of place-making. I take these patterns as examples of resilience, because, by a spatial expansion of their patterns of (im)mobilities, Kel Tamasheq adapt to historical and environmental contexts which are rapidly changing.

Before entering this analysis, though, I want to explain how I met Zeinabou in Ouagadougou in 2014, when I was researching the conditions of Kel Tamasheq refugees. Her contact was given to me by an English music producer with whom she had worked. She immediately became my hostess and translator, and our exchanges continued even after her return to Mali in 2015. In 2017, as I began fieldwork for my Phd thesis in Bamako, I relied on her assistance to start my research. My fieldwork lasted 10 months (between 2017 and 2019), for which I spent several nights per week at people's houses. Zeinabou's house was where I stayed first. She has been gravitating near Bamako since a young age to go to school, working in the humanitarian sector. After marrying her husband, a member of the military, she built a house in Bamako in the early 2000s. She then lived throughout Mali, and nurtured her social networks to continue her engagement in cultural and humanitarian/development projects. Becoming acquainted with Bamako and Kel Tamasheq families through her experience and networks has been extremely enriching for me. She provided access to her social network and, through this, I developed my own network of acquaintances, informants, and friends, and learnt to move autonomously within Kel Tamasheq's enclaves of different geographical, socio-economic or political origins (Gonzales 2020).

Practices of mobility by visiting

To avoid traffic jams on the way to the naming ceremony in Sirakoro, the car takes a dusty and bumpy road that that passes through the center of the neighbourhood. However, passing through Yorodjambougou, the neighbourhood before Sirakoro, the car makes an important detour to the house of Zeinabou's brother. The house is quiet, and the brother is relaxed in the courtyard taking his morning tea. In the saloon, a vibrant situation welcomes us. Zeinabou's sister, arrived from the other side of the city, and her sister-in-law are chatting lively, while, on their side, a plate of grilled meat, a left-over from the previous day's celebration of Zeinabou's brother's arrival, is assaulted by a teeming group of children enjoying this mouth-watering breakfast. Zeinabou arrives and, with a quick sweeping gesture, drives away the children and sits down, finishing the rest of the plate and joining the discussion. The women were waiting for Zeinabou to come and leave all together for the naming ceremony. Waiting, they spent their time chatting about Bamako's news, concerning another relative's arrival in Bamako: an old man from the Kel Ansar arrived from Mberra, the refugee camp in Mauritania. He has been hosted by his daughter, who lives on the other side of the street. If it had not been for his daughter Faata convincing him to come to Bamako to get medical treatment for his aggravating sickness, he would still be in the refugee camp and involved in schooling programs as a retired teacher, the women comment. Following Kel Tamasheq norms of hospitality, Zeinabou and her sister should have already greeted him in Bamako, but neither had yet managed to do so as: their houses are not close by, but are located in other neighbourhoods. Now, at their brother's house, just a street away from Faata's house, the two women decide it is the perfect time to strategically pay a visit. After the first glass of tea following the meat, the female convoy leaves the house, saluting Zeinabou's brother who will remain at home all day, receiving friends and relatives who will come to greet and welcome him back to Bamako.

Arriving in front of Faata's house, we see a parked pick-up. A child sneaks his head out of the gate, as he hears the engines of the arriving cars. In the court, Faata, the owner of the house, is making soured milk, mixing it by pouring it again and again from one plastic jar to another. She welcomes us, pointing at the living room, where her father is. In the semi-darkness of the room, two figures of men lie on the ground, leaning on some pillows. One must be a visitor, maybe the owner of the pick-up, because immediately after us, a child enters the room, bringing him a cup of soured milk, a custom of hospitality on someone's arrival. The visitor is also a Kel Ansar: he arrived this morning from Magnambougou, a neighbourhood on the right side of the river, to welcome the old man. Like Zeinabou and her sister, he had not yet managed to come all the way to Faata's house, because of other urban commitments. Our women's group, now composed by Zeinabou, her sister, her sister-in-law, Aishata and myself, sits down with the two men, forming a large circle. They exchange salutations, greetings, and some laughs.

After a few exchanges, the women stand up and leave, going outside to join Faata, who now serves us soured milk. She gives a small record of her father's state of health, expressing both her relief on having him there, and her latest hardship since his arrival. Indeed, his presence entails a continuous flux of visitors, and she, as the owner of the house, must welcome them with full hospitality (providing soured milk, tea, etc.). This is an intense activity, which prevents her from joining Zeinabou's caravan to the naming ceremony today. She must stay at home to receive guests. And this, she comments, negatively impacts her *petit commerce*. Like Aishata, Faata sells goods (mainly veils and beauty products) and her customers need to be continuously engaged with. Missing social events around the city can result in losing some old clients, and failing to attract newer ones. In Bamako, you need to move to make things happen (Gonzales, forthcoming). However, she acknowledges that having visitors all day long has triggered some sales. Furthermore, staying at home meant that she could leave her boutique (a small room with entrances outward to the road and inward to the courtyard) open, providing a more continuous service to the neighbours.

Our visit does not last long but, before leaving, Faata announces that, soon, her house will host another member of their extended family, a marabout, coming also from Mberra. His son, the child who snuck out from the door, is also Faata's nephew from his mother side. The marabout will join his son in Bamako, maybe seeking novel opportunities there. After Zeinabou's sister buys a hand-cream from Faata, we leave.

Our destination is only ten minutes away: from Yorodjambougou we take the only paved road of Sirakoro, stretching to the outskirts of the city, and, after a brief ride, we stop at a villa with many cars parked in front. We are sent to the living room where the women are concentrated, chatting and waiting for lunch. The new mother is lying in a corner of the room, tapping her baby to make her sleep, while chatting with peer-age women. We salute her and find a place to sit. Aishata sits close to the new mom and her young companions, taking out her merchandise which soon starts circulating into the women's hands. Zeinabou, instead, engages with an older woman, "the President", who is one of Zeinabou's aunts and the sister of the new-born's grandmother. Her nickname is due to her role as the organiser in the *tontine*, a money-collection group, generally based on kinship, that provides financial support to its members. They discuss some details for their *tontine* meeting the next day. After we are served tea on our arrival, the meal is served: many big plates of *ris gras* with a grilled fish on top, are positioned around. Silence falls and, after the meeting ends, we leave for other daily commitments.

Resilience is framed as a continuous shaping not only of livelihood practices but also of socialities and identities (Semplici, 2021). The social journeys narrated above show that forms of (im)mobilities, through practices of visiting, for example, generate continuities by reproducing patterns of socialities, practices (e.g. hospitality), and exchanges of information. For example, Zeinabou, her sister, and her sisters-in-law attended the naming ceremony as an expression of their sociocultural belonging to the same *tewsit*. The intention to fulfil social obligations is further emphasised by their decision to visit Faata's father, which they had neglected due to other urban commitments and distance. These daily circulations are expressions of reconsolidations of these networks, retracing patterns of visiting and hospitalities, reinforcing a sense of solidarity around identitarian continuity.

This sense of continuity is also articulated across the rural-urban spectrum, evidenced in this vignette by the different men visiting (Zeinabou's brother), or temporarily moving to Bamako (Faata's father, the marabout arriving). Moving between the rural-urban dimensions is not limited to men, but rather is a structural dimension of Kel Tamasheq's enclosure of Bamako within their regional networks.

By this rural-urban continuity, Bamako becomes a meeting hub for Kel Tamasheq arriving from different areas, with Sirakoro being where the majority of Kel Tamasheq's families are finding places to live, especially those from the Kidal region. Participating in these festivities offers possibilities to meet new people and craft socioeconomic relationships. In the movements of Zeinabou and other family members, these exchanges are clear. For example, by stopping in Yorodjambougou to pick up her sister and her sister-in-law, Zeinabou greets her brother's arrival in Bamako, and together with the women, they decide to go to Faata's house, to fulfil their obligations to greet Faata's father after his arrival from Mauritania. Practicing hospitality, by visiting and receiving visitors, maintains a continuity between rural and urban cultural customs and a sense of belonging. Like Zeinabou's brother and Faata's father, many visitors we met and greeted at the naming ceremony were not long-term residents of Bamako but came from Timbuktu, Kidal, Tamanrasset and other rural areas to visit families in the Malian capital and, potentially, to also find possibilities to improve their life circumstances.

Urban spatial knowledge goes hand-in-hand with socio-political proximities among families. Boilley (1999b) speaks of overlapping pastoralist geographical, socio-economic and political zones of influence. We can draw a parallel with Kel Tamasheq's patterns of dwelling in Bamako. As in Tamanrasset in the case of Kel Tamasheq migration (Dida 2012), or in

Zinder in case of Fulani migration (Kohler 2020), it has been noted that Saharan-Sahelian rural-urban migration tends to recreate socio-political enclaves. The same trend is observable in Bamako, where families of the same geographical, social status, or political origin tend to concentrate close to one another, forming sorts of urban enclaves with rural roots. Visiting practices are then shaped by these clustered patterns in Bamako: for example, spatial proximity makes it faster and cheaper to pay a visit to a relative's household, or to participate in a celebration, or even to make an appearance as a sign of courtesy. Bamako is not an easy city to navigate with public transport, taxis are not affordable to all, and times of travelling can be limited by traffic jams. Thus, clustering and proximity permit easier reproduction of already existing relationships.

Movement and invitations, then, create discreet knowledges of the urbanscapes, while reproducing, and transforming, patterns of social stratification based on family links, kinship or ethnic affiliation. For example, Zeinabou's convoy has been invited to the naming ceremony because of their family's relationship to the new mother. So, these encounters spark opportunities to discuss socio-cultural issues pertinent to the extended family, but also economic and political topics. For example, that is what Zeinabou and 'the President' do when engaging in discussions about the *tontine*, their system of internal wealth redistribution.

New contexts imply changes, and Bamako enhances different types of connections and reproduction of habits, depending, for example, on urban spatial transformations. Bamako is one of the fastest growing African cities (Pilling 2018), and new neighbourhoods mushroom continuously, opening possibilities to find cheaper places to rent or buy land (e.g., Sirakoro). It is not easy to track people's movements in the city. Visiting is frequent done without knowing the precise location and narratives of the difficulties in getting to someone's new place are common urban anecdotes. Collective visits, by which I mean a group of people visiting a place together, are common, as they facilitate getting to unfamiliar destinations, while they produce shared experiences of urban travel. This is the case of Zeinabou's sister who waited for her sister in Yorodjambougou to arrive at the ceremony together, because the sister had never been to that house, and so took advantage of Zeinabou's knowledge of the neighbourhood to plan her visit. Moreover, these collective visits make moving through Bamako less expensive (whether by private car or taxi), enabling people with less means to do their rounds of visiting.

Mobility is key in narratives of dealing with Bamako's arrangements. This is evident in the stories of women's *petit commerce* (Gonzales, forthcoming). As Faata clearly states, the failure to move around, to participate in visiting practices or in ceremonies taking place in the city, has a negative impact on her sales. Her current family's situation is now limiting her mobile practices around the neighbourhood and the city, fostering her feeling of losing lucrative opportunities. By the same reasoning, Aishata did not waste the possibility to jump into Zeinabou's car to go to the naming ceremony: it was an opportunity to go beyond her daily transactions and potentially meet new customers for her *petit commerce*. To be mobile means to diversify customers, a key practice that can be conceptualised in a continuum with wider pastoralist trends in the region, of increasing mobile patterns for the sake of diversifying livelihoods in the face of growing environmental, political, and socioeconomic varia-bilities (Grémont 2014: Kohler 2020: Oxby 2014). At the same time, fruitful circulation entails degrees of staying put, and of switching to immobility.

Practices of immobility by being visited

In this vignette, staying put in a place is a current instance of a key strategy for nomadic pastoralist groups claiming the right to belong and relate to their territories (Grémont 2011). Fixation is also an important variant in how resiliently nomadic populations respond to increasing variabilities, as immobility provides novel opportunities to be catalysed. For example, if Faata cannot move to do her *petit commerce*, she can nonetheless keep her boutique open, which would otherwise remain closed. Having the boutique open is a way to make it more available to neighbours, and indirectly, to potentially strengthen and also increase the number of bonds with them. Spending time at Faata's home, in the following months, allowed me to observe the wide range of neighbourhood interactions that the boutique afforded. Women on their way to or from the market would stop by to greet her; some would buy beauty products or send their children to do so; some would come to chat a bit with Faata, who would suggest the latest merchandise.

Moreover, staying put meant that Faata, as the manager of the house and the one who best knows the neighbourhood, could arrange sharing food, which according to Kohler (2020) is an important indicator of 'integration' in relationships of proximity. For her father's arrival, or the marabout (who requires frequent sheep' sacrifices), Faata would organise delivery of food to poor neighbours. Importantly, she prioritized spatial proximity (giving to the neighbours) over relational proximity (giving to members of her extended family in the city): It happened that a cousin, who lived just 10 minutes away by car in Sirakoro, complained about Faata's negligence in delivering her meat. In this instance, Faata facilitated exchanges with neighbours, of different ethnic and geographical origins over kin. This is a practice that can enhance novel relationships and contributes to processes of place-making. If kin connections provide a safety network to rely on in times of difficulty, lighter connections are also very useful because they can be pulled and pushed without deeper consequences. Moreover, with prolonged contact, these lighter relationships can sediment and become relevant bonds. By investing in them, Faata resiliently faces the situation she is plunged into.

Generational change plays a great role in crafting new relationships. Faata's children nurture friendships in the school, going beyond Kel Tamasheq's ethnic bonds. Faata's daughter fervently recalls her day at the Bamako zoo with her school companions. However, in contrast to the Wodaabe situation in eastern Niger in the early 2010s, where children invited school companions to their homes (ibid.), in Bamako I only rarely witnessed this practice. This might be due to the tensions and the stereotypes attached to the Kel Tamasheq community in Bamako, and their association with terrorist or bandit activities, narratives reinforced since the Malian geopolitical crisis.

Immobility also means to 'be reached', to receive guests and to perform hospitality. Faata's father or Zeinabou's brother are at home and will remain there to welcome friends and family members. Celebrations, such as the naming ceremony, are generally organised at home. Very rarely did I hear of weddings held in public places, and they were generally public figures who sought to increase their visibility in the capital's social environment. Holding celebrations at home recreates a familiar habitat, where continuities of socio-cultural practices can be performed. However, 'to be visited' does not work similarly for everyone, but is entangled with issues of social status, lineage, and the reason for visiting. In her day of rest, Zeinabou stays home and, most likely, receives visitors all day long, people arriving from Bamako or the regions who want to stay at her place when in Bamako, take advantage of the opportunities arising around her, or just want to salute her. Zeinabou has both the social and cultural capital (Bourdieu, 2000) that enables her to have people coming to her. If this is not necessarily always wanted (unpleasant visits can happen), she has become known as a point of reference for many female-centred affairs in Bamako, especially among her kin group. Immobility permits variability of movement, affairs and interactions: movements of the residents and guests, who visited who, circulating news, gatherings and invitations.

Conversely, when Aishata moves to live in a flat in Sirakoro by herself, peripheral to other family members' houses, she will not be easy to visit. However, she will not necessarily complain about that, but would instead reflect on her current 'failure' to visit Zeinabou, because of lack of means. Being in Sirakoro for her meant developing a more lucrative network with the Kel Tamasheq community, but to have fewer everyday interactions with members of her extended family (Gonzales, forthcoming). After a couple of months, she found suitable accommodation closer to one of her family's houses. This new positioning permits her to be included in family businesses: it implied that family members coming to Sirakoro would include her in their visiting practices because they would go just next door. But it also gave her more flexibility to join them in visiting around the city, to go with them in other neighbourhoods, to do some *tbushak* in other parts of the city, taking advantage of their rides.

To conclude, these practices of (im)mobility craft resilient patterns of Kel Tamasheq dwelling in Bamako, reconsolidating lineage bonds, while building new ones. Practices of (im)mobility also correspond to the heterogenous perceptions and representations of Kel Tamasheq in Bamako, which are sometimes positive, other times negative because of urban pollution, socio-cultural distance or socio-political discrimination. And it is in these narratives of Kel Tamasheq in Bamako that the concept of *tazidert* is sometime invoked, as a characteristic trait of Kel Tamasheq collective identity, to express an attitude of 'keep on going' despite their negative representations and experiences of Bamako.

Tazidert: **Resilience through patience**

Semplici argues that "it is by understanding how people construct their collective identities [...], and how these flexibly respond to changes in the surroundings, that one can understand resilience" (2021: 229). Feelings of belonging craft senses of solidarities which, she argues, enlighten understandings of resilience. Kel Tamasheq's urban dwelling indicate that patterns of (im)mobilities are structural in the making of socio-economic connections. Dwelling practices in Bamako are paired with the mainstream Kel Tamasheq's representations of the city as an alien place, politically, socio-culturally, and sensorially (e.g. a different type of hot weather, less spacious), boasting narratives of Kel Tamasheq in Bamako as being 'out-of-place' (Gonzales 2022). These narratives structure Kel Tamasheq experiences of the capital, contributing to a sense of continuity in Kel Tamasheq nomadic identity as markedly opposed to the city of Bamako and what it stands for, the Malian state. At the same time, reality is more complex, and concomitant to narratives of urban estrangement and continuity with the rural setting, adaptation to the urban space is invoked. In this context, the concept of *tazidert*, and the 'keep on going' attitude, underlines the importance of acknowledging the state of things – Kel Tamasheq currently in Bamako – so to actively foster embracing the situation. The value that *tazidert* expresses is associated religiously with the sphere of Islam, as well as with the sociocultural and environmental spheres rooted in dryland practices and knowledges.

Arriving in Bamako from Mberra, the marabout at Faata's house received many guests, and customers who came to him for recommendations and religious advice. He was not new to Bamako, but this time, due to personal reasons, he decides not to return directly to Mberra. He says that he takes his decision consciously; he wants to see if his marabout expertise will economically support him so that he might be able to re-marry. However, consciously deciding to remain in Bamako is not for everyone, he comments: many Kel Tamasheq do not accept being there. He traces the Kel Tamasheq's negative representations of Bamako: dislike for the urbanscape, its pollution and population density, cultural differences and the heritage of the politically contested relationship with Mali, the weather and ecological challenges (e.g., malaria), the labour market not being as expected, feelings of alienation. "They cannot accept to be here," the marabout continues, "but this will not help them. They should be patient and realise that, if they are here, they should embrace the situation".

This explanation refers to Islamic religious creed based in beliefs about predetermination (Oxby 2014.). Faith that God has placed you where you are should trigger an openness to life's events, which can be cultivated and endured with patience. Like praying, patience needs to be practiced, where practicing entails proactive stances and not passive toleration of a life situation. *Tazidert* is, then, invoked when a subject needs to acknowledge a situation, and it expresses its full conceptual potential when invoked in relation to events or historical processes that are described as 'overwhelming' (*dépasser*), beyond the subject's comprehension.

Patience inspires both endurance and proactivity at the same time. It is an important attitude that reflects the sociocultural and ecological background of Kel Tamasheq as nomadic and pastoralist people and serves to legitimise identarian continuity across the rural-urban divide. Indeed, representations of life in the drylands generally emphasise the harsh living environments. Patience is an important component of living without much suffering under these conditions. In collective narratives, it is associated with strength and endurance, characteristics which are associated with pastoralist practices and nomadic identities (e.g., in the case of Turkana, the *raiya* 'bushman', in Semplici 2021). According to a long-term resident man in Bamako, well-educated on cultural issues of nomadic communities, and involved in the Kel Tamasheq cultural scene in Bamako, patience is associated with resistance to difficult situations by surrendering to them. Indeed, if ecological variabilities are not embraced and dealt with, but countered, suffering will increase. This is true for life in general (a religious precept), but especially in the desert, he comments. His explanation is rooted in the culture/nature divide, which suggests that, if the desert is a place outside the human realm (it is generally associated with spirits and *djin*), people need to adapt to it by acknowledging its particularities. It cannot be the other way round, as nature does not adapt to humans. To emphasise this concept and place it in Kel Tamasheq pastoralism practice, he tells the following proverb: *tazidert atirua tiknéuen*, meaning that with patience, two sets of twin lambs can be achieved. More lambs translate into more milk and more flock, he explains. As animals represent the source of livelihood, as well as political, sociocultural, and economic status, having a growing number of animals is associated with improved quality of life. But this is to be gained by nurturing patience.

In this case then, ecological and sociocultural processes are entangled in the concept of patience and are used to read and filter new experiences, like migration to the urban landscapes of Bamako. Migration studies have underlined the frustrations related to forced immobility (Diallo 2018; Dudley 2010), while anthropological works have looked at youth waiting for opportunities in precarious labour markets (Hage 2009; Jeffrey 2010), as well as African pastoralist male migration to urban sites (Ancey et al. 2020). Patience here reveals alternative ways in which (im)mobile practices, and their expectations, are lived and conceived by subjects. In the Bamako context, these intersecting dimensions are dealt with by leaning on a common pastoralist background and references to the environment. In parallel, religious beliefs provide crucial references to sketch subjective and shared approaches to life and its unfolding. Patience is the acknowledgment of one's situation, and a voluntary submission to historical factors, which are devolved to the transcendent realm of God.

That does not mean embracing passivity, but rather the opposite. By invoking *tazidert*, Kel Tamasheq do not submit but actively face the situation, bluntly engaging with it. Nevertheless, although narratives of *tazidert* accompany (im)mobile practices and dwelling in Bamako, they should not be essentialised as a unique 'nomadic pastoralist characteristic', because, as the marabout underlined, many Kel Tamasheq in Bamako do not act accordingly. Rather, they show how cultural and religious dimensions are used flexibly to face and adapt to new situations, suggesting that to 'keep on going' is a feature of resilience.

Conclusion

Through this chapter, I analysed resilience by looking at two different but entangled dimensions of Kel Tamasheq life in Bamako in relation to the concept of patience: practices of (im)mobility, specifically visiting patterns, and narratives of collective identity. Resilience is here considered as the capacity to switch from being mobile to being immobile depending on opportunities, interests, emotional attachments, affections and socio-political and economic ties. Transformation is inherent to processes of switching from being mobile to immobile, and vice versa. Visiting processes, and *tbushak* are important vehiculating modalities of these switches. At the same time, resilience implies an ability to acknowledge that transformations in politics and the environment take place, and that patience is the concept that can be used for coping with that. Embedded in their collective nomadic identity as pastoralist and Muslim, patience provides a continuity in Kel Tamahseq collective identity representations despite switches from mobility to immobility, and across the urban-rural continuum. Patience embodies this ability to embrace opportunities and is inscribed in collective emic representations of a Kel Tamasheq identity. In this chapter, I wanted to think more broadly about resilience as the capacity to switch from mobility to immobility, and vice versa, and to integrate this idea of embracing change, through the concept of patience, as a trait of collective nomadic identity.

References

Ancey, V., C. Rangé, S.D. Magnani, and C. Patat. 2020. Pastoralist youth in towns and cities. Summary report. Supporting the economic and social integration of pastoralist youth in Chad and Burkina Faso. Rome: FAO. http://www.fao.org/3/ca7213fr/ca7213fr.pdf

Badi, D. 2012. 'Cultural interaction and the artisanal economy in Tamanrasset.' In J. McDougall and J. Scheele. (eds). *Saharan Frontiers: Space and mobility in Northwest Africa*. Bloomington: Indiana University Press. pp. 200–214.

Belalimat, N. 2019. 'Chapitre 7 – Réseaux sociaux, nouveaux médias et territoires au Sahara.' In S. Boulay and S. Fanchette. (eds). *La question des échelles en sciences humaines et sociales*. Éditions Quæ. Tiré de. http://books.openedition.org/quae/2057

Boesen, E., L. Marfaing and M. de Bruijn. 2014. 'Nomadism and mobility in the Sahara-Sahel'. *Canadian Journal of African Studies* 48(1): 1–12. https://doi.org/10.1080/00083968.2014.935101

Boilley, P. 1999a. *Les Touaregs Kel Adagh: dépendances et révoltes: du Soudan français au Mali contemporain*. Paris: KARTHALA Editions.

———— 1999b. 'Les Touaregs entre contraintes géographiques et constructions politiques.' *Études rurales* 151/152: 255–268.

Bonte, P. 1986. 'La tⵝwshet est-elle un groupe de filiation? Alliance, pouvoir et appartenance sociale chez les Kel Gⵝres'. In E. Bernus, P. Bonte, L. Brock and H. Claudot. (eds). *Le Fils et le neveu*. Paris: Editions de la Maison des Sciences de l'Homme. pp. 237–276

Bourdieu, P. 2000. *Esquisse d'une théorie de la pratique. Précédé de trois études d'ethnologie kabyle*. Paris: Le Seuil.

Bourgeot, A. 1990. 'Le désert quadrillé: des Touaregs au Niger'. *Politique Africaine* 38: 68–75.

Chauveau, J.-P. 2017. 'Introduction : Le nexus État, foncier, migrations, conflits comme champ social.' *Critique internationale* 75(2): 9–19. https://doi.Org/10.3917/crii.075.0009

Claudot-Hawad, H. 2002. 'Noces de vent: Épouser le vide ou l'art nomade de voyager'. In H. Claudot-Hawad. (ed). *Voyager d'un point de vue nomade*. Paris: Éditions Paris-Méditerranée. pp. 11–36.

Diallo, S. 2018. *The Truth about the Desert: Exile, memory, and the making of communities among Malian Tuareg refugees in Niger*. Cologne: Modern Academic Publishing.

Dida Badi Ag Khammadine. 2013. 'Les dynamiques politiques à l'œuvre dans l'Azawad (Nord-Mali)'. *El Watan*, 12 August. https://www.elwatan.com/archives/interna-tionale/les-dynamiques-politiques-a-loeuvre-dans-lazawad-nord-mali-12-08-2013 (accessed 10 December 2018).

Dudley, S.H. (2010). *Materialising Exile: Material culture and embodied experience among Karenni refugees in Thailand*. New York: Berghahn Books.

Geschiere, P. (2005). 'Autochthony and Citizenship: New modes in the struggle over belonging and exclusion in Africa. *Forum for Development Studies* 32(2): 371–384.

Giuffrida, A. 2010. 'Tuareg networks: An integrated approach to mobility and stasis.' In A. Fischer, and I. Kohl. (eds). *Tuareg Society within a Globalized World: Saharan life in transition*. London: Tauris Academic Studies. pp. 23–41.

Gonzales, G. 2020a. '*Niglá*: Methods and discontinuous (im)mobilities among Malian Kel Tamasheq in Bamako.' *Nomadic Peoples* 24(4): 195–209.

————— 2022. '"*Bamako war ikna*": Politics of wellbeing in Bamako.' *Le bienêtre au Nord et au Sud: Explorations*, In J. Tantchou, F. Louveau, and Marc-Éric Gruénais. (eds). Louvain-la-Neuve: Collection Investigations d'anthropologie prospective.

————— forthcoming. "« Il faut bouger » Mobilité, politique, et adaptation : parcours de deux femmes Kel Ansar à Bamako." *Ouest Saharien*, numerò thematique 2022.

Grémont, C. 2011, «Ancrage au sol et (nouvelles) mobilités dans l'espace saharo-sahélien: des expériences similaires et compatibles», *Année du Maghreb*, VII: 177–189. https://journals.openedition.org/anneemaghreb/1203

————— 2014. 'Mobility in pastoral societies of northern Mali: Perspectives on social and political rationales.' *Canadian Journal of African Studies* 48(1): 29–40. https://doi.org/10.1080/00083968.2014.91832

Hage, G. 2009. 'Introduction.' In G. Hage. (ed). *Waiting*. Melbourne: Melbourne University Press.

Homewood, K. 2008. *Ecology of African Pastoralist Societies*. Melton: James Currey.

Jeffrey, C. 2010. *Timepass: Youth, Class, and the Politics of Waiting in India*. Stanford: Stanford University Press.

Köhler, F. 2020. *Space, Place and Identity: Wodaabe of Niger in the 21st Century*. New York: Berghahn Books.

Konaka, S., and P.D. Little. 2021. 'Introduction: Rethinking resilience in the context of East African pastoralism'. *Nomadic Peoples* 25(2): 165–180.

Lecocq, B. 2011. *Disputed Desert*. Amsterdam: Brill.

Lecocq, B., G. Mann, B. Whitehouse, D. Badi, L. Pelckmans, N. Belalimat, B. Hall, and W. Lacher. 2013. 'One hippopotamus and eight blind analysts: A multivocal analysis of the 2012 political crisis in the divided Republic of Mali.' *Review of African Political Economy* 40(137): 343–357.

Lentz, C. 2013. *Land, Mobility, and Belonging in West Africa*. Bloomington: Indiana University Press

Lunacek, S. 2020. 'Sedentarisation, continuous mobility and gender among Tuaregs in the north of Niger.' In H. Horáková, S. Rudwick and M. Schmiedl. (eds). *Africa on the Move: Shifting Identities, Histories, Boundaries*. Münster: LIT Verlag.

McDougall, J. and J. Scheele. 2012. 'Introduction: Time and space in the Sahara.' In J. McDougall, and J. Scheele. (eds). *Saharan Frontiers: Space and mobility in northwest Africa*. Bloomington: Indiana University Press. pp. 1–24.

MNLA. 2012. 'Déclaration d'indépendance de l'Azawad'. https://s.france24.com/media/display/4482187a-14ed-11e9-8574-005056bff430/Declaration%20independance%20de%20l%27etat%20de%20l%27Azawad.pdf (accessed 16 November 2022)

Nori, M. 2022. 'Assessing the policy frame in pastoral areas of Sub-Saharan Africa (SSA).' *EUI RSC PP*, 03, Global Governance Programme [Global Economics].

Oxby, C. 2014. 'Social differentiation of risk: Perceptions of the future in drought-prone central Niger.' *Journal des Africanistes* 84(1): 106–129.

Pilling, D. 2018. 'African cities surge to top of global growth league'. *Financial Times*. 11 Sept. Accessed 11 Nov. 2019. https://www.ft.com/content/9d457d54-b272-11e8-8d14-6f049d06439c

Poudiougou, I., and G. Zanoletti. 2020. 'Fabriquer l'identité à la pointe de la kalache: violence et question foncière au Mali.' *Revue internationale des études du développement* 243: 37–65.

Rossi, B. 2015. 'Kinetocracy: The government of mobility at the desert's edge'. In D. Vigneswaran and J. Quirk. (eds). *Mobility Makes States: Migration and power in Africa*. Philadelphia: University of Pennsylvania Press. pp. 149–168.

Sèbe, B. 2015. 'A fragmented and forgotten decolonisation: The end of European empires in the Sahara and their legacy.' In T. Chafer, and A. Keese. (eds). *Francophone Africa at Fifty*. Manchester: Manchester University Press. pp. 204–219.

Scheele, J. (2012). *Smugglers and Saints of the Sahara: Regional connectivity in the twentieth century*. New York: Cambridge University Press.

Schlee, G. 2017. 'Introduction: Difference and Sameness as modes of transformation.' In G. Schlee and A. Horstmann. (eds). *Difference and Sameness as Modes of Integration: Anthropological perspectives on ethnicity and religion*. New York: Berghahn Books.

Semplici, G., 2021. 'Resilience and the mobility of identity: Belonging and change among Turkana herders in northern Kenya.' *Nomadic Peoples* 25(2): pp.226–252.

UNHCR (United Nation High Commission for Refugees). 2015. 'Mali: Refugees, IDPs and returnees'. https://reliefweb.int/sites/reliefweb.int/files/resources/20151031-UNHCR-Mali-Refugees-Returnees-IDPs_ENG.pdf (accessed 10 October 2019).

CHAPTER

11

Resilience under Strain: Spatial Dimensions of 'Farmer–Herder Conflict' in the Sahel

Takuto Sakamoto

Introduction

Deadly conflicts involving pastoralists in West Africa and the Sahel have been increasingly common in recent decades. In Nigeria, for example, numerous clashes between farmers and pastoralists have been reported not only in the traditionally pastoralist-dominated northern states, but also from the central 'Middle Belt' to the south of the country, resulting in the loss of many lives. The country's security has been seriously eroded by the activities of so-called 'bandits' (e.g., village raids and livestock rustling), which allegedly involve armed pastoralists (UNOWAS 2018: 74–75; Krätli and Toulmin 2020: 37–39). Other countries in the region are also experiencing pastoralist-related conflicts of varying intensity. In particular, the border areas of Mali, Niger, and Burkina Faso have seen a significant increase in large-scale armed clashes involving pastoralists in recent years (UNOWAS 2018: 63–65).

Although these conflicts, which often pit pastoralist herders against sedentary farmers, are not new (e.g. Basset 1986; Bukari, Sow and Scheffran 2018; Hussein, Sumberg and Seddon 1999; Moritz 2006; Turner 2004), it is notable that they have recently attracted significant interest with a strong policy focus from a broad array of stakeholders. For example, these conflicts have occasionally been mentioned and discussed in the Security

Council of the United Nations (UN), arguably the primary multilateral body in global security, as they have frequently been highlighted in the semi-annual reports that the UN Secretary-General submits to the organ (Report of the Secretary-General on the Activities of the United Nations Office for West Africa and the Sahel). The UN stresses a multi-layered approach to addressing these conflicts, urging coordinated policy responses from the countries of the region, regional organizations such as the Economic Community of West African States (ECOWAS), and other UN agencies (UN 2018; UNOWAS 2018).

Given such developments, these conflicts involving pastoralists might plausibly be considered to be in the process of 'securitization' (Buzan, Wæver and de Wilde 1998). However, caution is required before taking such a perspective. These conflicts, which have frequently been described as 'farmer-herder conflict' or 'pastoralism-related insecurity' (see Delsol 2020: 4), have far deeper roots than these representations might imply. That is, rather than being simply reduced to specific inter-group relationships or, even less, particular ways of livelihood, these conflicts are deeply embedded within wider social-ecological systems in the Sahel, in which pastoralists, farmers and numerous others have been involved in complex interactions under various natural and social constraints (see Galvin 2009; Reid, Fernández-Giménez and Galvin 2014; Robinson and Berkes 2010 for a similar perspective). Accordingly, the increasing and widening occurrence of 'farmer-herder conflicts' in the region cannot be fully understood, and therefore practically addressed, without an adequate account of the functioning of these social-ecological systems at large, whose resilience now seems to be under considerable strain due to multiple sources of shock and stress, including, among others, climate change and population pressure (UN 2018: 4; UNOWAS 2018: 13–15).

This chapter advances such an understanding of the contemporary farmer-herder conflict in the Sahel by addressing three broad questions concerning pastoral resilience: (1) how, and to what extent, does the evolving social-ecological system in the Sahel constrain the behavior of local pastoralists, especially in terms of their mobility for resource access?; (2) how, and to what extent, do the pastoralists adapt their spatial behavior to these systemic constraints on a daily basis?; and (3) what wider consequences does the daily adaptation of individual pastoralists add up to, especially regarding farmer-herder relationships?

In addressing these three questions, the chapter illuminates one microprocess that has been widely cited as a primary 'trigger' of such conflict, namely, crop damage (Basset 1986; Bukari, Sow, and Scheffran 2018; Krätli and Toulmin 2020: 55–56; Turner 2004; Turner et al. 2011). In the present context, crop damage refers to the destruction of crops caused by intentional or inadvertent invasion of farmland by a pastoral herd. It does not always lead to a violent clash. Most of the time, ensuing tensions are resolved peacefully through ad-hoc negotiations between the parties concerned, or through formal and informal dispute resolution procedures (Bukari, Sow, and Scheffran 2018; Oyama 2015). However, these negotiations and procedures often fail to function effectively, in which case deadly violence can erupt between herders and farmers.

Crop damage is a non-trivial subject for the resilience of the contemporary African pastoralism, which is broadly understood here as the capacity of African pastoralists to flexibly respond to the socio-ecological vagaries of drylands. Crop damage frequently takes place during daily herding, an essential pastoral practice constituting 'micro-mobility' (Niamir-Fuller and Turner 1999: 37). Controlling micro-mobility (daily movements around the village or camp), along with 'macro-mobility' (long-distance routes and seasonal grazing areas), has traditionally ensured pastoralists much-needed flexibility and thus resilience. As such, systematic (that is, not sporadic) crop damage in a given social-ecological system could indicate disruption of an important adaptive mechanism supporting pastoral resilience. Hence, this chapter's analytical focus is on pastoralists' daily herding and its wider spatial implications for crop-damage.

The following sections present a series of analyses that examine processes and patterns of crop damage involving pastoral Fulbe and Tuareg herders and Hausa farmers in southwestern Niger. Based on field observations as well as Global Positioning System (GPS) logs obtained on the ground, the study comprises three analytical components, which correspond to the three research questions raised above. The first part estimates a human ecological pattern of land cover (e.g., vegetation) and land use (e.g., cropland) in the study area with the combined use of field data and satellite images. It reveals the significant spatial constraints that herders now face with respect to resource access. Using the estimated land cover and land use (LCLU) pattern, then, the second part analyzes GPS records of Fulbe pastoralists' daily herding activities. It sheds light on how these herders are locally coping with the suggested spatial constraints. Employing the findings and outputs obtained from the preceding analyses, the third part constructs a simple computational model that simulates the daily herding practices of pastoral households. This model provides a scalable method for assessing and predicting the wider spatial consequences of herding behavior of numerous pastoralists interacting with a given landscape. The simulation seeded with the obtained empirical data produces a 'hazard map' of crop damage occurrence, which largely reveals the systematic nature of these incidents even in the presence of pastoralists' efforts to prevent their occurrence.

The study area

The study area is one anonymous village (hereafter called 'village D') and its surrounding area in the southwestern part of the Republic of Niger (Figure 11.1). The village is located approximately 7 km southeast of the city of Dogondoutchi in the Dosso Region. Ecologically, the area lies across the Sahel and savanna ecological zones. Annual rainfall averages about 500 mm in the vicinity of Dogondoutchi with considerable interannual fluctuation. The precipitation also shows marked seasonal variation. The precipitation is largely concentrated in the four months from June to September, which constitute the rainy season (*damana* in Hausa).

Demographically, the area is predominantly inhabited by Hausa people. Most Hausas in the area are smallholder crop cultivators who grow pearl millet and cowpea in fields around their settlements. In recent decades, the area's Hausa population has increased rapidly, driving the steady expansion of the total cropland area (Oyama 2015: 232–233). Meanwhile, the area also has a small pastoral population consisting primarily of Fulbe but also

Tuareg households. Most Fulbe and Tuareg pastoralists in the area settle in or around Hausa villages, often have their own small fields for growing pearl millet and cowpea, and take care of their own livestock (if they have any) as well as livestock entrusted to them by Hausa farmers. Village D had a total of 88 households at the time of visiting in September 2019. Among them, only one household was Fulbe and one was Tuareg, and the rest were all Hausa. According to Fulbe informants, there are around 110 Fulbe and Tuareg households in the surrounding area. The size and composition of livestock herds vary enormously among these households, but most take care of some mix of goats, sheep, and cattle. A sizable number of households also retain a small number of camels and donkeys.

The author visited the study area in September 2017 and again in September 2019. With the help of collaborators and villagers, an array of field data was obtained. First, in numerous locations in the area, LCLU conditions, such as vegetation, cropland, and surface water, were directly observed and recorded. The geographic coordinates (longitude, latitude and elevation) of these locations were also recorded using a GPS logger. The point data thus collected constitute the so-called 'ground truth' data, which were employed for the automated LCLU classification using satellite imagery (see the next section). Second, 45 herd camp sites were visited

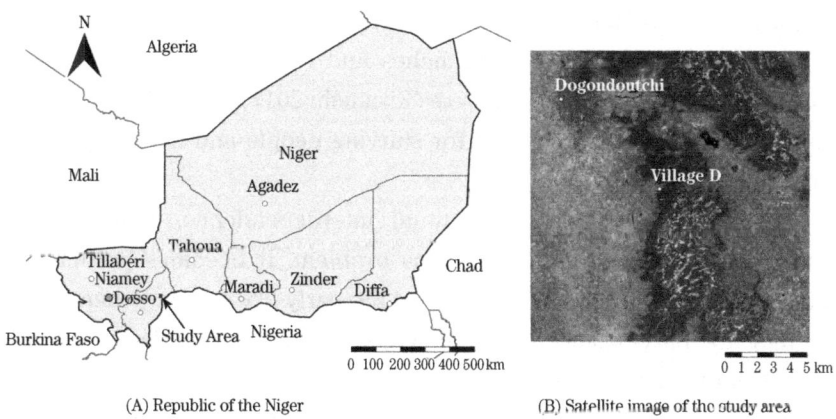

(A) Republic of the Niger   (B) Satellite image of the study area

Figure 11.1 The study area

and their geographic locations were recorded. These camps, which were scattered on the extensive plateau (inselberg) located east of village D, supposedly comprise a sizeable portion of the herd camps that existed in the study area at the time of visiting (September 2019). Third, the trajectories of four livestock herds were recorded over three consecutive days during the second visit. These herds – two cattle herds and two goat/sheep herds – were managed (not necessarily owned) by two Fulbe households who were camped on the plateau at the time of visiting. Herders from these households self-recorded their trajectories using GPS loggers during daily herding (from approximately 8 am to 5 pm).

Overall, Fulbe and Tuareg pastoralists, being an absolute numerical minority, have retained a more or less symbiotic relationship with their dominant Hausa neighbors. In addition to the livestock entrustment mentioned above, for example, there is a widespread practice of encampment contracts between herders and farmers in the dry season (Oyama 2017). In this contract, a farmer, most often a relatively wealthy family, invites a herder and his herd to set up a dry-season grazing camp in the middle of the former's field in exchange for cash or payment in kind (e.g., pearl millet). The livestock brought into the harvested field, then, feeds on the remaining stalks of pearl millet and drops dung, thus restoring soil fertility for the farming family. Moreover, in the middle of dry season (from March to May), during which the available pasture frequently dried up, pastoralists, along with other impoverished farmers, are often allowed to enter farmlands to collect branches and fruits from the trees (e.g., *Balanites aegyptiaca*) planted there (Kirikoshi 2017). These branches and fruits are valuable food sources for starving people and livestock in this difficult time of the year.

Despite these aspects of mutual interdependence, farmer-herder conflict remains a potentially serious problem. It becomes particularly salient during the late rainy season and the early dry season, when Hausa farmers are harvesting crops, the major output of months of intense labor. Conflict almost always arises over damage to crops caused by the accidental or intentional livestock invasion of a field. As mentioned, the mere incidence of crop damage seldom incites open violence, as the parties involved usually seek monetary settlements (broadly called *bana*) through ad-hoc negotiation with some communal engagement (Oyama 2014; 2015). Given the numerical and socio-economic dominance of famers vis-à-vis herders in the area, these negotiations often favor the former while suppressing the latter's dissenting voice concerning the determination of facts and the attribution of responsibility. Accordingly, large-scale clashes between farmers and herders at the community level are relatively rare.

Nevertheless, the incidence of crop damage is sufficiently perplexing that it can strain the generally cordial atmosphere among villagers. Against such a backdrop, sporadic violence occurs, such as throwing stones at herders. According to Oyama (2014: 113), the frequency of such incidents was increasing in the area surrounding village D, especially since 2010. In anticipation of such conflict, Fulbe and Tuareg herders usually distance themselves from their Hausa neighbors during this tense time of year by moving their livestock to temporary camps that they have set up on the nearby barren plateau (Oyama 2014: 108). Some undertake long-distance transhumance to the more arid north around the Sahara Desert, thus minimizing contact with farmers (Oyama 2015: 89–91).

Constraints on resource access: Land cover and land use estimates

Figure 11.2 visualizes the geospatial patterns of land cover and land use (LCLU) observed around the study area. These patterns were all derived from satellite images taken by the European Space Agency's (ESA) Sentinel-2 earth observation mission on different days of September 2019, the same month in which the author conducted field observation (see Sakamoto (2021: 78–81) for details). The Sentinel-2 images are composed of 13 spectral bands, with a maximum ground resolution of 10 meters. Therefore, in each image in Figure 11.2, one pixel corresponds to a land plot of approximately 10m × 10m. Since each image has 1734 and 1768 pixels in height and width, respectively, the entire image corresponds to an area surrounding village D that extends 17–18 km in both the north-south and east-west directions.

Of the four images shown in Figure 11.2, Panel (A) shows spatial variations in the value of Normalized Difference Vegetation Index (NDVI) calculated for each pixel (range: -1.0 to 1.0). NDVI can be computed from a simple interband operation on a multispectral image, and has been widely used as a proxy of vegetation density and activity (Tueller 1989). In this study, NDVI is used for crude assessment of forage availability to herders. In the figure, the more luminous pixels represent the more heavily vegetated plots.

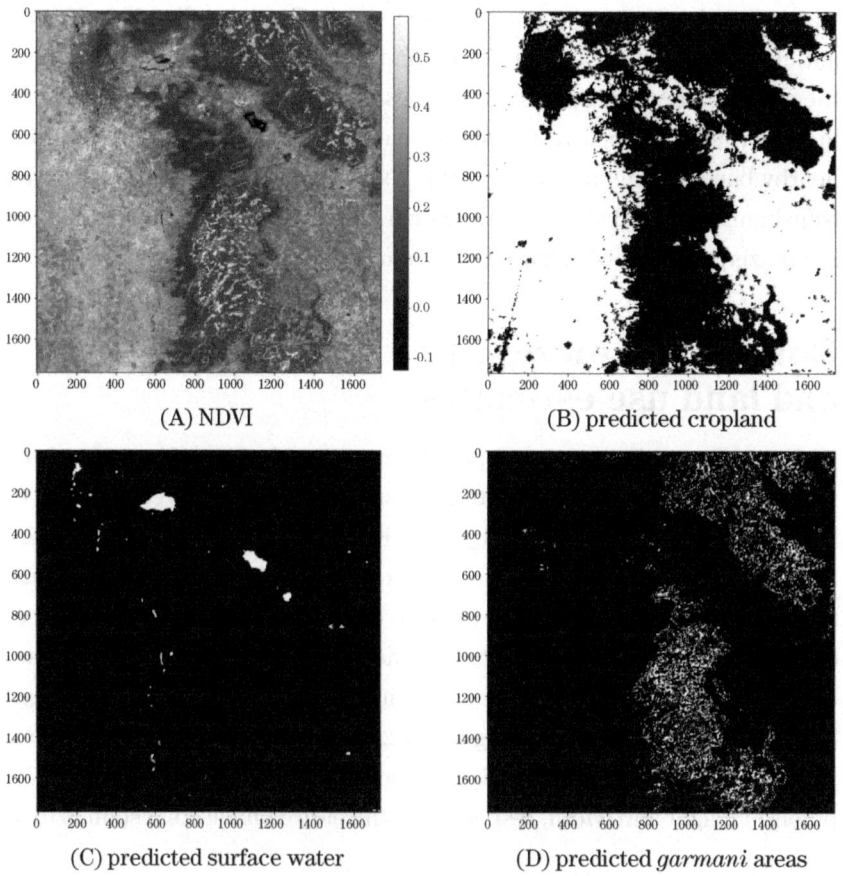

(A) NDVI

(B) predicted cropland

(C) predicted surface water

(D) predicted *garmani* areas

Figure 11.2 NDVI and LCLU classification

In contrast, the remaining three images in Figure 11.2 are products of more elaborate operations on the same Sentinel-2 images. These operations are essentially automated, pixel-by-pixel LCLU classification using deep learning (Ji et al. 2018; Waldner and Diakogiannis 2020). Combining the 13-band spectral information contained in a satellite image with the on-site observations obtained during the field visits, the classification algorithm estimates the corresponding ground surface condition for each of the pixels composing the image. Panels (B)–(D) in the figure depict estimation outcomes for several LCLU categories, that is, (B) cropland, (C) surface water such as lakes and ponds, and (D) a subshrub locally called *garmani* (scientific name: *Sida cordifolia*), respectively. In all these panels, pixels predicted to be dominated by the LCLU category concerned are colored white.[1]

Of these categories, *garmani*, shown in (D), is a perennial subshrub that is found extensively in the study area, especially on the barren plateau where many herders camp with their livestock during the rainy season. Although *garmani* can be valuable dry-season forage for livestock, livestock generally avoid grazing it in the rainy season for physiological reasons. Therefore, no matter how abundantly it thrives, *garmani* has almost no value as a grazing resource for pastoralists during this season.

All in all, the panels in Figure 11.2 confirm a significant degree of resource constraints faced by pastoralists around the study area. As Panel (A) shows, as of September 2019, the vegetation density as measured by NDVI was largely moderate throughout the area, even high in some locations, partly due to the relatively good rainfall that year. However, when Panels (B) and (D) are superimposed on this panel, it is clear that most of the areas with dense vegetation are either vast cultivated fields (55.3% of the total area on a pixel basis) planted with pearl millet and other crops, or *garmani* clusters (5.3% of the total area) found extensively on the plateau. That is, from the standpoint of local pastoralists, the vegetation to which

---

1 In addition to cropland, surface water, and *garmani*, the distributions of bare land, bush, and residential areas were also estimated and used in the analysis (Sakamoto 2021: 79).

they have effective access is quite meagre, despite it being the rainy season. To give concrete numbers, the sites where vegetation is relatively rich, for example, where NDVI is 0.25 or higher, account for rather impressive 55.4% of the total area on a pixel basis. However, excluding the land plots classified as croplands, *garmani*, and others from these resourceful sites, the percentage of the remaining sites – sites that presumably have a sizable amount of accessible grazing resource for pastoralists – drops to only 4.1%.

The 'micro-mobility' of pastoralists: Daily herding trajectories

The preceding geospatial analysis plainly demonstrated that local pastoralists and their livestock herds face considerable constraints on access to grazing resources in a difficult landscape spatially dominated by crop fields. How do pastoralists access these limited resources on a daily basis? In other words, what is the performance of their 'micro-mobility' (Niamir-Fuller and Turner 1999) for grazing their livestock? In order to address this pressing question, this section reports on a quantitative analysis of individual herders' access to grazing resources as well as the associated risk of cropland invasion.

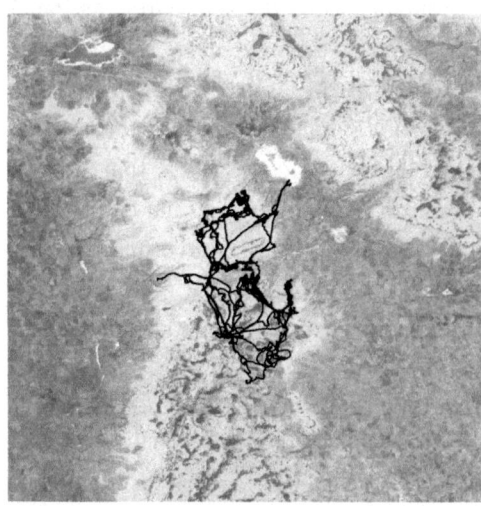

Figure 11.3 Herding trajectories

Figure 11.3 depicts the daily-herding trajectories of four livestock herds that Fulbe herders self-recorded on three consecutive days. Each day, each of the trajectories started from either of two closely located herding camps on the plateau. Crossing over extensive *garmani* clusters, each herd, accompanied by one or two Fulbe herders, went down a barren, steep slope of the plateau, and headed towards water sources while grazing grasses – which are often found on the margins of crop fields – along the way. After spending some time around water, the return journey began. The herd went uphill and came back to the herding camp in the late afternoon. The herding destinations and routes changed daily even for the same livestock herd, which indicates that the pastoralists' movements are flexibly determined, and do not strictly follow any pre-determined routes.

The GPS logs of these herding trajectories were projected onto the geospatial data presented in the previous section for a simple statistical analysis. Table 11.1 summarizes the results of this operation for each of the livestock herds, along with descriptive information on their composition and movements. On all three days during which their movements were tracked, the Fulbe herders grazed their livestock for less than nine hours, from just after 8:00 am to just before 5:00 pm. The 'maximum displacement' and 'total walking distance per day' in the table indicate that for this duration of time, the pastoralists and their herds walked over 15 km per day within a limited range of about 4 km from their respective camps.

Table 11.1 also presents another important set of quantities, namely, 'mean effective NDVI' and 'cropland frequency', which may indicate the degree to which herders have adapted to the surrounding human ecological environment. The former represents a crude proxy for the 'effective' forage availability along the corresponding herding track, that is, the potential per-pixel amount of grazing resources, as measured by NDVI, available to livestock animals, excluding those that should not or cannot be grazed, such as crops and *garmani*. The latter denotes the frequency with which a herd enters patches of land (each represented by a 10m × 10m pixel) that are classified as cropland during daily herding. In other words, mean effective NDVI measures pastoralists' access to grazing resources during daily herding, whereas cropland frequency quantifies the risk of crop damage. It should be noted, however, that the NDVI and land use classification values used to compute these quantities were derived at the scale of a 10m × 10m patch of land, since the resolution of the original satellite images is 10m per pixel. Therefore, even if a given herding trajectory overlaps a pixel that is classified as cropland in the estimated LCLU distribution, such an overlap happens at the relatively crude spatial accuracy of 10m, and thus one cannot say with any certainty that the tracked livestock herd indeed invaded a cultivated field and damaged the crops.

With this caveat in mind, Table 11.1 shows, for example, that the mean effective NDVI of Herd A, which consists of about 25 cows, is 0.053, and its cropland frequency is 0.120. It is not obvious whether these

Table 11.1 Statistical profile of herding trajectories

Herd	A	B	C	D
Composition (approx.)	25 cattle	55 goats/sheep	15 cattle	20 goats/sheep
Record freq. (approx. sec)	15	15	30	30
Recording dates	18–20 Sep. 2019	18–20 Sep. 2019	18–20 Sep. 2019	18 Sep. 2019
Recorded points	6,052	6,064	2,808	1,316
Total distance per day (m)	18,363	15,351	16,193	16,642
Max. displacement (m)	4,226	3,648	3,615	4,101
Mean velocity (m/sec)	0.732	0.650	0.578	1.176
Mean effective NDVI	0.053	0.044	0.050	0.028
Cropland freq.	0.120	0.029	0.144	0.036

Source: Produced by the author.
Notes: 'Record freq.': the frequency of recording the geographic coordinate of a herd's location in seconds; 'Recorded points': the total number of recorded locations pooled across the entire recording period (three days for herds A–C, one day for herd D); 'Total distance per day': the total walking distance of the herd per day in metres; 'Max. displacement': the maximum displacement from the herd camp in metres; 'Mean effective NDVI': the mean per-pixel NDVI value computed for the pixels where the herd's location was recorded, excluding NDVI values for pixels that were classified either as bare ground, surface water, cropland, *garmani*, bush, or residential area; 'Cropland freq.': the encountering rate of cropland, that is, the frequency with which the herd was on a pixel classified as cropland when its location was recorded.

values are high or low by themselves. In other words, some benchmarks are needed for quantitatively assessing the efficacy of pastoral micro-mobility. Figure 11.4 illustrates the results of such assessments for Herd A. For these assessments, 6052 (the exact number of recorded points of Herd A's trajectories) locations were first randomly selected from the land plots within a radius of 4.5 km (approximately corresponding to the herd's maximum displacement) centered on the camp of Herd A. The effective NDVI and land use category values were then sampled from the selected locations, and were aggregated to compute a set of benchmark values for the mean effective NDVI and the cropland frequency of Herd A. Finally, this procedure was iterated 1,000 times using different series of pseudorandom numbers to obtain the statistical distributions of these benchmarks. The two histograms in Figure 11.4 depict the obtained benchmark distributions for (A) mean effective NDVI and (B) cropland frequency, respectively. The solid vertical lines in the figure denote the empirical values derived from Herd A's actual trajectories.

As evident from these graphs, Herd A's mean effective NDVI, 0.053, is located far to the right of the entire distribution of benchmark values, whereas its cropland frequency, 0.120, is far to the left of the corresponding baseline distribution. That is, in comparison with the (admittedly not very efficient) theoretical baselines, Herd A clearly achieves a relatively high level of resource access in its daily herding, while markedly reducing

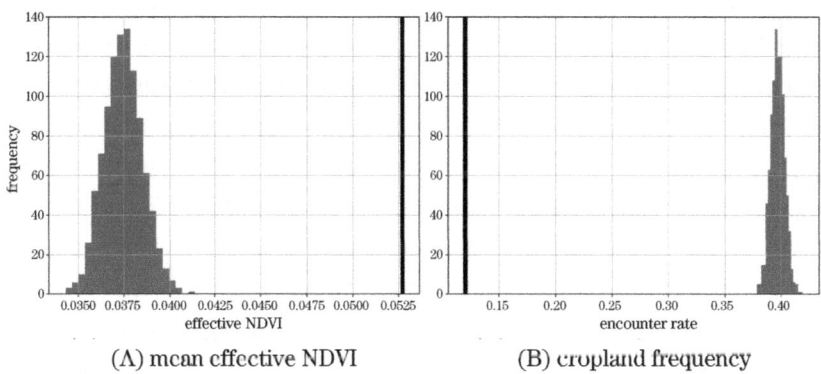

(A) mean effective NDVI   (B) cropland frequency

Figure 11.4 Evaluation of Herd A trajectories

the risk of cropland damage. Similar results emerge for the other live-stock herds. These results provide quantitative evidence that individual pastoralists have shown a considerable degree of local adaptation to their surrounding landscapes with limited resource availability, by strug-gling to strike a difficult balance between resource access and crop damage avoidance.

Limitations to pastoral mobility: Spatial simulation of crop damage

As indicated above, considering the current land use conditions in the study area, individual pastoralists are not unnecessarily damaging crop in their daily herding behavior. On the contrary, it appears that at least some of them are quite adeptly avoiding it. However, the preceding analysis also suggests that the risk of crop damage is never zero. As the numbers in Table 11.1 demonstrate, all the tracked livestock herds passed through (at a spatial precision of 10m × 10m) land parcels that were classified as cropland in some parts of their daily trajectories. Whether or not crops were damaged in these cases is of course another question, since such occurrences are influenced by a wide range of factors, including the composition and size of the herd and the herding skills of individual pastoralists. However, it would obviously be unrealistic to assume that no crops were damaged in these cases. Moreover, the findings presented in the previous section, which were derived from the GPS records of only three days of movements of the four livestock herds managed by two Fulbe households, are far from systematic, both spatially and temporally. In reality, there were more than 100 Fulbe and Tuareg households living in the surrounding area, many of whom similarly moved their livestock to rainy-season camps on the plateau and engaged in daily herding activities over an extended period of time.

To obtain a more systematic assessment of the occurrence of crop damage in the study area, this section investigates the wider spatial implications of micro-level pastoralist behavior by developing a compu-tational model that simulates the daily herding activities of numerous pastoralists in parallel (see Sakamoto 2021: 85–86 for more details). This model consists of virtual agents that represent pastoral households and their livestock herds residing at the 45 camps on the plateau the author visited. The virtual space that these agents inhabit was constructed using the empirical geospatial data (NDVI and LCLU categories) introduced above, whereas each agent's behavior was described by a simple stochastic process that reflects the actual tendencies observed in daily herding (e.g., orientation towards vegetation and water resources, cropland avoidance, etc.). Further, by employing a widely-used method called Approximate Bayesian Computation (ABC) as well as the GPS track records of daily herding, the model's parameters were extensively tuned so that the virtual agents' spatial behavior retains a sufficient correspondence to that of their real-world counterparts.

Figure 11.5 displays the results of simulation experiments wherein this model was run repeatedly using different seeds of pseudorandom numbers. That is, each of the 45 herding agents, interacting with its own local landscape, iterated daily herding from its base camp while slightly changing its route each day. Panel (A) in Figure 11.5 shows how often these agents passed each plot of the simulated landscape. The pixels with the highest luminance denote the land plots visited the most frequently, more specifically, those visited with a frequency of 10% or higher. Panel

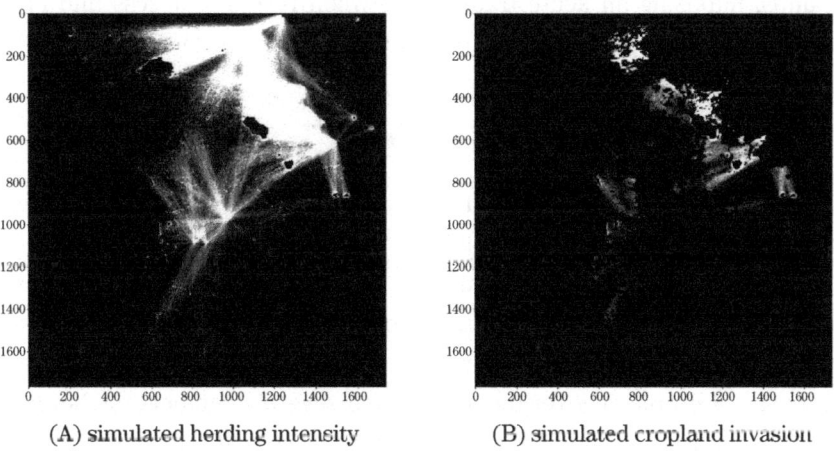

(A) simulated herding intensity   (B) simulated cropland invasion

Figure 11.5 Simulated herding intensity and cropland invasion

(B) is a similar output regarding cropland areas. That is, among the land plots classified as cropland, those visited the most often by the agents are colored the whitest. This latter image depicts a kind of 'hazard map' that visualizes spatial variations in the risk of crop damage incidence over the study area.

According to this map, the land patches predicted to have the highest risk of crop damage – that is, the cropland pixels that the herding agents visited in 10% or more of the entire iterations – constitute 0.877% of the total surface area on a pixel basis (26,911 pixels). Unsurprisingly, these patches are concentrated around several relatively large water sources in the northern half of the area, including two lakes (*Tabuki Dogondoutchi* and *Tabuki Ligido*). Being immediately below part of the plateau with a sizeable number of herding camps, these lakes and ponds are natural destinations for the herders camping there. The extensive spread of pearl millet fields in these agriculturally favorable locations, however, pose a serious challenge for reconciling conflicting land use requirements. Aside from these potential 'hotspots,' Panel (B) suggests other places associated with considerable risk of cropland trespassing, including those around the eastern edge of the extensive cropland that dominates the western half of the study area. There again, cultivated fields, scattered with water sources, are in close proximity to another part of the plateau hosting several herd camps, imposing an acute trade-off between resource access and crop damage avoidance for local pastoralists.

Discussion and conclusions

In the specific context of southwestern Niger, the foregoing analyses present unambiguous answers to the three research questions raised in Introduction. First, at least in this part of the Sahel, local pastoralists are indeed facing formidable constraints on their resource access. The estimated LCLU patterns in the study area have revealed the predominant expanse of croplands over the area, which severely hampers the herders' mobility and, in combination with the extensive spread of *garmani*, significantly reduces available grazing resources. Second, despite these obvious constraints and limitations, the quantitative analysis of the herders' daily trajectories indicates a considerable degree of adaptability in their spatial behavior. That is, their daily movements seem to be constantly adjusted, if not optimized, to ensure access to the patchy resources thinly spread over the area, while avoiding cropland invasion as much as they can. This is a clear sign of the resilience of local pastoralists, which is corroborated by field observations. At a micro level, local herders are indeed highly adept land users. Even when at a distance of just one or two meters from pearl millet fields, these herders often ingeniously keep their herds away from the fields, skillfully grazing their livestock nearby and then quickly leaving the site. The quantitative results obtained above suggest that these efforts on the part of pastoralists bear some fruit, potentially preventing unnecessary crop damage, and thus farmer-herder conflict, at a local level.

Third, nevertheless, on a wider spatial scale, there are definite structural limitations to these efforts. The analyses of the simulated daily herding trajectories of numerous herders over the study area indicate that there remains the likelihood – which might be alarmingly high in some 'hotspots' with a relative abundance of vegetation resources and surface water – that their livestock will enter cultivated fields, thus posing significant risk of crop damage. As mentioned, this does not mean that crop damage, much less open violence, inevitably happens. There are numerous factors and mechanisms working against these consequences. Furthermore, on a much larger scale, climatic and ecological conditions in the Sahel are notoriously variable both intra- and inter-annually. Accordingly, the spatial distribution of crop damage risk itself might considerably change depending on these temporal variations.

Despite all these considerations, however, there are strong reasons to believe that the patterns of risk outlined above are structurally stable and highly persistent. First, one cannot underestimate the spatial con-straints that local pastoralists now face in accessing resource (Figure 11.2). According to Oyama (2017), who manually inspected an aerial photograph of village D and the surrounding area in the colonial era, the spread of cropland was rather limited at the time, leaving a vast expanse of uncultivated land with abundant vegetation. Over recent decades, the landscape has radically changed. This change has largely been driven by massive social-ecological processes unfolding in the entire Sahel region, including, among others, rapid population growth and steady soil erosion (desertification). In particular, Niger has one of the fastest growing populations in the world. Its annual population growth rate, a staggering 3.82% (UN 2019), ensures that its population doubles in less than 19 years, putting enormous pressures on its already meagre resource base. Given their sheer scale and momentum, at least for the foreseeable future, these processes are unlikely to be contained, much less reversed. It is thus reasonable to expect that people's livelihoods, particularly those of the socially and politically disadvantaged such as pastoralists, will continue to be exposed to relentless social and ecological challenges.

In other words, the widespread risk of crop damage, a major trigger of 'farmer-herder conflict,' over the study area, especially around the 'hotspots' suggested above, has deep structural roots. It is inseparable from the resilient functioning, or lack thereof, of the entire social-ecological system that has long sustained the livelihoods and coexistence of people in the area, farmers and pastoralists alike. It is still unclear whether the system has already undergone a 'regime shift' (Walker and Salt 2006: chap. 3; Robinson and Berkes 2010) and transitioned to another 'stability domain' wherein a whole range of pastoral practices based on extensive mobility are no longer tenable. In any case, the foregoing analyses in the context of southwestern Niger at least suggest that pastoralists' traditional coping mechanism, namely mobility (Niamir-Fuller and Turner 1999), has largely been disrupted under the weight of enormous social and ecological pressures even at the most micro level (i.e., during daily herding), and the widespread nature of such disruption has clearly been driving the worrying trend towards the increased occurrence of pastoralist-involved conflicts with the accompanying losses of life. Mitigating and reversing this systemic trend obviously requires an extensive investigation of the resilience at much larger spatial and temporal scales, which will be addressed in future work.

Acknowledgements

I deeply appreciate the guidance and support of my collaborator, Shuichi Oyama. My thanks also go to friends and collaborators in Niger for their support and hospitality. I am very grateful to the editors of this book, Shinya Konaka and Greta Semplici, for their constructive and insightful comments on the manuscript. Finally, I acknowledge the financial support of the Japan Society for the Promotion of Science (JSPS) (JSPS KAKENHI; Grant Numbers: 17H04506 and 18H03606).

References

Bassett, T.J. 1986. 'Fulani herd movements.' *Geographical Review* 76 (3): 233–248. doi: 10.2307/214143.

Bukari, K.N., P. Sow, and J. Scheffran. 2018. 'Cooperation and co-existence between farmers and herders in the midst of violent farmer-herder conflicts in Ghana.' *African Studies Review* 61 (2):78–102. doi: 10.1017/asr.2017.124.

Buzan, B., O. Wæver, and J. de Wilde. 1998. *Security: A new framework for analysis.* Boulder, Colo.: Lynne Rienner Pub.

Delsol, G. 2020. *UN Peacekeeping Operations and Pastoralism-Related Insecurity: Adopting a coordinated approach for the Sahel.* New York: International Peace Institute.

Galvin, K.A. 2009. 'Transitions: Pastoralists living with change.' *Annual Review of Anthropology* 38 (1): 185–198. doi: doi:10.1146/annurev-anthro-091908-164442.

Hussein, K., J. Sumberg, and D. Seddon. 1999. 'Increasing violent conflict between herders and farmers in Africa: Claims and evidence.' *Development Policy Review* 17 (4): 397–418. doi: https://doi.org/10.1111/1467-7679.00094

Ji, S., C. Zhang, A. Xu, Y. Shi, and Y. Duan. 2018. '3D convolutional neural networks for crop classification with multi-temporal remote sensing images.' *Remote Sensing* 10 (1): 75.

Kirikoshi, H. 2017. 'Economic differentiation and safeguard against hunger in southern Niger of Sahel region: Agro-landscape and multi-purpose tree use of Hausa farmers.' *Japanese Journal of Human Geography* 69 (1): 43–56. doi: 10.4200/jjhg.69.01_043.

Krätli, S., and C. Toulmin. 2020. *Farmer-herder conflict in sub-Saharan Africa?* London: International Institute for Environment and Development (IIED).

Moritz, M. 2006. 'The politics of permanent conflict: Farmer-herder conflicts in northern Cameroon.' *Canadian Journal of African Studies/Revue canadienne des études africaines* 40 (1): 101–126. doi: 10.1080/00083968.2006.10751337.

Niamir-Fuller, M., and M.D. Turner. 1999. 'A review of recent literature on pastoralism and transhumance in Africa.' In Maryam Niamir-Fuller. (ed). *Managing Mobility in African Rangelands: The legitimization of transhumance.* London: Food and Agriculture Organization of the United Nations: Beijer International Institute of Ecological Economics; IT Publications.

Oyama, S. 2014. 'Farmer-herder conflict, land rehabilitation, and conflict prevention in the Sahel region of West Africa.' *African Study Monographs.* Supplementary issue. (50): 103–122. doi: info:doi/10.14989/189724.

———— 2015. *Tackling the Land Degradation in Sahel Region of West Africa: Trash input for land rehabilitation, food security and conflict prevention.* Kyoto: Showado (in Japanese).

———— 2017. 'Hunger, poverty and economic differentiation generated by traditional custom in villages in the Sahel, West Africa.' *Japanese Journal of Human Geography* 69 (1): 27–42. doi: 10.4200/jjhg.69.01_027.

Reid, R.S., M.E. Fernández-Giménez, and K.A. Galvin. 2014. 'Dynamics and resilience of rangelands and pastoral peoples around the globe.' *Annual Review of Environment and Resources* 39 (1): 217–242. doi: 10.1146/annurev-environ-020713-163329.

Robinson, L.W., and F. Berkes. 2010. 'Applying resilience thinking to questions of policy for pastoralist systems: Lessons from the Gabra of northern Kenya.' *Human Ecology* 38 (3): 335–350. doi: 10.1007/s10745-010-9327-1.

Sakamoto, T. 2021. 'Human ecological foundations of farmer-herder conflict in the Sahel: Combining field observation, remote sensing and computational modelling.' In Ochiai, T., M. Hirano-Nomoto and D. E. Agbiboa. (eds). *People, Predicaments and Potentials in Africa.* Bamenda: Langaa RPCIG. pp. 73–93.

Tueller, P.T. 1989. 'Remote sensing technology for rangeland management applications.' *Journal of Range Management* 42 (6): 442–453.

Turner, M.D. 2004. 'Political ecology and the moral dimensions of "resource conflicts": The case of farmer–herder conflicts in the Sahel.' *Political Geography* 23 (7): 863–889. doi: http://dx.doi.org/10.1016/j.polgeo.2004.05.009

Turner, M.D., A.A. Ayantunde, K.P. Patterson, and E.. Patterson. 2011. 'Livelihood transitions and the changing nature of farmer–herder conflict in Sahelian West Africa.' *The Journal of Development Studies* 47 (2): 183–206. doi: 10.1080/00220381003599352.

United Nations (UN). 2018. *Activities of the United Nations Office for West Africa and the Sahel: Report of the Secretary-General.* S/2018/1175 (28 December 2018).

United Nations (UN). 2019. *World Population Prospects 2019* (https://population. un.org/wpp/, accessed on April 30 in 2022).

United Nations Office for West Africa and the Sahel (UNOWAS). 2018. *Pastoralism and Security in West Africa and the Sahel: Towards Peaceful Coexistence.* UNOWAS Study. Dakar.

Waldner, F., and F.I. Diakogiannis. 2020. 'Deep learning on edge: Extracting field boundaries from satellite images with a convolutional neural network.' *Remote Sensing of Environment* 245: 111741. doi: https://doi.org/10.1016/j.rse.2020.111741

Walker, B.H., and D. Salt. 2006. *Resilience Thinking: Sustaining ecosystems and people in a changing world*. Washington, DC: Island Press.

The Resilience of Former Refugees in Rural Zambia

Rumiko Murao

Introduction

The importance of state leadership and national systems in protecting human rights and responding to shocks has been recognized and reinforced in the "2030 Agenda for Sustainable Development," and the 2016 "New York Declaration for Refugees and Migrants," as well as other recent initiatives. The United Nations High Commissioner for Refugees Executive Committee stated that "resilience refers to the ability of individuals, households, communities, national institutions, and systems to prevent, absorb, and recover from shocks, while continuing to function and adapt in a way that supports long-term prospects for sustainable development, peace, security, and human rights" (UNHCR ECOM 2017: 3). Even before these initiatives began, refugee studies critically examined arguments based on the assumption that refugees are vulnerable, demonstrating that refugees are fully capable of re-establishing their own livelihoods by acquiring land through the internal support systems within refugee communities (Hansen 1979). Though analyses of these support systems generally paint positive pictures of refugees' resilience, cohesion, and benevolence, they have also been the basis for the introduction of new measures in recent years (Omata 2012). The aim of these new measures is to reduce the burden of sharing for refugees, which has resulted in further impoverishing refugees in developing countries, such as those in Africa, who are in extreme conditions. However, the sites for refugee assistance are diverse, as are the changing forms of autonomy constructed by refugees. Without oversimplifying the resilience of refugees, which varies from period to period and place to place, this paper focuses on the livelihood activities of the Mbunda, an agrarian former Angolan refugee community in the context of forced resettlement under a local integration project in Zambia.

Local integration, one of three durable solutions for refugees, was emphasised in the Comprehensive Refugee Response Framework (CRRF) implemented in Zambia (United Nations High Commissioner for Refugees [UNHCR] 2018). Local integration allows refugees to permanently settle and rebuild their lives in the first country of asylum (Pillay 2011; Verduijn 2018). However, because many refugees stay as protracted refugees more than 5 years and have already rebuilt their lives, the concept of local integration has not been necessarily matched to the African cases such as Zambia (Omata 2017). For more than fifty years, Zambia has been a peaceful and humanitarian country that has hosted refugees from neighbouring countries, especially in regions close to the border. Some refugees were born in Zambia who have not been to the country of origin. After the cessation of refugee status was declared for Angolan refugees in 2012, and then for Rwandan refugees in 2013, many Angolans and Rwandans remained in Zambia after organized repatriation, refusing to return to their countries of origin. One such group was the Mbunda people who had fled from Angola. This prompted the Zambian government to implement local integration and resettlement projects supported by CRRF funding. Through these projects, the Mbunda people were granted legal status and land rights to live and cultivate land in Zambia from 2014; however, they had to set up their lives autonomously without humanitarian assistance for refugees (Murao 2020).

In recent years, other African states have also amended relevant legislation to promote private land ownership and grant land rights, with the aim of settling nomadic populations, intervening in resource management, accepting foreign investment in the agricultural sector, and boosting the economy. Cases of increased private ownership resulting from ambiguous land rights legislation have been reported. For example, in East African pastoral societies, the main factor contributing to socio-economic stability was mobility; comprising seasonal migration of herds and rotational grazing often across vast territories based on their residential and grazing unit (Niamir-Fuller 1998).

However, as overgrazing in East Africa became an issue, the Maasai – East African pastoralists – were divided into common landholdings, and government-led common land tenure based on collective ranching, introduced by the Land Act in Kenya. In recent years, land reform has led to further land division and the emergence of individuals who have set aside land as private property (Galaty 1993; Mwangi 2007a, 2007b; Ntiati 2002). In Ethiopia, direct foreign investment in the agricultural sector has led to large-scale land enclosure in pastoralist settlements in the south (Sagawa 2016). The land enclosure in Ethiopian pastoralist societies have resulted in the division of residential areas of the same lineage group with both the youth and the elders. Whereas, in agrarian societies, there are reports that after the land act changed to promote private land ownership and grant land rights with the aim of boosting the economy by the state, the strengthening of the role of traditional chiefs has helped to ensure that the village settlements remain unaffected by the land enclosures (Oyama 2015; Takeuchi 2017). These reports have attracted much attention because the agrarian societies tended to be focused on their chief's power when the enclosure of land occurred.

In Zambia, land law reform led to increased control of land by certain traditional authorities, such as chiefs enclosing land, etc. (Oyama 2015). The granting of land rights by the local integration projects to the Mbunda people incorporated elements of community development, enabling former refugees to achieve economic independence like the rest of the population and improved their livelihoods.

The Mbunda, like many farmers in Zambia and elsewhere in Africa, were aware of their rights of land use but did not have private ownership of land, since the chiefs and heads of kinship-based residential groups are in charge of land allocation (Murao 2012). After the liberation war in Angola, the Mbunda were hosted both as refugees by the UNHCR and as migrants by their kin, chiefs, and related ethnic groups from eastern Angola. The

Mbunda people lived together within a *limbo* (a matrilineal kinship group, discussed in more detail below) with relatives and friends from Zambia; in some cases, even without the Mbunda traditional chiefs' assistance (Murao 2012). Residents of the refugee settlement of Zambia lived under bureaucratic conditions. Particularly, in the Mayukuwayukwa Refugee Settlement in the Western Province of Zambia, residents were given land to cultivate and had to achieve food autonomy after they initially received humanitarian aid as newcomers as well as after they reorganized their social relationships like membership in *limbo* (Murao 2020).

Some studies have shown the former Angolan refugees in refugee settlements were unsuccessfully integrated after the implementation of the cessation clause in Zambia (Nyamazana et al. 2017; Development and Training Services 2014; Frischkorn 2013). These studies clarified the livelihood changes, strategies, and identities of former refugees in Zambia. Other studies feature Angolan refugees who have successfully settled in the local villages of Zambia, called self-settled refugees (Hansen 1979; Bakewell 2002; Murao 2012). Some studies suggest that these cases are examples of successful local integration because they were hosted and supported by chiefs and matrilineal kin and rarely identified them as refugees. The dangers of private land tenure settings leading to increased inequality and poverty have also been identified (cf. Rutten 1992). However, in the Mbunda of Zambia case, the legal status and resettlement projects involving the acquisition of land rights have not been identified, especially agriculture through local integration projects.

This chapter aims to clarify how the refugee peasants, the Mbunda people, reconstructed their livelihood based on agriculture with land rights after implementation of the local integration project. I focus on how they reorganized social relations both in the refugee settlements and in the resettlement schemes to examine the internal support that has aided their resilience.

The fieldwork was conducted by interviews with sixty-four people from the Mayukwayukwa Refugee Settlement (Mayukwayukwa) and the adjacent resettlement scheme, for a period of one month in 2007, and one and a half months between 2016 to 2020. This study examines the livelihood and society of the Mbunda people in Angola as well as their history of exile; the lives in Mayukwayukwa and the impact of regional integration on residential group formation; the livelihood changes in the refugee settlement and resettlement scheme through the local integration projects; the role of daily social relationships in their livelihood strategies.

The Mbunda people and Mayukwayukwa of Western Zambia

The Mbunda people are a Bantu-speaking people who live in the Democratic Republic of the Congo (DRC), Eastern Angola, and Western Zambia (Von Oppen 1996). Previous studies have noted that the Mbunda people originated from the Luba and Lunda Kingdoms of the DRC (Cheke 1994; McCulloch 1951) where the chiefdoms flourished under their king. After they arrived in Upper Zambezi in Angola, the ninth king expanded the territory further, trading materials such as arms, hoes, crops, and animal skins with other ethnic groups, including the Chokwe, Luchazi, Luvale, and Lozi. The groups sometimes fought over intermarriages and territory.

The Mbunda people lived along small rivers and engaged in agriculture in shifting cultivation fields and wetlands located near these rivers. They engaged in pearl millet farming, hunting, gathering, and fishing. After the Portuguese arrived in Angola, they also began to cultivate cassava and maize, which they ate as a thick porridge, and which remains a staple food in the region. They also boiled maize that was harvested early while it was still green.

A Mbunda *limbo* (pl. *membo*) comprised three to four generations of matrilineal kin who reside together. A *limbo* was traditionally a political and reciprocal unit with a headman. According to the Mbunda people in Angola, the headman of a *limbo* is responsible for deciding where the people belonging to that *limbo* will cultivate crops. The chief rules over a few *membo* and distributes their land as well as resolving problems and conflicts.

The Mbunda people were forced to migrate following the Portuguese invasion of the Moxico province in 1911. The Angolan War of Independence began in 1961, after Portugal and Britain divided Angola's territory in 1926. The Mbunda people joined the conflict as freedom fighters. During this time, the Portuguese colonizers set fire to villages, and more Mbunda people fled to Zambia, settling in local villages. In 1966, independent Zambia's first official refugee settlement, Mayukwayukwa was established in the Kaoma District, 85 km west of Kaoma (Fig. 12.1). The settlement is one of the oldest refugee settlements in Africa. By late 1967, the total population at Mayukwayukwa had reached 2,200 (Clark & Stein 1985).

The Mbunda people traditionally settled in local villages with chiefs, kin, and acquaintances and mainly cultivated cassava. However, in refugee settlements such as Mayukwayukwa, the Mbunda people resided with chiefs who had no authority over land distribution and lived under bureaucratic control. Because there was little humanitarian assistance in Mayukwayukwa in 1966, these people built their own houses. The Mbunda people started to form their *membo* with their kin and were able to cultivate

Table 12.1 History of the Mbunda in Zambia and Angola

By early 19th century	Mbunda people originated from the Luba and Lunda Kingdoms of the DRC Expanded the territory to eastern Angola, western Zambia and Namibia
1911	Portuguese invasion of Moxico province, Angola initiates forced migration to Zambia
1961	Angolan war of independence begins
1964	Zambia becomes independent
1966	Mayukwayukwa Refugee Settlement established in Zambia
1975	Angola becomes independent and civil war begins
2002	Peace accord in Angola
2003	Repatriation from Mayukwayukwa to Angola organized
~2007	Zambia Initiative
2012	Cessation clause declared from Angolan refugees
2014	"Strategic Framework for the Local Integration of Former Refugees" in Zambia was declared
~2016	Local Integration project begins
2016	Relocation of plots in the settlement scheme begins
	New York declaration for refugees and migrants Comprehensive Refugee Response Framework (CRRF)
~2022	Sustainable Resettlement Program begins

fields on their own. They also began to receive humanitarian aid in the form of maize, rice, beans, vegetables, and so on.

After the Mbunda settled in Mayukwayukwa, they severed ties with the former *limbo* inhabitants. This was not the case when the *limbo* inhabitants split in the Mbunda villages of Zambia and Angola. In Mayukwayukwa, the places of abode had been allocated by officers of the High Commissioner of Refugees. Residents could not easily relocate because they needed permission from the refugee officer. Some *limbo* residents moved to other places in Mayukwayukwa because of witchcraft or disagreements and in some cases women and children had come and gone because of marriage or divorce.

When Angola gained independence from Portuguese colonial rule in 1975, a civil war soon followed. Similar to other Angolan refugees in Mayukwayukwa, the Mbunda did not return to their country. More people, including the Mbunda, escaped from eastern Angola where the MPLA (People's Movement for the Liberation of Angola) was fighting the guerrilla

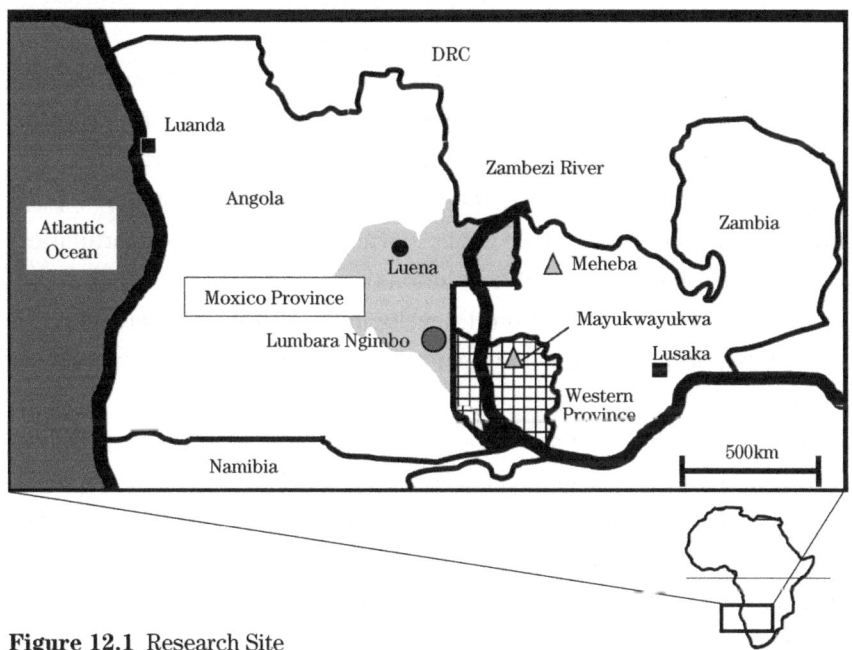

Figure 12.1 Research Site

forces of the UNITA (National Union for Total Independence of Angola) and became protracted refugees.

Further waves of refugees arrived following Zambia's independence in 1964, following the Angolan and Mozambican liberation struggles, the Zimbabwean war, and the Namibian struggle (Burdette 1984). Zambia needed to maintain a politically friendly stance with its neighbours, protect itself from any incursions of violent conflict, and at the same time, make provisions for the refugee populations expanding both in size and in length of stay (Murao 2020).

As the refugee population in Zambia expanded, reaching its peak of nearly 300,000 registered refugees in 2001, and the pressures of protracted refugee-hosting increased, the Government of the Republic of Zambia (GRZ) began seeking alternatives to its continued maintenance of refugee settlement schemes, including repatriation and local integration after the Angolan war ended in 2002 (Sumbwa 2012). Between 2003 and 2007, GRZ together with the UNHCR conducted 'organized repatriation' of tens of thousands of Angolans, many directly from Mayukwayukwa. The organized repatriation process returned approximately 74,000 refugees to Angola from 2003 to 2007, including some Mbunda.

At the same time, the Zambia Initiative (ZI) – a DLI Governmental holistic approach – was implemented from 2003 to 2007 for Angolan refugees including the Mbunda in Mayukwayukwa (UNHCR 2018). The main objective was to alleviate the combined effects of food deficit, poor infrastructure, limited access to public services and economic opportunities, and in the process find durable solutions for refugees. Through the ZI, a learning process for further rural development projects for refugees, the 'Local Integration' (LI) project was developed by international actors such as the UNHCR.

By 2010, there were approximately 9,000 Angolans remaining at Mayuk-wayukwa. Then, the cessation clause was declared for Angolan refugees in 2012 and for Rwandan refugees in 2013. As such, LI managed the Angolan and Rwandan people as former refugees in Zambia, including the Mbunda in Mayukwayukwa refugee settlements. GRZ pledged to integrate former Angolan refugees locally in late 2011, and the Minister of Home Affairs approved the expansion of the LI criteria in 2015 to include almost all remaining former Angolan refugees and 4,000 Rwandans whose status changed in the summer of 2013. As of January 2016, about 6,500 Angolan refugees and 140 Rwandan refugees were almost all eligible for resident permits and identity documents.

After the LI started, the former refugees with occupancy letters were entitled to receive a five- or ten-hectare plot to settle on and cultivate. The plots were vast (over 100 square kms in Mayukwayukwa and over 300 square kms in Meheba Refugee Settlements in Kalumbila District, North Western Province) and were expected ultimately to grow to around 8,000 to 10,000 households (approximately 30,000 to 40,000 persons). These new resettlement schemes were also opened to Zambian citizens to help integrate the new permanent residents. All these new resettlement schemes were near the refugee settlements of Mayukwayukwa and Meheba.

In 2014, the 'Strategic Framework for the Local Integration of Former Refugees' in Zambia was designed to assist the GRZ in advancing durable solutions for former refugees still in the country who wanted to be locally integrated. It brought together key programmatic, coordination, and advocacy objectives that the GRZ and its partners pursued to support the implementation of the LI initiative from 2013 to 2015. It was envisaged that the main activities would be implemented in two years, from 2014 to 2015, with an expected extension into 2016, particularly focused on the acquisition of formal land titles (deeds) and potential late relocations of a limited number of beneficiaries. As of January 2016, 5,412 refugees and forty-four asylum seekers were living in Zambia. Additionally, 6,561 former Angolan refugees and 141 former Rwandan refugees whose legal grants were completed through the LI project were in the process of obtaining residence permits and being resettled from the refugee settlements to the resettlement scheme plots.

Continuation of the LI strategy – a partnership between the office of the Vice-President, Department of Resettlement, and the United Nations Development Program (UNDP) – was identified as a sustainable resettlement program from 2017 to 2021, after CRRF was declared in 2016. Even when the LI project terminated, Mayukwayukwa and Meheba settlements lacked connections to local planning and administrative structures, adequate services, and facilities for agriculture as well as other livelihood options, and had insufficient health, education, social, transport, and communication infrastructure.

Lives in the Mayukwayukwa resettlement scheme

This section describes the impact of regional integration on livelihood changes in the refugee settlement and resettlement through the local integration projects. The resettlement scheme in Mayukwayukwa was established by expanding part of the southern plot of the Mayukwayukwa refugee resettlement area in 2014; a 150 km² resettlement area (Fig. 12.1). The Ministry of Lands arranged plots of five and ten ha each for redistribution to settlers by GPS (Fig. 12.2), and Habitat for Humanity (NGO) provided supporting poles and cement materials for the construction of houses in each plot. Land allocation by GRZ was initially done from the roadside of

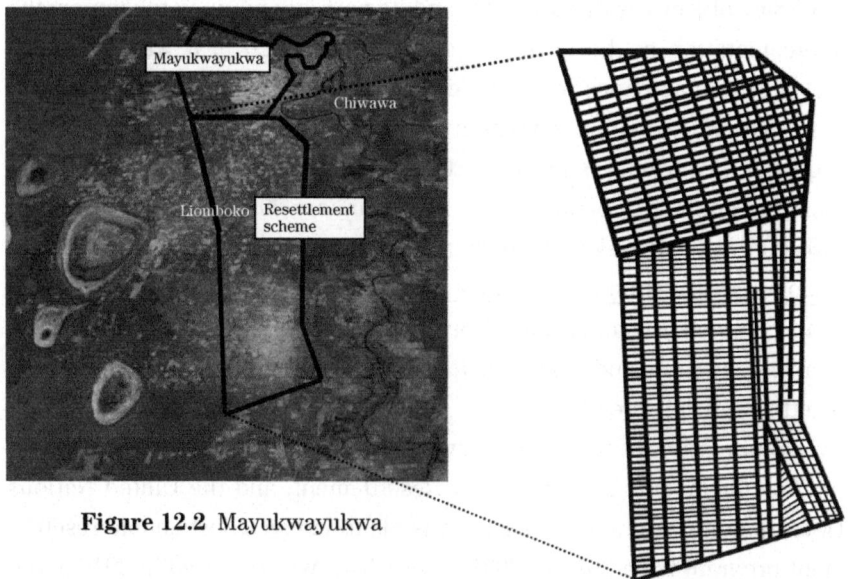

Figure 12.2 Mayukwayukwa

Figure 12.3 Resettlement scheme

the refugee settlement, with priority given to socially vulnerable groups such as the elderly, widowed households, and the physically challenged. In addition, the NGOs International Development Enterprises (iDE) and Concern provided agricultural materials, and a clinic was being constructed. In the resettlement plot, Zambian nationals and former Angolan and Rwandan refugee households had begun to settle in scattered locations and were cultivating the land around their houses.

As repatriation to Angola and resettlements occurred simultaneously, the Angolans were not necessarily refugees in the conventional understanding of the term. Having arrived in 1966 as young children, many of the elders in Mayukwayukwa had only vague memories of life in Angola and a limited sense of Angola as their place of origin. Some of the interviewees were born in Mayukwayukwa and had never known Angola, aside from stories told by their parents and grandparents. There were third and fourth generation Angolan refugees within Mayukwayukwa who had never crossed an international border or sought refuge. Additionally, many second and third generation Angolans were married to Zambians; however, even the children in those marriages were considered to be Angolan refugees.

The Angolan former refugees, including the Mbunda, earned their livelihoods by cultivating vegetables for sale (21.4%), household work (12.5%), and labour exchange, retail work, etc. (27.4%) (Nyamazana et al., 2017). They mixed up these activities in their daily lives. In particular, the Mbunda made a living from agriculture within Mayukwayukwa.

Mayukwayukwa is located along the river, and the Ministry of Agriculture, NGOs, and other organizations implemented large and small projects to support vegetable cultivation in the surrounding lowland fields (*miyaka*). Many people grew vegetables in this area in response to the market demand in Mayukwayukwa. In addition to cassava and maize, which were mainly grown in shifting cultivation fields called *lihya*, similar crops were grown for subsistence on *muhanga*, fields on the slopes of small hills in Mayukwayukwa (Fig. 12.4). Cowpeas and groundnuts were also grown in shifting cultivation fields and slope fields. Meanwhile, the resettlement plots were created on small hills that included shifting cultivation fields and abandoned lands that had already been cultivated by former refugees

(Fig. 12.5). The Mbunda residents tended to cluster in *limbo* dwellings next to each other in Mayukwayukwa by sector. The land was titled to the government and could be occupied and used by refugees but could not be bought or sold. Their shifting cultivation fields were loosely managed, with land near *limbo* available for use by anyone in the *limbo*. In addition, the shifting cultivation fields were managed loosely, with the land near the *limbo*.

The UNHCR and the Zambian government approached the chief, who oversaw land use in the area and bought the land for the refugees in the settlement. During the research in Mayukwayukwa in 2007, *membo* (sl. *limbo*) residents, particularly those living closer to Mayukwayukwa, who lived next to each other, individually cultivated the land between the forests where they employed shifting cultivation in the fields. When disputes arose, the Mbunda and others living in the refugee settlements usually resolved them through discussions between *limbo* residents, or between the parties concerned if they were from different *membo*. If the dispute remained unresolved, it was reported to the officer in charge of the Commissioner for Refugee, who ruled on it.

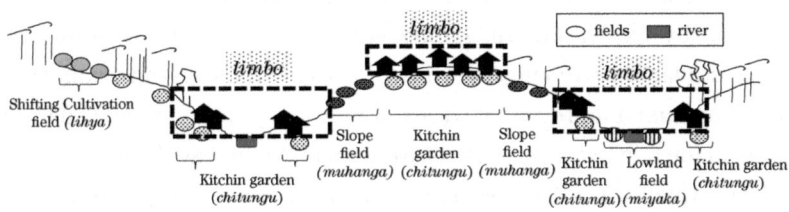

Figure 12.4 Land use in Mayukwayukwa (conceptual diagram)

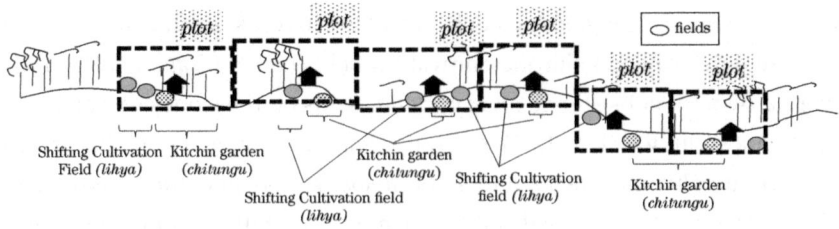

Figure 12.5 Land use in resettlement scheme (conceptual diagram)

Local integration projects began, and the resettlement was subdivided into numbered plots. Although there were delays in issuing land titles, some of the former refugees, such as the Mbunda, acquired five or ten hectares of new land, including others' shifting cultivation fields and abandoned fields. They had both housing and arable land within their plots and earned income by growing cassava, maize, cowpeas, pumpkins, and groundnuts, which they sold at the Mayukwayukwa markets that were two hours away by foot. Lowland fields, slope fields, and kitchen gardens, which had been cultivated in the refugee settlement, were abandoned when the former refugees moved to their own plots in the resettlement. The main drivers of change in the Mbunda livelihood after resettlement was that the soil of resettlement scheme was not fertile enough to grow vegetables. Thus, the resettlement scheme could be seen as a loss, especially for those who had cultivated vegetables in lowland fields that were only available in the refugee settlement.

Table 12.2 shows the crops grown by former Angolan refugee households including the Mbunda, Kwa-mashi and Chokwe people, resettled in 2017. Fourteen of the sixteen former refugees in the resettlement scheme grew crops for their own consumption and for sale, of which, cassava and maize were the most common. They generally produced for their own consumption and sold any surplus when they had enough for themselves. Their cash income changed from vegetable sales from the lowland fields to the sale of cassava and maize when they abandoned these fields. Though some had cultivated rice for a long time, near inland wet land (*dambo*) for cash income through negotiation with local Zambian chiefs, they began relying on the resettlement area through these integration programs. Some people maintained their livelihoods by simplifying the crops grown and increasing yields in the area. This was a change from their livelihoods in the refugee settlements, where people had a variety of arable land, cultivated a variety of crops, and sold vegetables with high commercial value. Thus, through the LI project, forestlands that had once been allocated semi-autonomously, with refugees sharing the land with each other, were subdivided, and promoted for cultivation in this arid environment.

Table 12.2 Crops grown by former refugee households in the resettlement scheme

No.	1		2		3		4		5	
	food	cash	food	cash	food	cash	food	cash	food	cash
	cassava	cassava	cassava		cassava	cassava	cassava	cassava	cassava	cassava
	maize	maize	maize		maize	ground nuts			maize	pearl millet
	rice	rice			ground nuts					maize
	ground nuts	ground nuts			pumpkin					cow pea
	cow pea									ground nuts
	pumpkin									chindambe

No.	6		7		8		9		10	
	food	cash	food	cash	food	cash	food	cash	food	cash
	cassava	rice	maize	cassava	maize	maize	maize	maize	maize	
	maize	maize	cassava	ground nuts	cassava	cow pea	cassava	cassava		
	ground nuts		ground nuts	rice	ground nuts		cow pea	cow pea		
	pumpkin		rice		cow pea		ground nuts			
	cow pea									
	sweet potato									

No.	11		12		13		14		15	
	food	cash	food	cash	food	cash	food	cash	food	cash
	maize	cassava	maize	maize	maize	maize	cassava	cow pea	cassava	cow pea
	cow pea	maize			cassava	cassava	maize		maize	
	cassava				ground nuts	ground nuts	ground nuts		cow pea	

Note: No. in the table shows households the number of former Angolan refugees in the resettlement scheme.

Land use by separated *limbo* residents

This section looks at how the Mbunda reorganized their social relationships according to their daily livelihood based on agriculture, both in the refugee settlement and the resettlement scheme, to examine their resilience. As noted in the previous sections, before LI and the resettlement started, they lived in *limbo* (pl. *membo*), where the principal ethnic group existed. In Mayukwayukwa in 2007, when ZI were underway, the remaining Angolan refugees were living in their own *membo* in their sector (Table 12.3). Table 12.4 shows an example of the composition of ethnic groups in the Sector 4 *membo*. At that time, there were seven Mbunda *membo*. Among them, No. 6 *limbo* had ten inhabitants when visited in 2007. However, after repatriation to Angola, the population decreased, then increased again by marriages and births.

After implementing LI in 2014 and resettlement in 2017, at the end of 2019, only 420 households (21.6% of total) of the Angolan former refugees moved to the resettlement scheme. The Mbunda both in Mayukwayukwa and in the resettlement scheme still lived in *limbo* reorganizing the memberships. Here I will focus on a few cases of *limbo* 6. In *limbo* 6, all of them stayed together before local integration projects started. The composition and

Table 12.3 Ethnic groups in each *limbo* of Sector 4

No.	Ethnic Group
1	Chokwe
2	Mbunda
3	Chokwe
4	Chokwe
5	Chokwe
6	Mbunda
7	Mbunda and Chokwe
8	Mbunda
9	Nyemba
10	Mbunda
11	Mbunda
12	Mbunda
13	Mbunda
14	Nyemba

Table 12.4 Population in each sector

Sector No.	households	population
1	92	520
2	140	401
3	92	276
4	130	578
5	105	387
6	157	552
7	116	473
9	144	592

population of the No. 6 *limbo* of Sector 4 in 2020 is shown in Figure 12.6. In the No. 6 *limbo* of Sector 4, there were five remaining people in 2020. Others, like M, had already moved to a resettlement plot (Fig. 12.7). The date in parentheses shows the issue dates of each identification or letters of the legal status and rights.

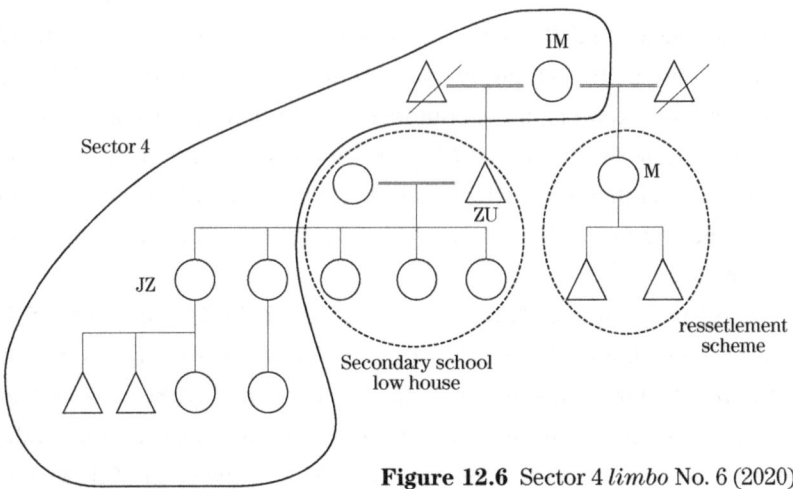

Figure 12.6 Sector 4 *limbo* No. 6 (2020)

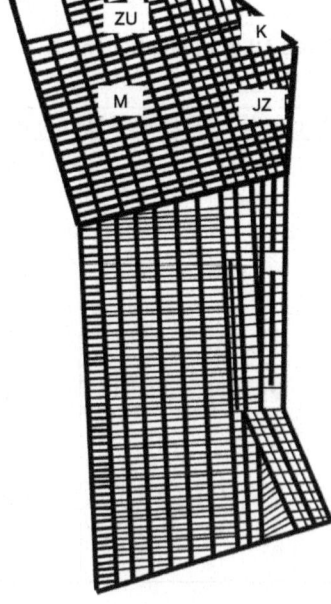

Figure 12.7
Plot sites of *limbo* No. 6 of Sector 4 and No. 15 of Sector 1

Case 1: M (born 1981, Mbunda woman)

I was born in a hospital in Mangango, nearby Mayukwayukwa. My parents came from Lumbara Ngimbo in Angola to Mayukwayukwa in 1966. I lived in Sector 4 when LI started, but decided I had to leave and moved to the resettlement area. Before I moved to the area, I sold maize. I cultivated around my plot before I was given a plot; therefore, the plot was my old fields. I abandoned the fields near *limbo* 6 when I moved to this plot. Habitat for Humanity (NGO) finished building the house in October 16, 2014. At that time, I was dissatisfied with the drinking water coloured in orange and the unknown bad conditions. In 2015, I received maize, groundnuts, cowpea, Chinese cabbage, onions, and African eggplants (*impwa*), as well as a hoe, an axe, cooking tools, and a machete from the NGO Concern and I moved to my plot. In 2016, I received only a few crop seeds and fertilizers. I did not participate in the agricultural cooperative societies under Ministry of Agriculture of Zambia. In 2016, I also got cassava stems from the uncle of my husband who stayed in the other sector of Mayukwayukwa and transplanted them in 2017. Again, I transplanted cassava stems from the field in 2017 and bought maize seeds at the market in 2018 to plant around the house. I grew maize, cowpeas, and pumpkins. After harvesting maize and cowpeas, I sold them at the market in Mayukwayukwa. In addition, I have earned cash by working as an agricultural labourer in Mayukuwayukwa and the resettlement scheme, where I have two fields. In 2016, I made a small field of vegetables and planted cabbage and tomatoes, but they were infested by insects. Last year, because it did not rain, people in Mayukwayukwa could not harvest enough crops like maize there. Sector 4 people including *limbo* 6 people came to visit us to buy cassava with peace work. I do not have a passport or residence permit, but a registration card issued by the Angolan government (January 13, 2014), alien card issued by the Zambian government (January 9, 2014), and occupancy letter (October 28, 2014).

Because M abandoned her fields in Mayukwayukwa when she moved, her fields were not fragmentated in 2020. M gained five ha of her plot, which is larger than what she had before she moved to the area and concentrated her efforts there. Rather than growing a diverse range of crops, both with the support of the NGO and her trials, she increased her cassava cultivation area for food with support from her relatives who stayed in another *limbo* of Sector 4 in Mayukwayukwa. She did this to sell maize and cowpeas as well if there was a surplus. She also had to travel more than two hours on foot to the market in Mayukwayukwa to sell her crops or was dependent on income from labour in Mayukwayukwa. However, Mayukwayukwa, which is lower in elevation than the resettlement scheme, has a poorer crop yield with less water content in the soil. A new relationship is developing with the growing number of former refugees who are trying to buy cassava by working in the fields at the resettlement scheme.

As mentioned above, M and her children were allocated their plot in 2015; however, other *limbo* residents were still living in Sector 4 and other places in Mayukwayukwa. M's half-brother ZU, lived in a row house for general workers of the Secondary School, having his plot in the resettlement area. The case of ZU shows the land he used were fragmentated both in Mayukwayukwa and in the resettlement scheme.

Case 2: ZU (born 1972, Mbunda man)

I live in a row house for general workers at Mayukwayukwa Secondary School. I started managing the water pumps at Peace Works and the Secondary School in 2014 when I moved from Sector 4. The house in my plot where I cultivated before has only a shelter, iron sheets, and bricks in the resettlement scheme, though I did not move there. However, I was given cassava stems by the NGO and my relatives in Sector 4. I continue to transplant cassava stems and cultivate cassava and maize in my plot. I have a resident alien identity card issued by the Zambian government on May 13, 2013. I haven't received other identifications; residence permit, aliens' card, and occupancy letter (August 12, 2014) are available.

> I cultivate maize and pumpkins near the school. I sell those in Sector
> 4. I also get cash through my peace work at the school. I can be on leave
> from work at school and may go to my plot after the rainy season.

ZU was earning cash by renting housing at a secondary school in Mayukwayukwa at a discount and working as a pump manager. He was also allowed to cultivate land near the school as an employee; he cultivated crops and sold it in Sector 4, including *limbo* 6 residents. Although he had moved out of *limbo* 6 in Sector 4 during LI, he was still able to maintain his slope fields in Sector 4 nearby *limbo* 6, lowland fields that he could cultivate near the secondary school where he worked, and the resettlement plot where he could do shifting cultivation fields, eat, and sell his crops. ZU wanted to continue to expand his subsistence by growing his own crops to eat and sell. In doing so, he was able to utilize his connection with Sector 4 as a source of assistance. ZU's case shows that through the local integration projects, the Mbunda could maintain the land of the *limbo* 6. The land of the *limbo* 6 people continued to be cultivated by the remaining people of *limbo* 6 in 2018 like one of ZU's daughter, JZ with her mother.

Case 3: JZ (born 1994, Mbunda woman)

> I have a plot in the resettlement scheme. My mother is aged. I have a
> registration card issued by the Angolan government (May 23, 2013)
> and a resident alien card issued by the Zambian government (May 23,
> 2013) but have not obtained a passport or residence permit. In my
> plot, in 2014, the NGO, Concern distributed maize, cowpeas, fertilizer,
> hoes, axes, and machetes; therefore, I started cultivating with them. In
> 2015, I sowed maize, cassava stalks given by a neighbour (AG's wife),
> and groundnuts purchased at the market. In 2016, I and my husband
> (at that time we were not married) sowed the seeds they were giving
> free of cost. In addition, a UNDP tractor ploughed 2,500m². In 2017, we
> did not plant anything and in 2018, we planted maize that we got from
> both of our parents when we got married and cassava from my father.
> However, we could not produce enough to sell and just used it for meals.

The land was not fertile. My father cultivated the land until 2018. I did not cultivate any crop this year because I just got married in 2019. I cultivate maize, cowpeas, and soybeans in slope fields around including where M abandoned when she moved to the resettlement scheme. At the waterfront location (*chitunkul*), I grow maize, sweet potatoes, soybeans, and *mutete* (*Hibiscus Spp.*). In lowland fields including where M abandoned when she moved to the resettlement scheme, maize, onion, tomatoes, and rape are planted. Cabbage and tomatoes are sold at the market.

My mother IM has no plot because she was in Kaoma, where her daughter was married for 3 years. Plots in the resettlement scheme were allocated during that time but she could not come back because she was sick. In Sector 4, she plants maize and pumpkin in the slope fields. There are no lowland fields for my mother. I am currently doing peace work with Congolese refugees in Mayukwayukwa. I would like to continue to live in Sector 4.

JZ depends on the sale of vegetables grown in lowland fields and other areas of Mayukwayukwa for her livelihood where M in the resettlement scheme abandoned her fields in Mayukwayukwa. JZ also tried to rely on the economic activity network in Mayukwayukwa, including engaging in agricultural labour in the Congolese refugee fields in recent years. Mayukwayukwa is a place where protracted refugees live, but there is also rapid change of people there. Nonetheless, it is possible to build economic

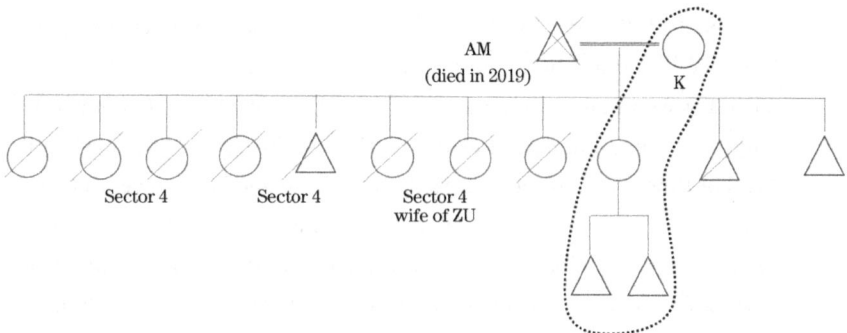

Figure 12.8 Sector 1 *limbo* No. 15 (2020)

relationships with them, allowing former refugees to utilize the advantage of being able to cultivate lands in both the resettlement and Mayukwayukwa. Mbunda like ZU are able to cultivate in Mayukwayukwa while also taking advantage of the resettlement. Thus, with the introduction of new capital and policies, they choose means of livelihood such as cultivation through both the former *limbo* residents where they resided and other kin, selling vegetables for cash to both kin and the market, engaging in agricultural labour and reorganizing new relationships.

Meanwhile, in Sector 1, No. 15 *limbo* K (Fig. 12.8) was disadvantaged by local integration projects.

Case 4: K (born in 1945, Mbunda woman, AM's wife)

I was born in the Hasabisalu village of Angola. My husband, AM, was born in Luena, the capital of Moxico Province in Angola. We fled from the Angolan war and came to Mayukwayukwa Sector 1 in 1966. I have an alien card issued by the Zambian government and national registration card from Angola. In 2014 when I was given a plot through LI, there was only the skeleton of a house; the pump from which to take water for daily use was far from there; and the colour of water was orange; therefore, I thought it was very difficult to migrate. My husband was blind and could not even finish building the house. I was doing peace work and UNHCR was providing money for the blind old man. The year 2014 was the first year for agriculture in the plot, and the NGO Concern gave me maize, cowpeas, groundnuts, fertilizer, and a watering can. I planted those seeds donated by Concern with cassava, maize, groundnuts, cowpea, and sorghum in shifting cultivation fields, and in the lowland fields, rape, cabbage, and sweet potatoes were planted. In the kitchen garden, maize and pumpkins were also planted. However, the plot was near Sector 28 and was ruined because the pigs of Sector 28 destroyed the crops. Hence, we decided not to plant anymore crops in that plot. Vegetables in lowland fields are sold in the market.

In 2019, Congolese refugees came to one of my fields in Mayuk-wayukwa. In 2019, UNHCR cleared the whole area and the government's Community Services Department started distributing the land to the Congolese refugees. At this time, we were not informed about the clearing. When we complained, the refugee officer explained to us that there was land in the resettlement, and we should move to there and cultivate it; but I am still here until now.

When I interviewed K in 2016, K's husband who was still alive then insisted that he could not move to the plot in the resettlement scheme because of his blindness. She was also adamant that her crops had been damaged by livestock. The location of the plot was close to Sector 28 in Mayukwayukwa, and she would rather live there and have the stability guaranteed by vegetable sales and agricultural labour than suffer any disadvantages from relocating. The case of K shows the local integration projects' privatization of resettlement land was not successful. However, the death of her husband and the occupation of her arable land by Congolese refugees could affect her stability.

Thus, the livelihood of Mbunda through local integration projects after the cessation clause was still in transition. However, livelihoods based on agriculture continued to be based on the land use relationships of separated *limbo* residents who had previously lived together, both in Mayukwayukwa and the resettlement schemes. Even though they were given land rights in the resettlement scheme and moved there, their rights to the land of the former *limbo* continued to be loosely recognized among the *limbo* members who had continued to reside in the refugee settlement. Thus, as the Mbunda's internal support for their resilience, without the strong influence of traditional authorities, *limbo*-residents' relationships based on land differs from the kin who do not live or did not live together, but helps food, agricultural labour and the plants to cultivate to maintain their livelihood despite the disruption of the local integration projects.

Conclusion

This chapter clarified how refugee peasants, the Mbunda people, reconstructed their livelihood based on agriculture with land rights after the implementation of the local integration project. I focused on how they reorganized their social relations both in refugee settlements and in resettlement schemes to examine the internal support for their resilience. In comparison with East African pastoral societies, in African agrarian societies chiefs are the traditional authorities who rule land distribution. However, the Mbunda in Mayukwayukwa suffered loose land management, mainly with *limbo* residents and former residents, because of the rule by the Zambian government.

Previous studies identified Angolan refugees who stayed in local villages of Zambia as local farmers (Hansen 1979; Bakewell 2002; Murao 2012, 2014) because they were hosted by chiefs and matrilineal kin who helped them; and they were rarely identified as refugees. These are considered to be cases of successful local integration. In addition, previous studies of highly mobile African farmers showed their residential members reorganisation with matrilineal kin and shifting cultivation strategies when they move (Oyama 2002; Kakeya & Sugiyama 1987). In the Mbunda people's case of Mayukwayukwa, the UN agency and GRZ protected them through the local integration projects, giving them land, first aid, food, and legal status. Specifically, after LI started in 2014, resettlement schemes were developed by subdividing land.

On the other hand, the *limbo* residents became fragmented, with some people, like ZU staying neither in the resettlement schemes nor in *limbo*, earning cash while holding a more dispersed landholding in the period of local integration. As of December 2019, only 21.6% of total households had moved to the resettlement scheme. The reasons for not moving varied. However, land used in Mayukwayukwa, where mutual assistance is routine and land use was easy to claim, is more likely to be held for environmental changes associated with the transition from refugee status to resident foreigner status, such as the ability to grow high value-added vegetables. This has resulted in people remaining in the refugee settlement areas even after becoming resident aliens, continuing to have dispersed land use.

In this way, the livelihoods of Mbunda were based on kin, who flexibly managed their cultivation lands through *limbo* residents' relationship both at present and in the past, staying both in Mayukwayukwa and the resettlement scheme. Under these circumstances, although the former refugees were granted land rights under the law, Mbunda people were not substantially dependent on the livelihoods based on the privatized land rights without chiefs' authority. Most Mbunda simply continued in their previous dispersed land use, rather than actively taking up the privatized land through the local integration projects. Even those who moved to the scheme, such as M, continued to cultivate her former *limbo* land in Mayukwayukwa.

Land reform in East African pastoral societies in recent years led to further land division and the emergence of individuals who have set aside land as private property (Galaty 1993; Mwangi 2007a, 2007b; Ntiati 2002). It was not the traditional authorities but the government-led system that exercised authority on land rights there.

In contrast, as previous studies of agrarian societies in Africa show, land privatization by the residents did not essentially occur by the government-led regulation changes; however, it tended to be strengthened by the traditional chief's authorities (Oyama 2015; Takeuchi 2017). The Mbunda farmer cases show, however, that they were not greatly dependent for their livelihoods on privatized land rights because of their mutual relationships based on *limbo* both in the Mayukwayukwa and in the resettlement scheme, independent of the chiefs or the other traditional authorities. Thus, the Mbunda's resilience developed through the internal relations of the former refugees under land privatization for refugees as the part of durable solution for refugee problem.

Further, the cases of M, ZU and JZ showed how the Mbunda share crops, they plant from their relatives around the former *limbo*. Even though M and ZU are away, they receive assistance from the *limbo* and their relatives in Mayukwayukwa. Not only such kin based internal helps, such as M, who moved to the resettlement scheme maintained the social relationship with kin to continue agriculture and daily cash income. Also, the increasing number of agricultural labourers from Mayukwayukwa to the resettlement scheme showed the reverse economical flow through kin relationships who lived in the same sector.

However, we should be cautious about extolling the resilience of Mbunda. As previous studies showed, if land is increasingly privatized with permanent residents in the years to come, the economic disparities among people are likely to increase. This chapter revealed the resilience of Mbunda people through the local project after 2014, when land privatization was introduced in Mayukwayukwa.

Acknowledgments

I would like to thank the people who participated in this survey for their cooperation. This work was supported by JSPS KAKENHI Grant Numbers 15KK0099, 18H03606, 20K12378.

References

Bakewell, O. 2002. 'Returning Refugees or Migrating Villagers? Voluntary Repatriation Programmes in Africa Reconsidered.' *Refugee Survey Quarterly* 21 (1): 42–73.

Burdette, M. 1984. 'Were the copper nationalizations worthwhile?' In K. Woldring. (ed.). *Beyond political independence. Zambia's development predicament in the 1980s*. New York: Mouton Publishers. pp. 23–71.

Cheke. 1994. *The History and Cultural Life of the Mbunda Speaking Peoples*. Lusaka: Cheke Cultural Writers Association.

Clark, L., and B. Stein. 1985. *Older refugee settlements in Africa: Final report*. Geneva: Refugee Policy Group.

Development and Training Services. 2014. *Evaluating the Effectiveness of Humanitarian Engagement and Programming in Promoting Local Integration of Refugees in Zambia, Tanzania, and Cameroon*.

Frischkorn, R. S. 2013. *We Just Aren't Free: Urban Refugees and Integration in Lusaka, Zambia*. Ph.D. Dissertation, American University, Washington, DC.

Galaty, J. 1993. 'The land is yours': Social and economic factors in the privatization, sub-division and sale of Maasai ranches, *Nomadic Peoples* 30: 26-40.

Hansen, A.G. 1979. 'Once the running stops: Assimilation of Angolan refugees into Zambian border villages.' *Disasters* 3 (4): 369–374.

Kakeya, M. and Y. Sugiyama 1987. 'Shifting cultivation of Bemba people in middle southern Africa.' In I. Ushijima. (ed.). *Ethnology of Symbol & Society.* Tokyo: Yuzankaku. pp.111–140.

McClluoch, M. 1951. *The Southern Lunda and Related Peoples: Ethnographic Survey of Africa. West Central Africa. Part I.* International Africa Institute, London.

Murao, R. 2012. *Creativity of African Farmers.* Kyoto: Showado.

———— 2014. 'Land use of Angolan immigrants in western Zambia: Rethinking the autonomy and coexisting of self-settled refugee communities in host countries.' In A. Takada and I. K. Nyamongo. (eds). *Mila Special Issue.* pp. 59–67.

———— 2020. 'Zambia.' In K. Makino and E. Iwasaki (eds). *Global Social Welfare: Africa and Middle East.* Tokyo: Jyunpousha. pp. 225–247. (in Japanese)

Mwangi, E. 2007a. 'Subdividing the commons: Distributional conflict in the transition from collective to individual property rights in Kenya's Maasailand.' *World Development* 35 (5): 815–834.

———— 2007b. 'The puzzle of group ranch subdivision in Kenya's Maasailand.' *Development and Change* 38 (5): 889–910.

Niamir-Fuller, M. 1998. 'The resilience of pastoral herding in Sahelian Africa.' In F. Berkes and C. Folke (eds). *Linking Social and Ecological Systems: Management practices for building resilience.* Cambridge: Cambridge University Press. pp. 250–284.

Ntiati, P. 2002. Group Ranch Subdivision Study in Loitokitok Division of Kajiado District, LUCID Working Paper 7. Nairobi: International Livestock Research Institute.

Nyamazana, M., G. Koyi, P. Funjika and E. Chibwili. 2017. *Zambia Refugees Economies: Livelihoods and Challenges.* Lusaka: UNHCR.

Omata, N. 2012. 'Community resilience or shared destitution?' Refugees' internal assistance in a deteriorating economic environment' *Community Development Journal* 48 (2): 264–279

Omata, N. 2017. *The Myth of Self-Reliance: Economic lives inside a Liberian refugee camp (Forced Migration, 36).* Oxford: Berghahn Books.

Oyama, S. 2015. 'Protector of customary land or abuser of power? The role of chiefs in land allocation under the Zambia 1995 Land Act.' In M. Kakeya. (ed.). *Special issue on 'Conflicts over land and traditional authority in contemporary Africa', Asian and African Area Studies.* Kyoto: Asian and African Area Studies, Kyoto University. pp. 3–49.

———— 2002. 'The market economy and transformation of shifting cultivation farmer society.' In M. Kakeya. (ed.). *African Farmers World: The custom and changes.* Kyoto: Kyoto University Press. pp. 3–49.

Pillay, K. 2011. *The effectiveness of local integration as a durable solution: The situation of Mauritanian refugees in Senegal.* MA thesis, Université Gaston Berger de Saint Louis. (Accessed: 10 April 2022).

Rutten, M. 1992. *Selling Wealth to Buy Poverty: The process of individualization of landownership among Maasai pastoralists of Kajiado District, Kenya 1890–1990*. Saarbrücken: Verlag Breitenbach Publishers.

Sagawa, T. 2016. 'Frontier Potential – Local range responses to land grabbing in Ethiopia.' In Mitsugu Endo. (ed.). *Transcending Armed Conflict in Contested Institutions and Strategies*, Kyoto University Press. pp. 119–149.

Sumbwa, A. G. 2012. 'Settlement, protracted displacement, and repatriation at Mayukwayukwa in western Zambia.' *African Geographical Review* 32 (1): 1–15. DOI:10.1080/19376812.2012.715989

Takeuchi, S. 2017. 'Chapter 8 Land reform and rural changes: Comparison between Rwanda, Brundi, and western DR Congo.' In S. Takeuchi. (ed.). *Land and Power in Africa: Understanding drastic rural change in the age of land reform*. The research institute of developing economies, JETRO. pp. 251–291.

UNHCR 2018. Durable solutions for refugees. https://www.unhcr.org/uk/solutions.html (Accessed: 10 April 2022).

UNHCR Executive Committee of the High Commissioner's Programme (UNHCR EXCOM). 2017. *Resilience and self-reliance from a protection and solutions perspective*. EC/68/SC/CRP.4. Geneva: UNHCR EXCOM.

Verduijn, S. 2018. 'Local Integration in the Context of Protracted Displacement Inside Syria.' http://blogs.lse.ac.uk/mec/2018/02/25/local-integration-in-the-context-of-protracted-displacement-inside-syria/ (Accessed: 11 December 2018).

Von Oppen, A. 1996. *Terms of Trade and Terms of Trust-The History and Contexts of Precolonial Market Production around the Upper Zambezi and Kasai*. Münster: LIT Verlag.

Epilogue

Resilience in the Drylands: Contested Meanings

Ian Scoones

Introduction

Scattered across the drylands of East Africa, there are countless signboards from NGOs, donor projects and government agencies, with the word 'resilience' on them. Resilient pastoralism, farming, women, youth, livelihoods and more. But what does the word mean? Does adding the word 'resilience' to development activities make a difference, or do standard, failed approaches persist? What new, critical thinking and practice might inject some alternative perspectives? Through contributions from across East Africa, this book offers some answers.

Development is prone to fads, and sadly lessons from past failures, perhaps especially in the drylands, are often not learned (cf. Krätli et al. 2015). Too often 'resilience' projects in pastoral areas look remarkably like those that existed before – livestock marketing groups to improve incomes, livelihood diversification for women, community rangeland management and so on. How can the idea of 'resilience' be made more relevant and meaningful for pastoral areas?

Whether through climate change, population growth, conflict, land fragmentation or large-scale investment, East Africa's pastoral areas are changing dramatically (Lind et al. 2020a). This makes new thinking around resilience and vulnerability crucial, if carefully adapted to these new contexts. Based on in-depth empirical studies from Ethiopia, Kenya and

Uganda, this book joins other reflections on these themes (e.g., Lind et al. 2020b; McPeak and Little 2017; Catley et al. 2013; Catley 2013; McPeak et al. 2011) in attempting to offer a more critical, contextual understanding of a massively over-used term. Given the omnipresence of the 'resilience' buzzword, this book is very timely.

Multiple pastoralisms: Context matters

With a focus on context, the basic questions of resilience of what, for whom and where? come to the fore. In thinking about how pastoral settings have changed in the Greater Horn of Africa, Catley et al. (2013) offer a simple framework (Figure E.1).

Based on two axes – more or less market and resource access – four trajectories of pastoralism are suggested: increasing commercialisation and (export) market orientation; sustaining traditional mobile pastoralism linked to local economies; diversification of livelihoods but linking to the pastoral economy; and finally exit and movement out of pastoralism, including away pastoral areas. Within any area, different combinations of these trajectories (and their multiple variants) are seen, resulting in highly differentiated pastoral settings.

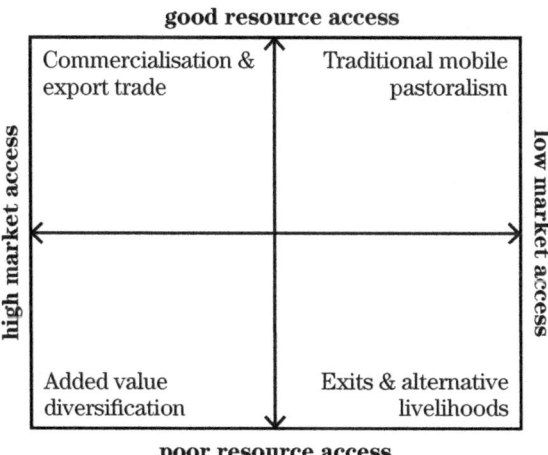

Figure E.1 Future trajectories of pastoralism (from Catley et al. 2013)

Extending the analysis to five different pastoral areas across the region, Lind et al. (2020b, 2016) examine the patterns of change in different settings. Those areas with good access to markets in the Gulf (such as the wider Somali region) are characterised by increasing patterns of market-orientation and commercialisation of livestock, with heightened social differentiation within the area, along the lines described by Catley and Aklilu (2013) as 'moving up, moving out'. By contrast, in other areas, such as in the Karamoja cluster, a process of 'de-pastoralisation' (Caravani 2019) is observed, where livestock numbers have massively declined in part through persistent conflict and in part through aid programmes encouraging 'diversification' away from pastoralism. The result is a highly impoverished setting with highly constrained opportunities (Catley et al. 2021).

Pastoral societies are thus increasingly differentiated, both within and between areas, with axes of class, gender, age and ethnicity all important. No longer are the pastoral settings of East Africa characterised by stable egalitarianism, if ever they were. Today class and social differentiation with varied patterns of accumulation are evident (Scoones 2021). Pastoralism therefore looks very different in different places and across time and for different people, and so 'resilience' must be looked at contextually too. This is the value of this book, offering multiple insights based on in-depth fieldwork.

What is resilience?

Across the chapters of this book, however, some quite different perspectives on resilience are offered. Accepting the dynamic, non-equilibrium nature of dryland environments (Behnke et al. 1993), all contributors reject the 'engineering' view of resilience, based on a stable system 'bouncing back' to its former state (Walker and Salt 2012; Holling 1996), as well as the frequently deployed 'coping with shocks' version of resilience, as seen in much development debate around social protection, for instance. Instead, a more social, cultural and political interpretation is offered, drawing on everything from socio-ecological systems understandings to perspectives from social and cultural anthropology. What should we make of this diversity, and what common threads emerge?

We see at least five different (although overlapping) definitions of resilience within the book – focusing on diversification, system reconfiguration, reliability generation, identity and belonging and citizenship practice – each linked to the pastoral contexts of East Africa.

- Resilience can be seen in terms of the persistence of diversified livelihoods, variously linked to, but not necessarily solely reliant on, livestock production. Responses to changes involve expanding scales of livelihood interaction and widening relationships over time. Resilience in this way can be seen in terms of 'pathway diversity' (Lade et al. 2019). This means forging links to different markets, and particularly to small towns in pastoral regions. It may even mean 'pastoralists' adopting totally new livelihoods, such as the Dassanech youth taking up fishing, as described by Sagawa. Resilience therefore emerges not just in response to short-term shocks, but in response to wider structural changes in contexts, such as improved communications, road building, urban expansion and so on. As in Little's reflection on Baringo in Kenya, a long durée perspective is necessary to understand how forms of production and livelihood reconfigure over time.

- Resilience can also be seen in terms of the ability to reconfigure elements of the socio-ecological system and, crucially, the connections between them. In this sense, resilience is very much about system change and transformation (e.g., Walker et al. 2004); not staying still, always being flexible and responding agilely to changing circumstances, thus drawing on Gunderson and Holling's (2002) 'panarchy' model. In many pastoral settings, livelihood strategies have shifted between intensification, extensification and diversification over time. Sometimes such strategies act to undermine extensive pastoralism – such as through the adoption of ranching or farming or wildlife use – with land fragmentation in particular undermining mobility. Thus, even though 'resilience' is achieved for some, it may undermine others' strategies, with impacts at the wider system level. The reconfigurations of socio-ecological systems may therefore come with costs and trade-offs, which require wider interventions, including compensation, targeted social protection and so on.

- Resilience can also be seen as centrally about assuring reliability – and the continued delivery of desired services from the system – in the face of high variability inputs. Such a perspective, following Roe (2020; also, Roe et al. 1998) and work on 'critical infrastructures' (Roe 2013), focuses on the ability of networks of 'high reliability' professionals, who are able to manage uncertainty and offset ignorance and danger in real-time. Such a perspective emphasises the active process of generating reliability (and so resilience) based on the skills, knowledge and agency of key actors. Seeing pastoralism as a 'critical infrastructure' and focusing on the practices of pastoralists that generate high reliability in the face of uncertainty (and ignorance) therefore highlights the importance of pastoralists' networks and social relations. Konaka looks at conflicts between the Samburu and the Pokot in Kenya and argues that 'reliability' (and so resilience) in the face of conflict emerges through the formation of clustered settlements (for protection, sharing knowledge and more easily mobilised collective action) and an inter-ethnic mobile phone network (for enhancing communication and negotiating peace). Key in these strategies are high reliability professionals at the centre of networks (Tasker and Scoones (2022) for an example around animal disease responses from Marsabit, Kenya).

- Another perspective sees resilience as embedded in cultures, belonging and solidarity, articulated around a shared identity. According to pastoralists' own definitions, resilience is therefore not a structural feature of 'the system', nor can it be created from outside through 'development' interventions but is central to how everyday life is conducted and understood, part-and-parcel of what others have termed 'vernacular resilience' (Wandji et al. 2021). Based on extended fieldwork in Turkana, Kenya, Semplici argues that resilience emerges from a sense of collective belonging, responding to a sense of isolation and marginalisation from wider society, combined with the self-identification as 'Turkana' in relation to other ethnic groups. This strong cultural- and identity-centred view of resilience resonates with the experiences of other pastoral groups, often cast aside by colonial rule, post-independence state-building and so-called 'development'. Identities are however not static, as pastoralists change their livelihoods – moving to town, working for others or farming in an irrigation scheme – their 'pastoral' identities are mutable, flexible to changing contexts. In such shifting conditions, resilience is always about solidarity, both in terms of group identity, but also in terms of the patterns of mutual support and moral economy with family groups, across age sets and among wider clans. These social relations, bound by a strong sense of identity and belonging, are central to pastoralists' ability to confront uncertainty and live with variability, whether in remote rangelands or in towns in pastoral areas. In this perspective, resilience is centred on relational identities, both within and between groups and with the wider social and political-economic system (Mohamed 2022).

- Finally, and relatedly, Hazama argues that resilience must be seen as part of a repertoire of citizenship practices, including human/non-human relationships. Work in Karamoja in Uganda shows how pastoralists have developed their own citizen practices, including relating to livestock as 'co-citizens', that have allowed a pastoral lifestyle to persist, at least for some. In the Karamoja case, this has been in the face of huge threats, including state impositions through peace and disarmament programmes, forced sedentarisation and long-run attempts by missionaries and development agencies to encourage pastoralists to farm, settle and adopt 'modern', more urbanised lifestyles. That such pastoral 'citizen practices' have allowed some to continue with pastoral livelihoods, despite sustained assault and victimisation, is testament to a particular, emergent and practised form of resilience embedded in Karamojong pastoral society.

Resilience as process

Reading across the chapters in this book, there is therefore a clear argument for a more disaggregated, context-specific view of resilience. As already noted, we must always ask, resilience of what in relation to what, and where? The chapters have offered diverse case studies and different conceptualisations, but there are three important, common threads.

First, resilience is not just a 'system property', able to be visualised in terms of balls and cups, as in the 'engineering view'. Just as with that other buzzword, 'sustainability', resilience is inevitably framed in different ways by different actors, and the process of (co-)construction of pathways to resilience is always a political act, negotiated among different players, and with power relations always central (cf. Scoones 2016; Leach et al. 2010). Resilience is therefore always embedded in social relations, culture and identity and wider political-economic relations.

Second, resilience is not just about responsiveness to and recovery from short-term, immediate shocks, but about longer-term transformations of livelihood systems and the relationships that make them up. 'Relational resilience' (e.g., West et al. 2020; Chandler 2014) highlights the importance of the reconfiguration of relationships, including human/non-human relationships, as people, labour and herds and flocks are restructured in the face of challenging events and changing contexts. A long-term perspective is called for as relationships emerge over time and in relation to wider structural features of society and political economy. Thus, for example, relations of age and gender are not fixed, nor are the locations where pastoral livelihoods are sought. A fixed view of pastoralism – the (male) pastoralist with his large herd on a distant rangeland coping with the shock of a drought – gives a narrow view of the challenges of contemporary pastoralism. What about the young, female smallstock owner living in town, also trading vegetables from her farm plot? What about the hired herder managing multiple animals owned by absentees? What about the displaced former pastoralist trying to survive off handouts, but gradually building the capacity to return to herding? And so on. Centred on diverse relationships at the centre of resilience building, identities and identifications, forms of solidarity and moral economy and social bonds forged through everything from age sets to religious congregations or WhatsApp groups become important. Relationships exist within networks, so how these are constructed – and the power relationships within them – become vital in thinking about how resilience emerges over time.

Third, resilience has to be seen as emerging out of people's everyday practices and their social relationships within networks. It is the process through which resilience emerges that is important, not the endpoint. Rethinking resilience as 'high reliability management' is helpful in this regard. Transforming high variance inputs into relatively low variance outputs via diverse practices and processes is central to generating reliability, and so a sustained provision of goods and services that support pastoral livelihoods (cf. Roe 2020). This may be around just managing a herd or flock, but it may be much else too, as pastoral livelihoods become more diversified and complex. But central to the process of generating reliability is the knowledge, skill and practice of so-called reliability professionals, embedded in networks. Thus, in order to ensure reliability in the marketing of pastoral products in the face of highly volatile markets, skilled, well-connected people with a good sense of the overall dynamic, as well as those who are able to adapt flexibly to changing circumstances in order to deliver milk or meat safely and on time, are crucial. They must be linked to each other in networks, and be able to communicate in real time, managing uncertainty and avoiding ignorance and potential disaster. All pastoral systems have such networks – for example, around the delivery of camel milk to consumer markets from remote pastures in the pastoral hinterlands or in relation to mobilising local responses to animal disease outbreaks. As a result, enhancing the capacity of reliability professionals and their networks is a sure way to increase resilience, but in ways that are very different to the standard, instrumental modalities of most development projects being promoted in pastoral areas in the name of 'resilience'.

Conclusion

Across the chapters and in relation to the three cross-cutting themes highlighted above, a strong conclusion emerges. Resilience is not a 'thing' – a definable system property offering stability in the face of variability and shocks – but a 'process' – something emergent out of relationships, connections, networks and practices, rooted in cultures and identities, and always in-the-making by people (and their relationships with the non-human world) in places. Resilience therefore should not be seen as noun, but a verb.

The chapters in this book offer important insights into such a contextual, processual perspective on resilience. The repeated emphasis on relationships, cultures, identities and practices highlights the importance of looking at particular settings and understanding their dynamics and the perspectives on 'resilience' of those who live there. This is a vitally important counter to the standardised, instrumentalised version of resilience promoted through development programmes across the drylands and promoted on the multiple project signboards littering the landscape.

Perhaps because of the disciplinary associations of most of the contributors, and their engagement with particular research sites over extended periods, there were perhaps some missing dimensions, however. As highlighted earlier, looking outwards from particular cases to the wider context, we see the way wider political economies – of investment, accumulation and class relations – structure – both constraining and enabling – the possibilities of building resilience, and indeed the wider potentials to transformations to more resilient futures (cf. Scoones et al. 2020). Implicitly, of course, all the contributions point to the fact that resilience is a political concept, but the politics of resilience is not especially highlighted. As a contested idea, being central to development activities across the drylands, the politics of resilience is however a vital concern, and needs to be at the forefront of any agenda for 'rethinking resilience' (e.g., Rigg and Oven 2015; Brown 2015). Even when stripped of its political meaning, the 'resilience programming' interventions of the aid agencies and NGOs in pastoral areas are always political – whether pushing particular types of market engagement, or encouraging livelihood diversification presented as 'alternative' livelihoods (to pastoralism) or urging other forms of 'sedentist' modernisation.[1]

Taking account of the increasingly differentiated nature of pastoral societies (and following Fraser 2013), resilience thinking and practice therefore must encompass a politics of redistribution, addressing increasingly important issues of class and social difference, including gender, age, ethnicity, and so take account of how resilience is framed by different social groups through a political process of deliberation and negotiation. This must go together with a politics of recognition, and so raise questions of identity and identification, and how resilience is associated with how people see themselves. And finally, this must connect with a politics of representation for pastoralists, often marginalised within wider political systems, and so questions of community, belonging and citizenship. As Fraser argues, in order to imagine a more progressive, emancipatory perspective in response to contemporary 'crises' all three political dimensions must be combined. This also applies to thinking about resilience, and so transforming development practice in the pastoral drylands.

Acknowledgements

This chapter was written thanks to support from the European Research Council through an Advanced Grant to PASTRES (Pastoralism, Uncertainty, Resilience: Global Lessons from the Margins (www.pastres.org); ERC grant number: 740342).

---

1   https://pastres.org/2022/03/18/how-sedentist-approaches-to-land-and-conservation-threaten-pastoralists/

References

Behnke, R.H. Jr., I. Scoones and C. Kerven. (eds.) 1993. *Range ecology at disequi-librium: New models of natural variability and pastoral adaptation in African savannas.* London: Overseas Development Institute.

Brown, K., 2015. *Resilience, development and global change.* London: Routledge.

Caravani, M., 2019. '"De-pastoralisation" in Uganda's Northeast: from livelihoods diversification to social differentiation'. *Journal of Peasant Studies* 46: 1323–1346. https://doi.org/10.1080/03066150.2018.1517118

Catley, A., 2013. *Pathways to resilience in pastoralist areas: A synthesis of research in the Horn of Africa.* Boston: Feinstein Center, Tufts.

Catley, A., and Y. Aklilu, 2013. 'Moving up or moving out? Commercialization, growth and destitution in pastoralist areas'. In A. Catley, J. Lind and I. Scoones. (eds.), *Pastoralism and development in Africa: Dynamic change at the margins*, Abingdon and New York: Routledge. pp. 85-97.

Catley, A., J. Lind, and I. Scoones, (eds.). 2013. *Pastoralism and development in Africa: Dynamic change at the margins.* London: Routledge

Catley, A., E. Stites, M. Ayele and R.L. Arasio, 2021. 'Introducing pathways to resilience in the Karamoja Cluster'. *Pastoralism* 11: 1–5. https://doi.org/10.1186/s13570-021-00214-4

Chandler, D. 2014. *Resilience: The governance of complexity.* London: Routledge.

Fraser, N., 2017. 'A triple movement? Parsing the politics of crisis after Polanyi'. In Burchardt, M. and G. Kirn. (eds). Beyond neoliberalism: Social analysis after 1989. Berlin: Springer. pp. 29-42.:

Gunderson, L. and C.S. Holling. (eds). 2002. *Panarchy: Understanding transfor-mations in human and natural systems.* Washington, DC: Island Press.1996. 'Engineering resilience versus ecological resilience'. In National Academy of Engineering (ed.). *Engineering within ecological constraints.* Washington, DC: The National Academies Press. pp. 31-44. https://doi.org/10.17226/4919.

Krätli, S., B. Kaufmann, H. Roba, P. Hiernaux, W. Li, M. Easdale and C. Hülsebusch. 2015. *A house full of trap doors. Identifying barriers to resilient drylands in development.* IIED Discussion Paper. London and Edinburgh: IIED.

Lade, S.J., B.H. Walker, and L.J.Haider. 2019. Resilience as pathway diversity: Linking systems, individual and temporal perspectives on resilience. *arXiv* preprint arXiv:1911.02294.

Leach, M., A.C. Stirling and I. Scoones. 2010. *Dynamic sustainabilities: Technology, environment, social justice.* London: Routledge.

Lind, J., D. Okenwa, and I. Scoones (eds.) 2020a. *The politics of land, resources and investment in Eastern Africa's pastoral drylands.* Woodbridge: James Currey.

Lind, J., R. Sabates-Wheeler, M. Caravani, L. B.D.Kuol, D.M. Nightingale. 2020b. 'Newly evolving pastoral and post-pastoral rangelands of Eastern Africa'. *Pastoralism* 10: 1–14. https://doi.org/10.1186/s13570-020-00179-w

Lind, J., R. Sabates-Wheeler, S. Kohnstamm, M. Caravani, A. Eid, D.M.Nightingale, C. Oringa, 2016. *Changes in the drylands of eastern Africa: case studies of pastoralist systems in the region*. Nairobi: DFID East Africa Research Hub.

McPeak, J.G. and P.D. Little. 2017. 'Applying the concept of resilience to pastoralist household data'. *Pastoralism* 7 (14): 1–18. https://doi.org/10.1186/s13570-017-0082-4

McPeak, J.G., P.D. Little and C.R. Doss. 2011. *Risk and social change in an African rural economy: livelihoods in pastoralist communities*. London: Routledge.

Mohamed, T., 2022. *The role of moral economy in response to uncertainty among pastoralists in Northern Kenya, 1975-2020*. PhD thesis, IDS, University of Sussex.

Rigg, J. and K. Oven, 2015. Building liberal resilience? A critical review from developing rural Asia. *Global Environmental Change* 32: 175-186. https://doi.org/10.1016/j.gloenvcha.2015.03.007

Roe, E., 2013. *Making the most of mess: reliability and policy in today's management challenges*. Durham NC: Duke University Press.

Roe, E., 2020. *A new policy narrative for pastoralism? Pastoralists as reliability professionals and pastoralist systems as infrastructure. STEPS Centre Working Paper*. Brighton: STEPS Centre.

Roe, E., L. Huntsinger, K. Labnow, 1998. 'High reliability pastoralism'. *Journal of Arid Environments* 39: 39–55. https://doi.org/10.1006/jare.1998.0375

Scoones, I., 2016. 'The politics of sustainability and development'. *Annual Review of Environment and Resources* 41: 293–319. https://doi.org/10.1146/annurev-environ-110615-090039

Scoones, I., 2021. 'Pastoralists and peasants: perspectives on agrarian change'. *Journal of Peasant Studies* 48: 1–47. https://doi.org/10.1080/03066150.2020.1802249

Scoones, I., A. Stirling, D. Abrol, J. Atela, L. Charli-Joseph, H. Eakin, A. Ely, P. Olsson, L. Pereira, R. Priya and P. van Zwanenberg. Scoones, I., A. Stirling, D. Abrol, J. Atela, L. Charli-Joseph, H. Eakin, A. Ely, P. Olsson, L. Pereira, R. Priya and P. van Zwanenberg. 2020. 'Transformations to sustainability: combining structural, systemic and enabling approaches'. *Current Opinion in Environmental Sustainability* 42: 65-75.

Tasker, A. and I. Scoones, 2022. 'High reliability knowledge networks: Responding to animal diseases in a pastoral area of Northern Kenya'. *Journal of Development Studies* 58(5): 968-988. https://www.tandfonline.com/doi/full/10.1080/00220388.2021.2013469

Walker, B., C.S. Holling, S.R. Carpenter and A. Kinzig. 2004. 'Resilience, adaptability and transformability in social-ecological systems'. *Ecology and Society* 9 (2): 5. https://doi.org/10.5751/ES-00650-090205

Walker, B. and D. Salt, 2012. *Resilience thinking: sustaining ecosystems and people in a changing world*. Washington DC: Island Press.

Wandji, D., J. Allouche, and G. Marchais. 2021. *Vernacular resilience: An approach to studying long-term social practices and cultural repertoires of resilience in Côte d'Ivoire and the democratic Republic of Congo. STEPS Working Paper.* Brighton: STEPS Centre.

West, S., L.J. Haider, S. Stalhammar, and S. Woroniecki. 2020. 'A relational turn for sustainability science? Relational thinking, leverage points and transformations'. *Ecosystems and People* 16: 304–325. https://doi.org/10.1080/26395916.2020.1814417

Index

Personal Names

Subjects